NUTRITION AND BIOTECHNOLOGY IN HEART DISEASE AND CANCER

ADVANCES IN EXPERIMENTAL MEDICINE AND BIOLOGY

Recent Volumes in this Series

Volume 365
MECHANISMS OF LYMPHOCYTE ACTIVATION AND IMMUNE REGULATION V:
Molecular Basis of Signal Transduction
Edited by Sudhir Gupta, William E. Paul, Anthony DeFranco, and Roger Perlmutter

Volume 366
FREE RADICALS IN DIAGNOSTIC MEDICINE: A Systems Approach to Laboratory
Technology, Clinical Correlations, and Antioxidant Therapy
Edited by Donald Armstrong

Volume 367
CHEMISTRY OF STRUCTURE–FUNCTION RELATIONSHIPS IN CHEESE
Edited by Edyth L. Malin and Michael H. Tunick

Volume 368
HEPATIC ENCEPHALOPATHY, HYPERAMMONEMIA, AND AMMONIA TOXICITY
Edited by Vicente Felipo and Santiago Grisolia

Volume 369
NUTRITION AND BIOTECHNOLOGY IN HEART DISEASE AND CANCER
Edited by John B. Longenecker, David Kritchevsky, and Marc K. Drezner

Volume 370
PURINE AND PYRIMIDINE METABOLISM IN MAN VIII
Edited by Amrik Sahota and Milton W. Taylor

Volume 371A
RECENT ADVANCES IN MUCOSAL IMMUNOLOGY, Part A: Cellular Interactions
Edited by Jiri Mestecky, Michael W. Russell, Susan Jackson, Suzanne M. Michalek,
Helena Tlaskalová, and Jaroslav Sterzl

Volume 371B
RECENT ADVANCES IN MUCOSAL IMMUNOLOGY, Part B: Effector Functions
Edited by Jiri Mestecky, Michael W. Russell, Susan Jackson, Suzanne M. Michalek,
Helena Tlaskalová, and Jaroslav Sterzl

Volume 372
ENZYMOLOGY AND MOLECULAR BIOLOGY OF CARBONYL METABOLISM 5
Edited by Henry Weiner, Roger S. Holmes, and Bendicht Wermuth

A Continuation Order Plan is available for this series. A continuation order will bring delivery of each new volume
immediately upon publication. Volumes are billed only upon actual shipment. For further information please contact
the publisher.

NUTRITION AND BIOTECHNOLOGY IN HEART DISEASE AND CANCER

Edited by

John B. Longenecker
University of North Carolina
Chapel Hill, North Carolina

David Kritchevsky
Wistar Institute of Anatomy and Biology
Philadelphia, Pennsylvania

and

Marc K. Drezner
Duke University Medical Center
Durham, North Carolina

SPRINGER SCIENCE+BUSINESS MEDIA, LLC

Library of Congress Cataloging in Publication Data

On file

Proceedings of a conference on Nutrition and Biotechnology in Heart Disease and Cancer,
held December 5–7, 1993, in Research Triangle Park, North Carolina

ISBN 978-1-4613-5804-6   ISBN 978-1-4615-1957-7 (eBook)
DOI 10.1007/978-1-4615-1957-7

© 1995 Springer Science+Business Media New York
Originally published by Plenum Press, New York in 1995
Softcover reprint of the hardcover 1st edition 1995

10 9 8 7 6 5 4 3 2 1

PREFACE

There is a unique nutritional commonality developing in research relating to *coronary heart disease* and *cancer*. The primary aim of this conference was to provide a forum for the leading researchers, clinicians, educators and administrators in these two fields to present a program on heart disease and cancer which included a) the major historical milestones, b) the present areas of greatest interest in research and therapy, c) the latest nutritional, molecular, and biotechnological advances, and d) a perspective on the most promising areas for future research and therapy.

Scientists have long contended that research marches on the feet of methodology. Thus there are numerous examples of research fields opening secondary to methodological advances. Some examples are: 1) thin layer and gas-liquid chromatography which, along with high pressure liquid chromatography have broadened the line of advances in lipid research and 2) PCR and the resultant impact on molecular biological approaches to several fields of science. The organizers of this conference thought the time was propitious for bringing together knowledge on newer aspects of molecular biological research with current advances in the two major areas of degenerative disease--*coronary heart disease* and *cancer*.

Our knowledge of these *"killer diseases"* has expanded greatly in the past few years and the advance has been catalyzed by use of an array of molecular biological techniques. Thanks to these, medical thinking in these areas is changing from considerations of treatment to strategies for prevention.

This conference has been successful in presenting the most advanced knowledge to an audience of interested participants who will be able to use what they learned for the advancement of their own research. Undoubtedly the same will be true for those who read these proceedings. It is our hope that future advances will make it possible to find the critical answers in the fight against these two perplexing diseases.

The Editors

ACKNOWLEDGMENT

We wish to sincerely thank the Institute of Nutrition of the University of North Carolina, the Sarah W. Stedman Center for Nutritional Studies of Duke University Medical Center and the North Carolina Biotechnology Center for co-sponsoring the conference and the Nabisco Biscuit Company, Coca Cola Company, Johnson & Johnson, Procter & Gamble, Hoffman-LaRoche Inc., and Glaxco Inc. Research Institute for their financial support.

We gratefully acknowledge the efforts of the members of the Program Committee and the dedication and cooperation of the staff members of the Institute of Nutrition of the University of North Carolina, the Sarah W. Stedman Center for Nutritional Studies, and the University of North Carolina Continuing Education Center.

PROGRAM COMMITTEE

Chair	*John Longenecker*	**Institute of Nutrition, University of North Carolina**
Co-chair	*Connie Bales*	**The Sarah W. Stedman Center for Nutritional Studies, Duke University Medical Center**
	Betsy Baker	**North Carolina American Cancer Society**
	Terry Bazzare	**American Heart Association**
	Mark Failla	**University of North Carolina at Greensboro**
	Charles Hamner	**North Carolina Biotechnology Center**
	David Kritchevsky	**The Wistar Institute**
	Gilbert Leveille	**Nabisco Biscuit Company**
	David Libby	**North Carolina A&T State University**
	Robert Maier	**East Carolina University**
	Donald McCormick	**Emory University Medical School**
	Bob Sanders	**The University of Texas at Austin**
	Jason Shih	**North Carolina State University**
	Tom Slaga	**M.D. Anderson Cancer Research Center**
	Steven Zeisel	**The University of North Carolina at Chapel Hill**

CONTENTS

HEART DISEASE

CANCER

HEART DISEASE AND CANCER

NIH INITIATIVE

APPENDIX

HISTORICAL REVIEW OF RESEARCH ON
ATHEROSCLEROSIS

Gardner C. McMillan

5305 Burling Terrace
Bethesda, MD 20814

INTRODUCTION

Atherosclerosis in humans can have severe clinical sequellae such as heart attack, stroke and peripheral vascular disease. These diseases are common and they have their own research histories, but it is the history of the underlying atherosclerosis or hardening of the arteries about which I shall speak. I have listed several sources for this paper and they have extensive bibliographies for specific points of reference.

It is possible to comment on several interwoven histories depending on whether the research viewpoint emphasizes the pathological, the biochemical, the metabolic, the epidemiological, the nutritional and recently the mechanistic and molecular biological.

PATHOLOGICAL RESEARCH AND DEFINITIONS

I shall begin with the pathological because it is by far the oldest, but touch on the others as needed to develop an outline that I hope will be useful background for you. Even today, the definition of atherosclerosis as a space occupying lesion or plaque of the inner coat of larger arteries that is focal, has a pattern of occurrence, is composed of an excess of fat, of an increased number of artery wall and inflammatory cells and their connective tissue products, and that may show calcification and ulceration, narrows the arterial lumen, may obstruct blood flow through the artery and may be associated with a local thrombus is a definition that uses the terms of classical pathology. In view of what we know today, a more etiological and mechanistic definition could be formulated, but, in fact we generally use a descriptive definition like the foregoing and add to it etiological associations and mechanistic cartoons as suits our purpose.

The first formal description of atherosclerosis was due to Fallopius in 1575 who described a degeneration of arteries into bone. Four hundred years ago educated physicians were aware of ossification of arteries. In 1695, that is three hundred years ago, we find the aorta of a 75 year old physician Johann Jakob Wepfer pictured and described. "The internal coat in several places was ruptured, lacerated and rotten like fruit and hurt the fingers when thrust into it, from the roughness of the bone". That the

Nutrition and Biotechnology in Heart Disease and Cancer
Edited by J.B. Longenecker *et al.*, Plenum Press, New York, 1995

1

disease is much older has been shown by autopsies on Egyptian mummies dating from 1580 B. C. to 523 A. D. and reported before World War I. One specimen was from Menephtah who was the Pharaoh of the Exodus.

During the 1700's descriptions improved and there were even some mechanistic discussions. Joseph Hodgson in 1815 gave an excellent gross description and included a chemical analysis of the arterial calcification and concluded it was not true bone. Most importantly he thought atherosclerosis was a disease process and not an effect of ageing.

Jean-Fredrich Martin Lobstein coined the term arteriosclerosis in Strassbourg in 1829. Our preferred term of atherosclerosis was proposed and justified in 1904 by Felix Marchand of Leipzig. Meanwhile, mechanistic ideas were proposed including focal inflammation and deposition of material from the blood onto the lining as an encrustation that was changed into plaque. The first detailed microscopic studies were those of Virchow (1858). He thought that the plaque began *in* rather than *on* the intima as a proliferation of cells with formation of cell products. Fatty changes followed. There were some elements of inflammation and lesions tended to locate where there was pressure from the blood stream. He noted that cholesterol was present but considered it secondary.

The descriptions of atherosclerosis became more refined and precise. A major addition was the so-called fatty flecks or streaks of the arteries of children which were described in detail in a large autopsy series in 1925 and suggested as the origins of the more advanced plaques ~ a thesis still under review to-day. Great improvements in technique and methodology have advanced us into the modern era when we can apply electromicroscopy, microchemistry, immunohistology and protein identification to follow cell lineage, cell kinetics, biochemical composition and apocrine and paracrine controlled cellular activity and metabolism in plaque formation.

RESEARCH ON CHOLESTEROL AND EXPERIMENTAL ANIMAL MODELS

But there are other aspects of the history of research on atherosclerosis. One of these deals with cholesterol and through it leads to experimental animal work, nutrition and metabolism.

The first recognition of cholesterol in plaques was by Vogel in 1843 and a suggestion of some connection between blood and tissue cholesterol was made two years later. Its prominence in plaques was emphasized by Aschoff in 1906 and in 1910 Windaus, a chemist working with Aschoff compared the cholesterol content of plaque and nonplaque tissue. Windaus found striking increases of free and esterified cholesterol in plaque. That blood cholesterol and clinical disease might be related was reported in 1913 when 13 patients with atherosclerotic disease were found to have elevated blood cholesterol

This state of knowledge coincided with nutritional animal experiments in which a rabbit model of atherosclerosis was developed in Russia between 1908 and 1913. This began with an experiment by Ignatowski who fed rabbits a diet rich in animal protein in the form of milk, eggs and meat. The idea, derived from Metchnikov, was that protein breakdown products would be toxic. The diet was atherogenic. The experimental result was confirmed by Fahr and in 1912-13 Stuckey and also Wesselkin showed that the atherogenic effect resided in the fatty aspect of the diet rather than its animal protein components. It remained for Anitschkow and Chalatow to produce in 1912-13 atherosclerosis in rabbits by feeding cholesterol dissolved in olive oil.

A huge literature of animal experimentation takes its origin from Anitschkow's finding. Other animal models were developed including dogs, chickens, pigs, nonhuman primates and even cell and tissue culture surrogates. Investigators used the model to study plaque evolution, the association or reaction to the cholesterol insult, and the influence of some third experimental circumstance such as a drug, hormone, nutritional deficiency or dietary modification on atherogenesis. Genetic questions were raised as well as questions about resistance to dietary cholesterol and regression of plaques over time.

The impact of Anitschkow and Chalatow's model was enormous and as experimental techniques were improved and refined many experiments of an earlier decade were repeated using the enhanced and more exact and informative new methods. For example the formulation and analysis of experimental diets was greatly refined; simple chemical measurements of blood cholesterol gave way to lipoprotein analysis, fractionation and quantitation. Other fats and phospholipids were studied, plaques were analyzed and quantitated in biochemical terms and the pathology of plaques was quantified in gross and microscopic detail. The emerging fields of statistical mathematics and biometry were applied to bring experimental rigor to the field and to aid interpretation. For the first time numerous reasonable hypotheses could be generated and mechanistic ideas tested. It was and is an intellectually exciting journey in which however, we often march to a technological drummer.

One effect of these animal experiments was to turn research to investigations in humans of the hypotheses generated by the experimental model. A huge clinical literature has developed since World War II that emphasizes the roles of cholesterol and fats. We must also note another very important aspect of Anitschkow's model. It and its derivative animal models did not display thrombosis or thrombotic processes. Consequently, and particularly in America, relatively little work was done on atherogenesis as a process involving or resulting from blood coagulation mechanisms. Even to-day we lack useful and convenient animal models to study such hypotheses and the literature is sparse although there is clear evidence of such processes in human atherosclerosis.

CHOLESTEROL AND ITS RELATION TO ATHEROSCLEROSIS

As we have seen cholesterol became of central interest in atherosclerosis both in the plaques and in the diet of experimental animals. Kritchevsky published a classic book on cholesterol in 1958 in which he also provided an historical review of the identification of this molecule and proof of its structure. Originally studied in gall stones as early as 1733 it was given its name of cholesterin--a name still used in German--by Chevreul in 1816. It was identified in atheroma in 1843 by Vogel and as we have heard was firmly established in the chemistry of atherosclerosis by Windaus in 1910. He found seven times more free cholesterol and twenty times more esterified cholesterol in plaques from aortic specimens than in more normal arterial tissue. Cholesterol has attracted much interest from biochemists and Brown and Goldstein remark in their Nobel Laureate lecture of 1985 that it has been involved in the award of thirteen Nobel prizes over the years. One of the early questions was whether the animal body could make cholesterol or whether it must always come from food. It was shown in 1913 that rats growing on a cholesterol-free diet increased their total body cholesterol. It was also shown that infants excreted two to four times the amount of cholesterol that they were fed and, with an experiment of exquisite simplicity and directness, Thannhauser showed in 1923 that incubated eggs contained more cholesterol than eggs that were not incubated.

The organism can make and excrete cholesterol. However, it was dietary cholesterol that became the dominant interest and here our history turns to animal and clinical investigation and to epidemiology. I shall not attempt to deal with animal and clinical investigation now although it is critical to our current understanding of metabolic pathways and the molecular biochemistry of lipids, their transport in the blood and their handling by various kinds of cells in the body. Perhaps two circumstances that speak to the role of elevated blood cholesterol in atherogenesis should be mentioned here. One is the very strong association between homozygous familial hypercholesterolemia, a genetic LDL-receptor disorder disease, and severe atherosclerosis in childhood and young adult life. The other is the currently emerging autopsy evidence that post-mortem serum lipoprotein and cholesterol levels are positively correlated with atherosclerosis in young men and women as seen in the Bogolusa Heart Study and the current PDAY study.

EPIDEMIOLOGICAL RESEARCH

I want to spend the remaining time on the intersection of epidemiology with our understanding of atherosclerosis. This is relatively recent history and it reflects an interplay between basic research in animals and clinical research with findings in populations.

An important study was published in 1968. It was an autopsy study conducted at Lousiana State University of some 21,000 cases drawn from one European and 13 American countries with very differing rates of heart attack. Two thirds were male and one half were white. It considered atherosclerosis itself rather than heart attack. It found that atherosclerosis was always similar, the gross and microscopic appearance being the same in all the countries and cases studied, differing only in average extent or severity and it found that when a heart attack occurred the severity of coronary atherosclerosis was about the same regardless of sex, race or the country of origin.

In 1983 professor Ancel Keys, one of the great figures in the field of cardiovascular epidemiology published an informal review of studies that had occupied him for several decades. He wrote about the historical background for his work and I have drawn from his paper which was titled "From Naples to Seven Countries--A Sentimental Journey". In 1916 De Langen had reported that the Javanese had less atherosclerosis than the people in Holland. He attributed this finding to differences in diet and in blood cholesterol. Javanese stewards on Dutch steamships eating Dutch diets had serum cholesterols like those in Holland. Snapper (1941) reported from China that atherosclerosis was minor and cholesterol levels were low. Aschoff had said in 1924 that there was a decrease in atherosclerosis in Germany following the food deprivation of World War I. Experience in Scandinavia during World War II suggested a fall in mortality from atherosclerotic disease during a period of food deprivation and an increase in mortality when normal amounts of food became available again. Keys was also aware of then unpublished data from the Netherlands that were similar. Keys had given low fat diets to hypercholesterolemic patients in the 1940's and obtained a fall in blood cholesterol. Restricting dietary cholesterol had had only a minor effect. By 1947 a conceptual framework was emerging and after some preliminary work in Minneapolis, in Naples and elsewhere, the full Seven Country Study was underway in 1957. He used dietary measures, cholesterol levels and other risk factor observations with coronary heart disease as an endpoint as well as a surrogate for atherosclerosis. These studies were able to show in geographically diverse cohorts that the fat, particularly saturated fat content of the diet was positively associated with the serum cholesterol level and this in turn was prospectively associated with heart attack rates.

Another noteworthy study has been the Framingham Heart Study. Planning for this population study began in 1947. Little was known about the epidemiology of arteriosclerotic heart disease or of hypertension but it was thought best to study them over time in an American population of normal or usual composition. It was also assumed from the available basic science and clinical research information that there were multiple causes for the diseases which worked slowly over time and also that precise methods to diagnose clinically the early stages were lacking. A tentative protocol involved as complete a periodic clinical examination as possible of otherwise free living men and women in a community. They were, for practical reasons to number about 6,000 and to be between ages 30 and 59 at entry. Follow-on was to be 20 years and the participants records would be searched retrospectively for characteristics that might relate to emergent or prospective events of disease. Because of interest on the part of the State Health Commissioner for Massachusetts who offered to cooperate with the USPHS Framingham was eventually selected as site for the study. Framingham included both rural and urban areas. It had a town meeting form of government with community participation and it had been the site of a successful six year tuberculosis study about 25 years earlier. The citizens were willing to participate and the study began in 1947. It became the responsibility of the newly created National Heart Institute on July 1, 1948.

The Framingham Heart Study, unlike the Seven Country Study, was conducted in a single community at a single site. It contributed little to dietary issues but it provided an immense amount of information about medical characteristics antecedent to clinical cardiovascular disease. Thus emerged the concept of risk factors for heart disease and indeed the Framingham Heart Study was able to quantify these associations so well that the study measurements could be transposed into probabilities of the occurrence of a clinical event over time ~ that is to say the risk of clinical cardiovascular disease (or its absence) in the patient's future.

The Keys' and Framingham studies were prospective and took many years before they provided data and, of course, they were not alone. Many other and various epidemiological studies also contributed along the way. These studies contributed very powerfully to our sense of risk factors for heart disease and the role they might play in clinical trials, in disease prevention in public health measures, patient care and for dietary and nutritional advice ~ at least for macronutrients since there are inadequate data on micronutrients.

Our history began with descriptive findings made with the unaided eye, the sense of touch and clinical history assisted after about three hundred years by the microscope and analytical chemistry. These phenomenological reports have recorded the remarkable number and variety of arterial changes that occur in human atherosclerosis. Until recently such studies have had little to say about the prevalence of such changes. Nevertheless, such changes are the essential and inescapable criteria of validity to which all subsequent findings must conform.

As phenomenological findings appeared, certain ones became salient and recognized as having special importance. Cholesterol and lipids were two of these. The dominance was due to the coincidence over the period of years from about 1900 to 1940 of an emphasis by pathologists, the availability of convenient and dependable analytical and quantitative chemistry for blood and tissue, and the development of cholesterol-fed hypercholesterolemic animal models. An additional important finding of the late 1800's and early 1900's was the recognition that myocardial infarction was associated with coronary artery occlusion by atherosclerosis and thrombosis.

Research then became more experimental. Both observations and questions to be tested were greatly refined and the approach became more and more reductionist and directed to causation.

Inevitably, and especially after the clinical diagnostic recognition of the signs and symptoms of coronary occlusion and myocardial infarction there was a strong impetus to study atherosclerosis and cholesterol and lipid metabolism in healthy humans and those sick with cardiovascular disease.

Clinical interest spread to the study of cardiovascular disease in the community with the methods of epidemiology. An epidemic of coronary heart disease morbidity and mortality was identified in the 1950's and 1960's in this country. Thus, an important public health impact of atherosclerosis was documented, the concept of risk factors emerged and a major role for nutrition was defined.

Since about 1970 research in atherosclerosis has emphasized molecular biology and the intimate mechanisms of atherogenesis and metabolism ~ evolving fields that I have not attempted to describe. But I hope that this brief review will let you appreciate what has gone before.

As stated earlier major sources for this paper are cited which have extensive bibliographies for specific points of reference.

REFERENCES

Anitschkow, N. N., 1967, A history of experimentation on arterial atherosclerosis in animals, in: "Cowdry's Arteriosclerosis", 2nd ed., H. T. Blumenthal, ed., Charles C.Thomas, Springfield.

Epstein, F. H., 1990, Die historische Entwicklung des Cholesterin-Atherosklerose-Konzepts, Therapeutishe Umshau 47: No. 6: 435-442.

Gallagher, G. L., 1990, Evolutions: Atherosclerosis, Journal of NIH Research 5: May 90.

Keys, A., 1983, From Naples to seven countries--a sentimental journey, Prog. Biochem.Pharmacol. 19: 1-30

Kritchevsky, D., 1958, Cholesterol, John Wiley & Sons, New York.

Long, E. S., 1967, Development of our knowledge of arteriosclerosis, in "Cowdry'sArteriosclerosis", 2nd ed., H. T. Blumenthal, ed. Charles C. Thomas, Springfield.

McGill, H. C., et al, 1968, The Geographic Pathology of Atherosclerosis, Lab. Invest. 18: No. 5: 463.

NUTRITION AND CORONARY HEART DISEASE EPIDEMIOLOGY

Herman A. Tyroler

Department of Epidemiology
School of Public Health
University of North Carolina at Chapel Hill
Chapel Hill, NC 27599-7400

INTRODUCTION

Epidemiologic studies of the etiology of Coronary Heart Disease (CHD), and clinical and population based approaches to its prevention and control, have evolved with recognition of the importance of many risk factors. However, despite acceptance of the concept of multifactorial causality, and continuing identification of an increasingly larger number of contributing factors extending from the molecular genetic to the societal level, the role of nutrition and the diet-lipid-CHD theory have remained central to CHD epidemiology. For this presentation, I have assumed that the theory, positing causal links between diet, serum cholesterol, atherosclerosis and coronary heart disease, is valid without presenting and evaluating all the massive evidence related to this assumption. Rather, I shall illustrate the theory's explanatory and predictive properties and its application in public health preventive and control programs as seen from a population oriented epidemiological perspective.

Early support for the theory came from animal experimental and human metabolic and clinical physiologic studies. The theory has subsequently been found predictive and explanatory of human, cross population and time trend, coronary heart disease epidemiological observations. Further confirmation presently is arising from studies of the relation of diet to pre-clinical, asymptomatic atherosclerosis. Human studies of atherosclerosis previously were possible only at time of surgery or autopsy. Studies of the presence and change over time of arterial lesions in clinical case series, clinical trials and in community-based large epidemiological studies are now feasible as a result of technologic advances in quantitative angiography and ultrasound imaging. Thus, a wide range of observational studies and clinical trials provides results which have been tested against predictions from the theory. The important role of the diet-lipid-CHD theory, the emergence of new methodological techniques in assessing CHD risk and in performing and evaluating clinical trials of CHD prevention and treatment, the increasing public health application of their results, and community surveillance of their impact will be highlighted to illustrate the relationships between nutrition and coronary heart disease epidemiology in this

Nutrition and Biotechnology in Heart Disease and Cancer
Edited by J.B. Longenecker *et al.*, Plenum Press, New York, 1995

7

conference on biotechnology, nutrition and health. This presentation is not an attempt at a comprehensive review of the thousands of reports of empirical research addressing these issues, nor even an attempt to synthesize and evaluate the numerous extant summaries (Stamler, 1979; Willett,1990; Stamler, 1992), interpretations, controversies and policy recommendations regarding the subject. Rather, this occasion provides the opportunity to illustrate findings related to the diet-lipid-CHD theory as derived from different types of epidemiologic investigations. What follows is an essay, selective in the studies chosen, the illustrative examples and the data presented.

The material has been organized in relation to stages in the study of the relations among diet, serum cholesterol levels and coronary heart disease and control measures in populations. A central role has been identified for clinical trials, as these provide human experimental tests of the theory, as well as evidence guiding clinical practice and community level policy decisions. As set out in the idealized schema of **Figure1**, observational studies, in principle, provide the descriptive background and hypotheses which lead to experimental testing by clinical trials; successful experimental results justify recommendations for medical practice, public health programs and population behavioral changes; community surveillance then permits assessment of population effects over time, with results and new hypotheses feeding modifications into the loop again, if goals have not been achieved. The schema implies successive steps with the results of one stage determining the next in series; in actuality, many of the studies and practices have been carried out concurrently, or in some instances advanced stages or applications have been introduced without the requisite antecedent evidence.

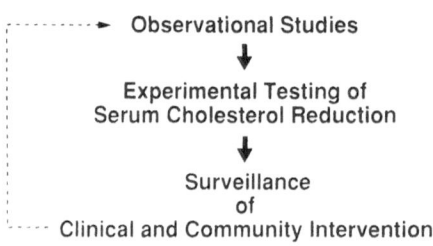

Figure 1. Diet. serum cholesterol. and stages in the epidemiological study and control of CHD

OBSERVATIONAL STUDIES

The specific observational studies chosen for presentation are illustrative examples only; the number of each type of such studies reported in the world's literature is extremely large and no attempt was made to provide a balanced, representative selection of such studies, nor an overview of them. The findings of each chosen study are generally congruent and consistent with the major tenets of the theory and provide a coherent "story", but it is not the purpose of this presentation to offer the study results as proof of the theory. Rather, in this context, they are confirmatory examples, chosen to illustrate the range of epidemiologic studies testing the theory, from the molecular to the community level.

Epidemiological observational studies confirmatory of the relation of dietary factors to serum lipid levels and CHD have been predominantly at the aggregate level (sometimes called ecologic studies). These are studies of population attributes, analyzing the association between average population indices of dietary practices , such as nutrient intake levels based on food balance or disappearance measures and average population levels of serum lipids, and population levels of CHD morbidity or mortality. In contrast, epidemiological studies at the level of individuals, relating the presence or absence of CHD in each person to her dietary and serum lipid attributes, have been more limited in number due to the technical difficulties in assessing usual dietary practices in large numbers of individuals. Investigations at the aggregate level of population attributes have studied co-variations of nutrients, serum lipids and CHD over time, place and population subgroups; they also have varied with the assessment of CHD as clinical manifestation, cause of death, or assessment at autopsy.

Among the earliest of the aggregate level studies was Keys' comparison of cross-national mortality rates with estimated nutrient consumption (Keys,1953). This disclosed increasing levels of mortality rates attributed to deaths classified as "Degenerative Heart Disease" in 1948-49 in association with increasing percentage of total calories consumed as fat. As an aside, it is noteworthy that atherosclerosis, now known as a metabolically active, modifiable process leading to end organ sequalae including coronary heart disease, fifty years ago was considered an inevitable, degenerative consequence of aging . Mortality increased monotonically with increasing fat consumption; moreover, the increase in degenerative heart disease mortality with increasing levels of fat consumption now appears in retrospect to be more than linear at higher levels of fat consumption. The possible departure from a linear relation, in this seminal work, was not originally a major consideration; it will be considered later in relation to subsequent studies. Succeeding analyses, as summarized by Stamler, of the relation between nutrition data and mortality patterns among nations, based on reports of the Food and Agricultural Organization and the World Health Organization, (Stamler,1979) indicated that ten such data sets all showed statistically significant positive associations of dietary total calories, total fat, animal fat, and saturated fat with CHD mortality rates and significantly negative correlations for vegetable protein, vegetable fat and total carbohydrate with CHD mortality rates. The magnitude of the correlation coefficients expressing the aggregate level associations between the dietary components identified above and CHD mortality rates were generally larger for men then women. Dietary cholesterol intake analyzed in the same mode also yielded positive simple correlation coefficients with CHD mortality rates for middle aged men. More recent work has combined saturated fat and cholesterol consumption and expressed their intake as a single index (Connor et al., 1986). When arrayed as a scattergram and summarized by functional relation, the association of CHD mortality rates with the cholesterol/saturated fat index is not dissimilar to the shape of the relation reported by Keys decades earlier.

Similar results have emerged from aggregate analyses of cohorts studied from 16 populations in seven countries relating the mean intake of dietary fat to the incidence to heart disease in these same groups (Keys, 1980). Although positive, aggregate level associations have consistently been found between CHD mortality and some nutrients (particularly saturated fat and cholesterol) and food groups (animal and dairy products), there has been controversy regarding the interpretation of the findings. Limitations in interpreting these findings in relation to the diet-lipid-CHD theory have been pointed out (Willet, 1990). For example, the observed association between the percent of dietary calories from saturated fat in relation to the ten-year coronary death rate was stronger than expected, if the effect of saturated fat were mediated only through the effect of raised serum cholesterol. This type of ecologic correlational study produces associations among aggregate level attributes of population groups and is limited in permitting inferences regarding cause and effect associations at the level of individuals. However, the consistency of results across the multiple types of epidemiologic studies and the additional types of investigations to be presented enhances the plausability of the detected associations as causal.

A relationship between percentage of calories from total fat in the habitual diets of populations and the presence and extent of atherosclerosis, determined at autopsy, (similar in kind to the findings relating diet to CHD mortality rates) has been reported from the International Atherosclerosis Project studies across 15 cities and countries (McGill, 1968). This study quantified the amount and extent of atherosclerosis of the aorta and coronary arteries and also reported significant aggregate level correlations between mean levels of serum cholesterol and severity of atherosclerosis of the populations studied.

The use of autopsy materials for epidemiologic investigations has the obvious and marked limitations imposed by the selective nature of the subjects available for study and analysis, and the pathologic end stage of the study material. Technologic development of methods for ultrasound interrogation of the arterial wall and computer assisted measurements of the intima-media has made possible the assesment of atherosclerosis in relation to candidate risk factors and determinants in large, representative population-based studies. For example, The Atherosclerosis Risk in Communities Study has reported on the association of habitual dietary intake with indices of atherosclerosis among 15,800 men and women aged 45-64 years sampled from four US communities. After adjusting for the possible confounding effects of age, energy intake, smoking and hypertension, the reported consumption of saturated fat, animal fat, monounsaturated fat, cholesterol and Key's score were positively related to carotid artery wall thickness (an indicator of atheroclerosis), while reported consumption of vegetable fat and polyunsaturated fat were inversely associated with carotid artery wall thickness (Tell et al., 1994). Studies of this type extend the range of findings consistent with, and predicted by, the diet-lipid-heart disease theory; the range of findings related to nutrient consumption now extends from mortality, morbidity, and autopsy evidence to earlier stages in pathogenesis, reflected in preclinical atherosclerosis.

Migration studies offer the opportunity to assess the impact of behavioral, life style and environmental changes upon population levels of disease; they have contributed important evidence regarding the diet- lipid- CHD theory. The comparison of coronary heart disease rates among Japanese living in Japan, Hawaii, and San Francisco is a prominent example of this type of study (Robertson et al., 1977). Aggregate level vital statistics studies indicated that coronary heart disease mortality rates for Japanese men were lowest among those resident in Japan and progressively higher among those resident in Hawaii and San Francisco. Studies performed on samples of Japanese men resident in these locations indicated that CHD prevalence, serum cholesterol levels, and dietary intake of cholesterol, total fat and saturated fat generally paralleled the mortality trends of higher levels in Hawaii and San Francisco than Japan. Follow-up mortality rates of individuals paralleled the aggregate level mortality findings and the individual level incidence findings. The characteristics studied as explanatory of the variations in findings by location of residence included adherence to Japanese, contrasted with western, life styles. Numerous social practice and behavioral changes are associated with westernization; among the characteristics in the migrants reflective of retention of Japanese life style and associated with relative CHD protection were dietary practices.

Serum cholesterol plays a central role in the diet- lipid-CHD theory, both conceptually as the intervening link in the postulated causal chain and also empirically as one of the strongest and most consistently observed correlates and predictors of coronary heart disease. Metabolic studies have defined the effect of nutrients on serum cholesterol levels. The effect of dietary cholesterol and fatty acids on blood cholesterol levels has been studied extensively and quantitative relations have been derived from metabolic and clinical research investigations in man (Keys, 1965). It has been difficult to confirm the quantitative funcional relations derived from these metabolic studies in population studies. One explanation for the divergent results often obtained in earlier population studies was the reliance on single day measurements to obtain estimates of usual dietary practices. The large magnitude of day to day intraindividual variation, compared with interindividual variation in dietary practices, made these studies ineffective tests of the effect of nutrients in studies with individuals as the unit of study. The difficulty in obtaining estimates of usual and cumulative dietary intake, given the limited number of diet relevent observations usually obtainable from each participant, remains a problem in large epidemiololgic studies.

In contrast, one of the most consistent findings across epidemiologic case-control, cross sectional and cohort incidence studies at the individual level has been the strong relationship consistently observed between total serum cholesterol levels and CHD incidence (Jacobs et al., 1992). This has been found among members of diverse populations from different cultures with varying average levels of CHD and at diferent periods of time. The largest of these studies has been the MRFIT screening follow-up study of 350,977 men. An exponentially increasing risk of CHD mortality in relation to serum cholesterol across a wide range of cholesterol levels was found in the MRFIT screenees and in the pooling of observations from a large number of other cohort studies (Tyroler, 1987); the relationship observed was not dissimilar in form to the type of functional relation observed in the aggregate level studies described previously. The association of total serum cholesterol with CHD is predominantly attributable to the low density lipoprotein-cholesterol component of total serum cholesterol (Tyroler, 1989). Several quantitative overviews in addition to those cited , summarizing the relation of total serum cholesterol to the incidence of CHD across observational epidemiologic studies, have been reported. One popular estimate is that each 1% change in serum cholesterol is associated with a 2% change in CHD. We shall return to this estimate, and suggestied modifications of it, after consideration of the evidence derived from experimental epidemiologic studies of interventions to modify serum cholesterol levels, i.e. the clinical trials evidence of the effect of lowering of serum cholesterol on CHD.

CLINICAL TRIALS OF SERUM CHOLESTEROL LOWERING WITH MORBIDITY AND MORTALITY END POINTS

Clinical trials provide human experimental evidence relevant to the diet-lipid-CHD theory. In contrast to the highly selective choice of a limited number of observational studies for this presentation, it is possible to discuss the outcome of virtually all the clinical trials designed to assess the experimental effect of lipid lowering on coronary heart disease. This is feasible as there have been far fewer clinical trials than observational studies. Additionally, the trial results have been summarized successively over time, using quantitative methods, as new trial results have accumulated. Despite the identification of a common set of trials for potential analysis, there have been differences among analysts in their choice of trials for overview analysis. Among the criteria relevant to this presentation is whether or not multiple intervention trials, in which diet was one among other interventions, were included in the overview of trials. This has resulted in controversial interpretations of the cumulative trials experience, and we shall refer briefly to the subject later.

An additional problem in evaluating the results of the diet trials is that there have been fewer dietary than pharmacologic trials which have tested the effect of reducing serum cholesterol on the risk of coronary heart disease. This has resulted in less stable estimates for the dietary effects. Among those that have been performed, the nature of the dietary interventions used have varied, for example from high polyunsaturated fatty acid diets (Dayton et al., 1969) to low total fat diets (Research Committee, 1965). Different interpretations have emerged among investigators and meta-analysts, reviewing the cumulative experience. However, the aggregate evidence derived from the experience of all trials and the comparison of dietary trials with trials using non-dietary means of reducing serum cholesterol does permit useful comparisons .

The effect of the amount of lowering of serum cholesterol, both by dietary and non-dietary means, on the frequency of coronary heart disease events has been quantitatively summarized by regression metaanalysis. One of the earliest quantitative overviews was attributed to Peto (Mann and Marr, 1981). In this early (five of the six trials were performed in the 1960's) quantitative overview of diet intervention trials, relatively little serum cholesterol reduction was achieved- serum cholesterol reduction in the diet trials ranged from 8 to 16%. Nevertheless, Peto reported an overall statistically significant association between the amount of cholesterol reduction achieved and the reduction in CHD in each trial . The meta-regression analysis indicated a 1.5% reduction in CHD for each 1% reduction in serum cholesterol in the diet trials. The same type of regression analysis of five early drug intervention trials, completed in the 1970s, in which the amount and range of serum cholesterol lowerering was small, varying only from 8 to 12%, indicated a 2.1% reduction in CHD for each 1% reduction in cholesterol . The combined estimates from these early clinical trials of both diet and drug interventions was an approximately 2% reduction in CHD for each 1% reduction in cholesterol, a value similar to that obtained from the quantitative overview of epidemiologic observational studies. The combined quantitative estimate of the effect of serum cholesterol lowering on CHD from these early diet and drug trials was also quantitatively predictive of the result of the Lipids Research Clinics Coronary Primary Prevention Trial (LRC-CPPT) (Tyroler,1987). There was an 8 1/2% greater reduction in serum cholesterol in the actively treated than the plecebo group in the LRC-CPPT, resulting in a predicted 17% reduction in primary end points based on the earlier meta-analyses; a 19% reduction was observed. Participants for the earlier trials were not selected based on hypercholesterolemia per se; by contrast, hypercholesterolemia due to elevation of plasma low density lipoprotein cholesterol , the atherogenic component of serum cholesterol, was a selection criterion for the LRC CPPT, which was a large, primary prevention trial.(Lipid Research Clinics Program, 1984a, 1984b). Following completion of the LRC CPPT and integration of its evidence of the efficacy of reducing CHD incidence by reducing LDL-cholesterol with the results of other trials and the totality of other forms of evidence regarding the causal nature of the cholesterol-CHD relation, the deliberations of an NIH Consensus Development Conference and recommendations of the National Cholesterol Education Program (NCEP) (Expert Panel on Detection, Evaluation and Treatment of High Blood Cholesterol in Adults, 1988,1993) lead to the goal of reducing CHD incidence and mortality in the US by lowering serum cholesterol. Dietary strategies are central to the NCEP recommendations, both for the high-risk approach targeted at individuals with marked elevations of serum cholesterol, and for the population approach of shifting the entire distribution of serum cholesterol to lower levels. The population approach is focused on the the large number of individuals with slight to moderate elevation of serum who contribute the largest population burden of CHD, and is designed to reduce risk among them primarily by dietary means.

A more recent meta-analysis of the aggregate experience of a larger number of individuals, studied in 22 trials, resulted in an estimate of a 2.5% reduction in coronary heart disease incidence associated with each 1% reduction in serum cholesterol (Holme, 1993). In this overview, the quantitative estimate of efficacy of dietary intervention in secondary prevention trials (in which participants with clinical evidence of pre-existent CHD were studied to prevent recurrent CHD events) was as great as for drug intervention. However, dietary intervention appeared less efficacious than drug interventions in primary intervention trials (in which partipants free of pre-existent CHD were studied to prevent the occurrence of first CHD events). In this context, it is important to note that the choice of primary versus secondary prevention trials for study and analysis has a major effect on the magnitude of the expected coronary event rate in the control group, on the expected mortality rate and on the proportion of fatal events which will be attributable to coronary heat disease. This follows from the study of higher risk individuals with prior evidence of CHD in secondary prevention trials, compared to the expected lower incidence of future CHD events among individuals free of clinical evidence of CHD in primary prevention trials.

A lag or latency period of approximately 18 months to two years before there is evidence of efficacy of serum cholesterol lowering in decreasing the incidence of CHD events had been observed in the LRC-CPPT (Lipid Research Clinics Program, 1984a) and in the Helsinki Heart Study (Frick et al., 1987) primary prevention drug trials. A recent meta-analysis (Law et al., 1994) indicates a positive quantitative relation between the duration of serum cholesterol reduction and its efficacy, i.e. an increase in efficacy with increasing duration of serum cholesterol reduction. When the diet trials were stratified by duration of cholesterol reduction during the trials into the intervals of less than two years, 2.1 to 5 years, and from 5 1/2 to 12 years, the percentage reduction in CHD events for each one percent reduction in serum cholesterol increased from 0.9% to 1.4% to 3.7% with increasing duration of cholesterol lowering. Efficacy and its relation to duration of cholesterol lowering was similar in the diet and the drug trials; in the drug trials the reduction in CHD associated with each one percent reduction in serum cholesterol increased from 1.0% to 2.1% to 2.2% as the duration of serum cholesterol increased. In this meta-analysis, the investigators compared the clinical trials results with the observational study findings, correcting for systematic causes of underestimation of the association between serum cholesterol lowering and coronary heart disease; this was achieved by adjustment for regression dilution and low density lipoprotein cholesterol effects. This overview of the clinical trials results indicated little effect of cholesterol lowering in the first two years, but that full reduction in risk was achieved within five years. The summary estimate of a 2.5% reduction in CHD risk for each 1% reduction in serum choleserol was similar to the estimate of an expected 2.7% effect, derived from the adjusted associations in the observational epidemiologic cohort studies.

There is consistent evidence from these meta-analyses that CHD outcomes are reduced by amounts proportionate to the amount of serum cholesterol reduction achieved during the trials of both dietary and non-dietary interventions, for both primary and secondary prevention of CHD events. As noted above, there is disagreement among meta-analysts about whether the efficacy of serum cholesterol lowering is greater for pharmacologic than for dietary intervention; the analyses of Holme suggest there is (Holme, 1993), those of Law et al suggest similarity of effect, controlling for duration of lipid lowering (Law et al., 1994). Given the differences in study inclusion criteria, differences in analytic approaches, the relatively few trials performed with diet as the single intervention and the small number of their study participants, the ability to draw firm conclusions is limited.

A significant decrease in all cause mortality has not been reported in the unadjusted meta-analyses of the aggregate experience of clinical trials of lipid lowering to date; rather, various analysts have reported either no change or slight increases in all cause mortality. The attained benefit of CHD reduction in the trials has been offset by an observed increase in non-cardiovascular disease deaths, which has been attributed by some to chance occurrence (Collins et al., 1992; Stamler et al., 1993; LaRosa,1993), or by others to possible iatrogenesis. An example of the latter was the reported excess of violent deaths, including the aggregate of homocides, suicides and accidental deaths in the primary prevention trials (Muldoon et al., 1990). Any increase of non-cardiovascular deaths has been most evident in the primary prevention trials. There is less evidence of excesses in non-cardiovascular events in secondary prevention trials, in which most of the deaths are cardiovascular in nature, as the baseline risk of deaths associated with CVD events is much higher. These findings of potential iatrogenesis, as controversial as they are overall, are less evident in diet trials. There also is no evidence of excesses of non-cardiovascular deaths in non-pharmacologic, multiple intervention studies, which include dietary intervention and often are excluded from meta-analyses to determine the effects of lipid lowering. (Stamler et al., 1993). Thus, it would appear from the clinical trials experience that, if there are harmful effects associated with blood cholesterol lowering, iatrogenesis is more likely atributable to the means of achieving the lowering, such as specific drug toxicity, rather than the effects of low cholesterol levels, per se.

CLINICAL TRIALS OF SERUM CHOLESTEROL LOWERING WITH ANGIOGRAPHIC END POINTS

The clinical trials just discussed were designed to assess the effect of serum cholesterol lowering upon the clinical manifestations of coronary heart disease as their primary end points. Study of the relationship between serum cholesterol lowering and progression or regression of atherosclerosis has become possible in clinical trials by the use of repeat angiography. The angiographic studies have focused primarily on the coronary arteries, although changes in other vascular territories, such as the femoral and carotid arteries, have also been explored. The angiographic trials also provide surrogate measures, as putative indicators of the ultimate effect on clinical events. The endpoints in these studies can be quantitative, based on computer assisted measures of arterial lumen parameters. Among the advantages of the quantitative endpoints of angiographic trials are the much smaller samples of individuals and shorter duration of time required to observe changes, contrasted with the much larger number of participants and longer periods of observation required for trial outcomes based on morbid and fatal endpoints.

Advances in computer assisted coronary angiography have permitted more objective, sensitive, and reliable quantitative assessments of vessel changes than previously were possible based on subjective, qualitative evaluations. Further, the angiographic studies initiated more recently have been able to test the effects of greater blood lipid lowering, given the current availability of potent pharmacologic agents and, in at least one trial, the marked lipid lowering achieved by ileal bypass surgery. In contrast to the average of approximately 10% serum cholesterol lowering achieved in the earlier trials, 25 to 40% lowering is presently reported. Among the disadvantages of the angiographic trial, in addition to its invasive nature, is the surrogate index nature of its end point. The estimate of coronary artery atherosclerosis, although correlated with extant coronary heart disease and predictive of future coronary heart disease, is not a direct measure of myocardial disease. Additionally, quantitative angiographic measures assess lumen encroaching lesions; changes within the vessel wall per se cannot be assessed by these methods, only the interference with complete filling of the lumen of the artery by the visualized dye.

Despite there limitations, the cumulative experience with clinical trials using coronary angiography has shown unequivacal evidence of stablization of lesions due to serum cholesterol lowering and in some instances suggestive evidence of slight regression of lesions(Brown et al., 1993a). Of particular interest has been the surprising emergence of statistically significant reductions in clinical events, which were secondary endpoints in the angiographic trials. Among the probable explanations for the remarkable reductions of clinical events, reaching statistical significance in trials of short duration with small numbers of participants, is the fact that most of these have been secondary prevention trials of patients at high risk of repeat events. Additionally, the amount of lipid lowering achieved in the angiographic trials has been considerably greater than that achieved in the earlier trials, which had morbidity and mortality as outcomes.

Considerable interest also has been stimulated by the finding that the reduction in clinical coronary events in the angiographic trials has been of magnitude considerably greater than the amount of change quantitated in the coronary arteries. It has been hypothesized that marked serum cholesterol reduction results in stabilization of relatively small, lipid-rich plaques, which otherwise would have been at risk of surface ulceration giving rise to thrombotic potential for catastophic clinical sequelae (Brown et al., 1993b).

The number of completed angiographic trials assesing diet therapy and the number of subjects exposed to this intervention have been few (Brown et al., 1993a). However, the favorable angiographic findings in the diet trials are similar to those resulting from non-dietary intervention.

The use of clinical trials methodology has permitted a rigorous, ethical means of obtaining experimental evidence, both qualitative and quantitative, of the efficacy of both dietary and non-dietary means of reduction of serum cholesterol to favorably modify the evolution of coronary atherosclerosis and reduce the risk of clinical CHD events. This is of obvious importance and utility for clinical practice and public health programs. In addition to its pragmatic value, the derivation of experimental evidence from clinical trials that lowering serum cholesterol modifies the course of coronary atherosclerosis and lowers the incidence of CHD also adds convincing evidence of the causal role of serum cholesterol in the etiology and pathogenesis of CHD. This is particularly compelling as the evidence of efficacy in the trials derives from markedly different means of cholesterol lowering including different diets, drugs of different modes of action, and illeal bypass surgery. Placing the results of clinical trials of serum cholesterol lowering in their epidemiologic context rounds out the evidence mandating clinical and community control programs (Peto et al., 1985).

SURVEILLANCE OF CLINICAL PRACTICE, POPULATION BEHAVIOR AND HEALTH MEASURES RELATED TO DIET AND CHOLESTEROL

As indicated above, review and evaluation of the totality of scientific evidence regarding the relation of diet, serum cholesterol and coronary heart disease has led to consensus statements regarding the causal nature of the associations and recommendations for treatment and prevention. Diet therapy in the treatment of individuals with elevated serum cholesterol and dietary recommendations for all individuals to achieve more favorable levlels in the blood cholesterol distribution of the total population remain the cornerstones of the programs (Cholesterol Adult Treatment Panel, 1988; NCEP Adult Treatment Panel 2 Report, 1993). The general dietary approach is to reduce total fat intake by reduction of saturated fatty acids, to reduce intake of dietary cholesterol, to reduce total caloric intake and increase physical activity leading to weight loss in those who are overweight. These recommendations are a codification and extension of similar proposals previously advocated by groups such as the American Heart Association Council on Nutrition and extend decades-long efforts in this and other countries. Epidemiologic surveillance permits an assessment of the impact of these programs on clinical practice, general population behavior and health status.

The acceptance of clinical trials results by academic cardiologists and their recommendations for practitioners regarding lipid lowering treatment lagged behind the available scientific evidence, until recently. An illustration of this was evident on comparison of the results of meta-analyses of randomized control trials with the published recommendations of clinical experts (Antman et al., 1992). Cumulative meta-analyses were applied for this purpose. Cumulative meta-analyses were performed by carrying out updated meta-analyses over time, as the results of each new clinical trial were published (Lau et al., 1992). This provided a measure of the strength of the cumulative evidence regarding the efficacy of serum cholesterol lowering, as the data accumulated at successive points in time, from the 1960s to the early 1990s. There was suggestive, but not statistically significant, evidence of efficacy of treatment effect as the results of secondary prevention trials accumulated until 1985. In 1985 there was highly statistically significant evidence of efficacy of serum cholesterol lowering in the secondary prevention of CHD . However, there was little evidence of treatment recommendations by experts in medical textbooks and reviews through 1990; this indicated a more than five-year lag between highly statistically significant evidence of efficacy of cholesterol reduction in the secondary prevention of coronary disease by dietary or pharmacologic means of cholesterol lowering and its mention in textbooks, manuals, or review articles related to therapies following myocardial infarction.

In contrast, however, comparison of national surveys of practicing physicians and the adult public indicated a more general increasing awareness of the issues related to serum cholesterol and professed actions in relation to it (Schucker et al., 1991). Both medical practice and the general public's health behavior were reported as increasingly following the recommendations for control of elevated cholesterol. The surveys were performed in 1983, 1986 and 1990. At those times, the physician reported median ranges of blood cholesterol levels chosen to initiate treatment decreased from 340-359 to 300-319 to 240-259 mg/dl. Between 1983 and 1990 the percentage of adults reporting a physician diagnosis of high serum cholesterol increased from 7% to 16% and the number reporting a prescribed cholesterol-lowering diet increased from 3% to 9%. Reports of self-initiated diets increased from 1% to 15%. It was reported that: "By 1990, over 90% of physicians reported awareness and use of the National Cholesterol Education Program Expert Panel on Detection, Evaluation, and Treatment of High Blood Cholesterol in Adults and the public reported marked increases in awareness of dietary methods to lower serum cholesterol."

Long term trends in the population's estimated food consumption are of obvious epidemiologic importance, however there have been methodologic problems in obtaining consistent estimates of them. For example, trends in individual consumption of dietary fat in the U.S. over the period 1920-1984 were estimated based on a quantitative regression analysis review of 171 studies (Stephen and Wald, 1990). This approach indicated that fat intake rose from 34% of energy in the 1930s to 40-42% in the late 1950s to mid 1960s, and then fell steadily to approximately 36% of energy in 1984. Among the components of fatty acids, saturated and monounsaturated fatty acid intake fell from 18 to 20% of energy in the early 1950s to 12-13% of energy in 1984, while polyunsaturated fatty acid intake rose from 2-4% energy to 7.5%. These trends are favorable in relation to the known physiologic relationship between type and amount of fat consumed and serum cholesterol concentration, and the relationship of serum cholesterol to CHD. Interestingly, the attempt to estimate individual food consumption in these analyses resulted in findings differing from estimates based on food supply figures, discussed earlier. The food supply figures indicated that consumption of total fat changed little since the 1960s but that there was a decline in the consumption of animal fat and a rise in the consumption of vegetable fat; the latter findings similar to those derived from the individual consumption estimates. It was postulated that one source of the difference in estimated trends was that the food balance estimates do not account for changes in wastage occurring over time, at all levels of the food supply from production to sale to wastage in the home. The analyses of time trend data derived from estimates of individual consumption are of interest in relation to the time trend data for serum cholesterol and coronary heart disease mortality levels.

There have been declines both in the mean serum cholesterol levels in the US adult population since 1960 (Johnson et al., 1993) and since 1988 in the proportion of individuals with levels of cholesterol classified as high based on the guidelines of the National Cholesserol Education Program Adult Treatment Panel (Sempos et al., 1993). Between 1988 and 1991 the proportion of adults with high serum choleterol, defined as (>240 mg/dl) declined 6% and the proportion requiring diet therapy decreased from 36% to 29%. Despite this improvement, it was estimated that approximately 52 million adults were candidates for dietary therapy. Comparisons based on the nationally representative surveys performed in 1960-62,1971-1974, 1976-1980, and 1988-1991 indicated consistent declines across the entire distribution of serum cholesterol levels and in all age-sex groups. Indirect evidence suggested that the decline in total serum cholesterol was due to decline in low density lipoprotein cholesterol, the atherogenic component.

The long term curvilinear, rising and then falling, pattern of dietary fat consumption based on surveys of individuals is similar to the long-term epidemic curve of coronary heart disease mortality in the United States. There also is consistency of the mortality trends in the US with the decreases in mean and elevated levels of serum cholesterol since 1960.

CHD mortality rose in the US until the mid 1960s, followed by plateauing, and has declined approximately 2% per year since then. This general pattern was shared across race, gender, and age groups. However, it is important to note that there have been differences in the timing of the onset of the decline and its subsequent magnitude among demographic subgroups in the United States (Davis et al.,1985). Since the onset of the decline, there has been the emergence of an inverse relation of CHD mortality to indices of socioeconomic status; the higher the socioeconomic status, the lower the CHD rates (Wing et al., 1987; Tyroler et al., 1993). The inverse gradient has been observed in cohort studies of incidence of CHD as well as in vital statistics studies of CHD mortality. The inverse gradient of CHD with socioeconomic status during the decline contrasts with a positive association during the ascending limb of the long term CHD epidemic curve. The gradient of CHD risk in relation to socioeconomic status is associated with differences in the traditional risk factors. However, analyses of their contribution suggests that the major, strong, individual risk factors, including serum cholesterol, explain only a portion of the socioeconomic gradient.

CONCLUSIONS

The original formulation of the diet-lipid-CHD theory postulated that elevated total serum cholesterol was the single most important link between diet and atherosclerosis. Developments in molecular and genetic lipidology and epidemiology have led to recognition of the importance of the distribution of the components of total serum cholesterol among the lipoprotens, as well as the apolipoproteins themselves and lipids other than cholesterol. The complex relationships of dietary components such as the different fatty acids, antioxidants, and fibers to these lipids and lipoproteins, and their possible independent relation to atherosclerosis, thrombosis, and clinical coronary events can not be addressed here. As stated at the outset of this presentation, CHD epidemiology has developed a multi-causal model, however with the diet-lipid-heart disease theory as its centerpiece. Perhaps the most succint summary and formulation of the importance of diet to the epidemiology of coronary heart disease was that of Stamler, when he characterized the causal diet as "rich" : "Rich diet is habitual fare high in animal products and processed animal products high in cholesterol and saturatd fat. High in refined and processed sugars. High in salt. High in alchohol for many in the population. High in caloric density and empty calories and in ratio of calories to essential nutrients. Low in potassium, fiber and often other essential nutrients and high in total calories for a low level of energy expenditure in the era of the automobile, television and mechanized work" (Stamler, 1979). This quotation vividly illuminates the interrelated, multifactorial nature of the dietary and nutrient components determining serum cholesterol levels, atherosclerosis and clinical cardiovascular events in populations. The metaphor of a "rich diet" was highly descriptive during the ascending limb of the CHD epidemic: cross-national studies indicated that affluent, developed societies had highest CHD mortality rates; individual level studies indicated that CHD rates were higher the higher the socioeconomic status. By contrast, with the onset of the decline in CHD mortality an inverse relationship of CHD mortality with socioeconomic status has emerged.

Stamler's summary also implies dietary effects on CHD risk additional to blood cholesterol effects, and the interrelatedness of diet to a multitude of risk factors other than blood cholesterol, such as hypertension, obesity and diabetes. Epidemiologic studies, however, indicate that other strong risk factors such as smoking do not lead to high population levels of CHD, if blood cholesterol levels are low in that population. Shift of the full distribution of blood cholesterol values towards higher values is a necessary population condition for epidemic coronary heart disease, and diet is one of the major determinants of that distribution.

REFERENCES

Antman, E.M., Lau, J., Kupelnick, B., Mosteller F. and Chalmers, T.C., 1992, A comparison of results of meta-analyses of randomized control trials and recommendations of clinical experts: treatments for myocardial infarction. JAMA. 268:240-248.

Brown, B.G., Zhao, X.Q., Sacco, D.E. and Albers, J.J., 1993a, Arteriographic view of treatment to achieve regression of coronary atherosclerosis and to prevent plaque disruption and clinical cardiovascular events. Br. Heart J. 69(Suppl.) S48-S53.

Brown, B.G., Zhao, X.Q., Sacco, D.E. and Albers, J.J., 1993b, Lipid lowering and plaque regression. New insights into prevention of plaque disruption and clinical events in coronary disease, Circulation 87:1781-91.

Cholesterol Adult Treatment Panel, 1988, Report of the National Cholesterol Education Program Expert Panel on Detection, Evaluation, and Treatment of High Blood Cholesterol in Adults, Arch. Intern. Med. 148:36-69.

Collins, R., Keech, A., Peto, R. and Sleight, P., 1992, Cholesterol and total mortality: need for larger trials, B. Med. J. 304:1689.

Connor, S.L., Artaud-Wild, S.M., Classick-Kohn, C.J., Connor, W.E., Gustafson, J.R., Flavell, D.P. and Hatcher, L.F., 1986, The cholesterol/saturated fat index: an indication of the hypercholesterolemic and atherogenic potential of food. Lancet 2:1229-32.

Consensus Conference: Lowering blood cholesterol to prevent heart disease, 1985, JAMA 253:2080-6.

Davis, W.B., Hayes, C.G., Knowles M., Riggan, W.B., VanBruggen and J. Tyroler, H.A., 1985, Geographic variation in declining Ischemic Heart Disease mortality in the United States, 1968-1978, AJE 122:657-672.

Dayton, S., Pearce, M.L., Hashimoto, S., Dixon, W.J. and Tomiyasu, U., 1969, A controlled clinical trial of a diet high in unsaturated fat in preventing complications of atherosclerosis. Circulation 39 and 40 (Suppl. 2) 1-63.

Frick, M.E., Elo, O. and Haapa, K., 1987, Helsinki heart study. Primary prevention of trial with gemfibrozil in middle-aged men with dyslipidemia. Safety of treatment, changes of risk factors and incidence of coronary heart disease. N. Eng. J. Med. 317:1237-45.

Holme I., 1993, Relation of coronary heart disease incidence and total mortality to plasma cholesterol reduction in randomized trials: Use of meta-analysis. Br. Heart J. 69 (Suppl.) S42-S47.

Jacobs, D., Blackburn, H., Higgins, M., D. Reed, Iso, H., McMillian, G., Neaton, J., Nelson, J., Potter, J., Rifkind, B., Rossouw, J., Shekelle, R. and DPhil, S.Y., Report of the conference on low blood cholesterol: Mortality associations, 1992, Circulation 86:1046-60.

Johnson, C. L., Rifkind, Basil M., Sempos, C. T., Carroll, M.D., Bachorik, P.S., Briefel, R., Gordon, D.J., Burt, V.L., Brown, C.D., Lippel, K. and Cleeman, J.I., 1993, Declining serum total cholesterol levels among US adults. The National Health and Nutrition Examination Surveys, JAMA 269:3002-08.

Keys, A., 1953, Atherosclerosis: a problem in newer public health, J. Mt. Sinai Hosp. 20:118-139.

Keys, A., Anderson, J.T. and Grande F., 1965, Serum cholesterol response to changes in the diet II. The effect of cholesterol in the diet. Metabolism 14:759-65.

Keys, A., 1980, Seven Countries. A Multivariate Analysis of Death and Coronary Disease, Harvard University Press. Cambridge.

LaRosa, J.C., 1993, Cholesterol Lowering, Low Cholesterol, and Mortality, Am. J. Cardiology 72:776-86.

Lau, L., Antman, E.M., Jimenez-Silva, J., Kupelnick, B.A., Mosteller, F. and Chalmers, T.C., 1992, Cumulative meta-analysis of therapeutic trials for myocardial infarction, New Engl. J. Med. 327:248-254.

Law, M.R., Wald, N.J. and Thompson, S.G., 1994, By how much and how quickly does reduction in serum cholesterol concentration lower risk of ischaemic heart disease, Br. Med. J. 308:367-68.

Lipids Research Clinics Program, 1984a, The lipids research clinics coronary primary prevention trial results. 1. Reduction in incidence of coronary heart disease. JAMA 251:351-64.

Lipids Research Clinics Program, 1984b, The lipids research clinics coronary primary prevention trial results. 2. The relationship of reduction in incidence of coronary heart disease to cholesterol lowering. JAMA 251:365-79.

Mann, J.J., and Marr, J.W., 1981, Coronary heart disease prevention trials of diet to control hyperlipidemia, in "Lipoprotein Atherosclerosis and Coronary Heart Disease", N.E. Miller and B. Lewis, eds., Elsevier/North Holland Biomedical Press, Amsterdam.

McGill, Jr., H.C., 1968, Geographic Pathology of Atherosclerosis, Williams and Wilkins, Baltimore.

Muldoon, F., Manuck, S.B., Mathews, K.A., 1990, Lowering cholesterol concentrations and mortality: a quantitative review of primary prevention trials. Br. Med. J. 301:309-14.

18

NCEP Adult Treatment Panel II Report. 1993. Summary of the Second Report of the National Cholesterol Education Program (NCEP) Expert Panel on Detection, Evaluation, and Treatment of High Blood Cholesterol in Adults (Adult Treatment Panel II). 1993. JAMA 269:3015-23.

Peto, R., Yusuf S. and Collins, R.. 1985. Cholesterol-lowering trial results in their epidemiologic context Circulation 72 (Suppl. III) 451.

Ravnskov, U.. 1992. Cholesterol lowering trials in coronary heart disease: frequency of citation and outcome Br. Med. J. 305:15-19.

Research Committee. 1965. Low fat diet in myocardial infarction - A controlled trial. Lancet 2:501-04.

Robertson, T.L., Kato, H., Rhoads, G.G, Kagan, A., Marmot, M., Syme, S. L., Gordon, T., Worth, R. M., Belsky, J. L., Dock, D. S., Miyanishi, M. and Kawamoto, S.. 1977. Epidemiologic studies of coronary heart disease and stroke in Japanese men living in Japan, Hawaii, and California. Incidence of myocardial infarction and death from coronary heart disease. Am. J.Cardiology 39:239-43.

Sempos, C.T., Cleeman, J.I., Carroll, M.D., Johnson, C.L., Bachorik, P.S., Gordon, D.J., Burt, V.L., Briefel, R.R., Brown, C.D., Lippel, K. and Rifkind, B.M., 1993. Prevalence of high blood cholesterol among US adults. JAMA 269:3009-14.

Schucker, B., Wittes, J.T., Santanello, N.C., Weber, S.J. McGoldrick, D., Donato, K., Levy, A. and Rifkind, B.M., 1991. Change in cholesterol awareness and action. Results from national physician and public surveys. Arch. Intern. Med. 151:666-73.

Smith, G.D., Song, F. and Sheldon, T.R., 1993. Cholesterol lowering and mortality: the importance of considering initial level of risk. Br. Med. J. 306:1367-73.

Stamler, J., 1979. Population Studies in "Nutrition, Lipids, and Coronary Heart Disease". R. Levy, B. Rifkind, B. Dennis and N. Ernst, eds., Raven Press, New York.

Stamler, J., 1992. Established major coronary risk factors in "Coronary Heart Disease Epidemiology", M. Marmot and P. Elliott, eds., Oxford University Press, Oxford.

Stamler, J., Stamler, Rose, Brown, V., Gotto, A.M., Greenland, P., Grundy, S., Hegsted, D.M., Luepker, R.V., Neaton, J.D., Steinberg, D., Stone, N., Van Horn, L., Wissler, R.W., 1993. Serum cholesterol, doing the right thing. Circulation 88:1954-60.

Stephen, A.M., Wald, N.J., 1990. Trends in individual consumption of dietary fat in the United States, 1920-1984 Am. J Clin. Nutr. 52:457-69.

Tell, G.S., Evans, G.W., Folsom A.R., Shimakawa, T., Carpenter, M.A., Heiss, G., 1994. Dietary fat intake and carotid artery wall thickness: The atherosclerosis risk in communities (ARIC) study, AJE 139:979-89.

Tyroler, H.A., 1987. Review of lipid-lowering clinical trials in relation to observational epidemiologic studies. Circulation 76:515-22.

Tyroler, H.A., 1989. Overview of clinical trials of cholesterol lowering in relationship to epidemiologic studies. Am. J. Med. 87 (Suppl. 4A) 14S-19S.

Tyroler, H.A., Wing, S., Knowles, M., 1993. Increasing inequality in coronary heart disease mortality in relation to educational achievement. Ann. Epid. 3:S51-S54.

Ulbright, T.L., Southgate, D.A.T., 1991. Coronary heart disease: seven dietary factors. Lancet 338:985-92.

Willet, W., 1990. Nutritional Epidemiology. Oxford University Press, Oxford.

Wing, S., Casper, M., Riggan, W., Hayes, G., Tyroler, H.A., 1988. Socioenvironmental characteristics associated with the onset of decline in ischemic heart disease mortality in the United States, AJPH 78:923-926.

OBESITY, FAT PATTERNING AND CARDIOVASCULAR RISK

June Stevens

Departments of Epidemiology and Nutrition
School of Public Health and School of Medicine
University of North Carolina at Chapel Hill
Chapel Hill, NC 27599-7400

INTRODUCTION

The hypothesis that where adipose tissue is located on the body is an important predictor of health has been the subject of intense research over the past 10 to 15 years. The concept of an "apple shaped" body being associated with higher risk than a "pear shaped" body has appeared in the lay literature, and as a result, questions concerning fat distribution are being directed to physicians and other health care professionals. The purpose of this paper is to summarize the history and major findings in the area of fat distribution, and to point out current controversies and future directions of research.

HISTORY OF RESEARCH ON FAT DISTRIBUTION

In 1956 a French physician at the University of Marseille, Dr. John Vague, published his observation that patients with greater upper body, as compared to lower body obesity, were more prone to certain disorders or diseases. Vague's work is often cited as the study to show relationships between fat distribution and morbid outcomes. To measure body types he developed the "index of masculine differentiation", calculated as the sum of ratios using adipose tissue and muscle thickness measures at the nap of the neck, the sacrum, the arm and thigh.(Vague, 1956). Vague termed subjects with relatively more upper body fat "android" and those with relatively more lower body fat "gynoid". He observed that men were more android than women, that the obese were more android than the nonobese, that diabetics were more android than non-diabetics and that older individuals are more android than younger.

It was approximately 25 years later that fat distribution again became of interest to the scientific community. Researchers in Gothenburg, Sweden and Milwaukee, Wisconsin lead the way to the "second wave" of research in fat distribution (Seidell, 1992). They, and other have confirmed Vague's original observations and other relationships have been uncovered.

Nutrition and Biotechnology in Heart Disease and Cancer
Edited by J.B. Longenecker *et al.*, Plenum Press, New York, 1995

21

CORRELATES OF FAT DISTRIBUTION

It is now clear that fat distribution is correlated with several cardiovascular disease (CVD) risk factors both physiologic and behavioral (Bjorntorp, 1993; Gillum, 1987; Kissebah et al., 1989). Fat distribution is directly correlated with diastolic and systolic blood pressure, plasma triglycerides, plasma cholesterol, plasma glucose and insulin, and inversely correlated with HDL cholesterol levels. Fat patterning is more adroid in those who smoke cigarettes (Shimokata et al., 1989) and in those who engage in less activity (Seidell et al., 1991; Tremblay et al., 1990). The upper or abdominal fat pattern has also been associated with alcohol intake and with stress (Larsson et al., 1989; Lundgren et al., 1989).

Vague pointed out cross-sectional associations between fat distribution and age in his early studies. Now there is evidence from longitudinal studies that the relative amount of abdominal fat increases with aging. In the Charleston Heart Study (Stevens et al., 1991) we found that with 25 years of aging and no change in body weight that men had an increase in abdominal girth of 6.7 cm, while women increased 4.6 cm.

Other results from the Charleston Heart Study indicate that there may be differences in body shape in black and white women (Stevens et al., 1993). On the average, black women are heavier than white women and have larger circumference measurements. In the 557 women studied in the 1963 examination of the Charleston Heart Study, the mean BMI of black women was 27.8 kg/m2 compared to 24.7 kg/m2 in white women ($p < 0.001$). Abdominal circumference was greater in black women as would be predicted (92.3 cm for black women vs 88.9 cm, $p < 0.001$ for white). However, after correcting for BMI, abdominal girth was actually larger in black women than in white women (88.9 for black women vs 92.2 for white, $p < 0.05$). Further, the ratio of abdomen to midarm circumference was larger in white women than in black women ($p < 0.05$), and could be interpreted to indicate a less central fat pattern in black women. This finding was unexpected, but very strongly illustrated in separate analyses of women who smoked and non-smokers. These results indicate that the increased risk for cardiovascular mortality in black versus white women (Keil et al., 1993) cannot be attributed to differences in relative distribution of adipose tissue. Nevertheless, black women may have a greater absolute amount of abdominal adipose tissue due to their increased levels of overall adiposity.

LONGITUDINAL STUDIES OF REGIONAL FAT DISTRIBUTION

Several cohort studies that have examined fat patterning as a predictor of mortality and morbidity over substantial time periods. The first longitudinal studies on this topic came from cohorts in Gothenburg, Sweden. A series of papers showed that waist-to-hip ratio was a risk factor for CVD (Lapidus et al., 1984; Larsson et al., 1984), stroke (Larsson et al., 1984), and non-insulin dependent diabetes (Lundgren et al., 1989; Ohlsson et al., 1985). The predictive power of waist-to-hip ratio was comparable to that of other established risk factors and was greater than that of BMI. These observations have subsequently been confirmed in analyses of the Framingham study (Stokes and Garrison, 1985), Honolulu Heart Study (Donahue et al., 1987), policemen in Paris (Ducimetiere et al., 1986), Mexican Americans (Stern et al., 1990), American Veterans (Terry et al., 1992) and Japanese-American men (Bergstrom et al., 1990).

Few studies have examined fat distribution as a predictor of disease in blacks (Stevens et al., 1992a, 1992b; Terry et al., 1992). To our knowledge, there has been only one study to examine fat patterning as a predictor of mortality in black women (Stevens et al., 1992a). We recently reported results from a 25-year follow-up of black and white women in the Charleston Heart Study. After controlling for the effects of BMI, the risk of all cause mortality was greater in white women with larger abdominal girths, while larger midarm girths were protective. The hazard at the 85th relative to the 15th percentile of abdomen/midarm ratio was 1.44 in models that included BMI, education and smoking as covariates. The girth measurements were not significantly predictive of CHD mortality in the white women. In the black women, the girths were not predictive of either all cause or CHD mortality, before or after controlling for BMI and the hazard estimate was close to 1.0. The failure of BMI and fat patterning to predict mortality in black women challenges assumptions regarding the role of fat patterning in the higher mortality rates experienced by black women.

More research is needed to define the role of fat patterning in disease, and to examine the consequences to health of racial or ethnic differences in body build and shape. Further, the relationship between total fatness and fat patterning to mortality in minority groups needs further exploration and explication.

MEASUREMENT OF FAT DISTRIBUTION

In epidemiologic studies, body fat distribution has been measured using a variety of indicies. Several ratios have been investigated including waist/hip circumference (Haffner et al., 1987; Hartz et al., 1984; Kalkhoff et al., 1983; Krotkiewski et al., 1983; Lapidus et al., 1984; Larsson et al., 1984; Ohlson et al., 1985), waist/thigh diameter (Ashwell et al., 1985) subscapular/triceps skinfold (Haffner et al., 1987) and abdominal circumference/height (Kannel and Gordon, 1979). In addition, principle components analysis has been used to construct fat distribution indices (Mueller et al., 1986; Reichley et al., 1987). Saggital diameter, the height of the abdomen at the umbilicus or at L3 or L4 with the subject supine has been used in some studies (Kvist et al., 1988).

Of all the measurements used, waist-to-hip ratio has been the most popular index of fat distribution. However, standardization of the anatomical location of the measurements in the index is lacking. The hip measurement used by one investigator may be referred to as the waist by another.

At present there is no consensus on how the anthropometric index of fat patterning is best constructed. New approaches to this problem have been made possible by the introduction of computed tomography and magnetic resonance imaging which can differentiate between subcutaneous fat and visceral or intra-abdominal fat. Although circumferences and ratios provide convenient methods to evaluate the abdominal distribution of body fat, it is the mass of intra-abdominal fat that appears to be most closely associated statistically to diabetes and CVD risk factors (Bjorntorp, 1990; Kissebah et al., 1982). Several attempts have been made to systematically explore the use of anthropometric measures to develop a multiple regression equations to estimate intra-abdominal fat (Baumgartner et al., 1988; Busetto et al., 1992; Despres et al., 1991; Ferland et al., 1989; Koester et al., 1992; Seidell et al., 1988; Seidell et al., 1987; Weits et al., 1988). One of the more recent studies is that by Keoster et al. (1992) in which intra-abdominal fat and a series of more easily obtained anthropometric measurements were investigated in a sample of 61 male Caucasian subjects. They found that waist circumference and log chest ratio (the logarithm of the chest circumference divided by the chest skinfold thickness) accounted for 67% of the variance in intra-abdominal fat. A validation of the regression equation in a separate sample of 21 similar men resulted in an even higher R^2 measurement of 0.79.

Although estimation of intra-abdominal fat by easily obtained measurements remains flawed, future work may produce better prediction equations. Also, despite the lack of very high correlations with intra-abdominal fat, simple anthropometric measurements of fat distribution have revealed numerous relationships with risk factors and disease.

MECHANISM OF FAT PATTERNING

Intra-abdominal adipose tissue has metabolic characteristics that are different from that of adipose tissue from other cites. These differences seem to be most pronounced in the regions that are drained by the portal circulation (Bjorntorp, 1990). These "portal adipose tissues" have a sensitive system for the mobilization of free fatty acids due to a preponderance of beta-adrenergic receptors and little alpha-adrenergic inhibition. This is seen in normal men and in abdominally obese women, but not in normal women or obese women with less upper body compared to lower body fat (Rebuffé-Scrive et al., 1989). This observation has served as the cornerstone of a proposed mechanism of the action of fat patterning (Bjorntorp, 1991; Kissebah et al., 1989).

The hypothesis has been advanced (Bjorntorp, 1991; Kissebah et al., 1989) that the responsiveness of intra-abdominal fat to lipolytic agents results in increased lipolysis with venous drainage of the released free fatty acids directly to the liver. These fatty acids may contribute to increases in triglyceride synthesis, and hyper-insulinemia secondary to decrements in insulin degradation. The hyper-insulinemia could eventually produce non-insulin dependent diabetes in susceptible individuals. It has been proposed that the hyper-insulinemia could also lead to increased blood pressure through increased sympathetic stimulation of the vessels, heart and kidneys. In addition, insulin resistance along with a relative increase in androgenic activity could lead to an unfavorable lipid profile. Observational research has supported the notion that a relative increase in androgenic activity is associated with a decrease in the size and number of gluteal-femoral adipocytes relative to abdominal adipocytes (Kirschner et al., 1990; Peiris et al., 1989).

The proposed mechanism for the action of intra-abdominal fat may need to be revised in response to recent studies which question some of the basic tenets of the theory (Seidell, 1992). First, Jensen et al. (1989) have shown in *in vivo* studies that women with high waist-to-hip ratios had lower whole body epinephrine-stimulated lipolysis compared to women with low waist-to-hip ratios. These results are in direct opposition to what was expected given the *in vitro* results of others (Rebuffe-Scrive et al., 1985, 1989). Also, intervention studies have shown that testosterone administration decreases visceral fat accumulation in men (Marin et al., 1992; Seidell et al., 1990). This is not what would be expected given the results of studies in women showing increases in upper body fat in women with higher testosterone levels (Kirschner et al., 1990; Peiris et al., 1989). These data challenge the hypothesis that an androgenic hormone profile contributes to a more "male type" fat pattern and the associated metabolic sequelae.

Hypotheses for the mechanism of action of fat distribution on morbidity and mortality will need to be reevaluated with these and other new findings in mind.

CLINICAL IMPLICATIONS

Obesity and regional fat distribution are among the risk factors for cardiovascular disease that can be changed. It is well known that general adiposity can be reduced by changes in diet and exercise, which are much more easily enumerated than they are accomplished. Fat pattern is also affected by caloric level and exercise, however the changes may be small. A weight reduction of 1 kg is associated with a change in waist-to-hip ratio of 0.001 to 0.003 (Bouchard and Tremblay, 1990). Nevertheless, longitudinal exercise training studies have indicated that small changes in the regional distribution of adipose tissue were associated with considerable improvement in plasma insulin and lipid metabolism, and there is some evidence to indicate that weight reductions may be more effective in alleviating insulin resistance and other cardiovascular risk factors in patients with relatively more visceral than peripheral fat (Despres et al., 1991; Krotkiewski and Bjorntorp, 1986).

Correlations between fat distribution and other cardiovascular risk factors deserve consideration. Smoking cessation and stress reduction could improve cardiovascular health through several mechanisms, one of which could be a more favorable fat distribution. Cut-points for a clinical definition of an upper body fat pattern have been proposed as a waist-to-hip ratio of 0.95 in men and 0.80 in women (Seidell, 1992). However, current research indicates that cut-points may need to be not only gender specific; but also specific to age, degrees of obesity and race.

FUTURE RESEARCH

Much has been learned concerning the role of fat distribution over the past 10 to 15 years, however, much remains unknown. It is apparent that more work needs to be done to elucidate the mechanism by which the regional distribution of fat affects metabolism and disease. Therapies, including drug and hormonal interventions, should be investigated, and the effects of changing fat pattern need further evaluation. More work is needed on methods and measurement technologies. In addition, normative values and relationships to disease need to be carefully examined in different populations and racial groups.

SUMMARY

Investigators have refined the study of the relationship of adiposity to disease by differentiating body shapes on the basis of the physical location of adipose tissue on the body. Different combinations of anthropometric measurements have been investigated as indicators of fat distribution. Recently combinations of easily obtained anthropometric measurements have been studied as correlates of visceral fat as measured by more complex measures such as computed tomography and magnetic resonance imaging.

Despite difficulties in the measurement of the regional distribution of adipose tissue, several studies have shown that an upper body, centralized or abdominal fat pattern is correlated with increased CVD risk factors and CVD risk. The majority of studies have been cross-sectional, although several prospective cohort studies have examined relationships between fat distribution and mortality or incidence of morbidity over rather long time intervals. Results from the Charleston Heart Study indicate these relationships may differ by race and gender.

More research is needed to define the role of fat patterning and changes in fat patterning on disease. Also, the relationship between total fatness and fat distribution to morbidity and mortality in specific populations and ethnic groups needs further exploration.

REFERENCES

Ashwell, M., Cole, T., and Dixon, A., 1985, Obesity: new insight into the anthropometric classification of fat distribution shown by computed tomography, Br. Med. J. 290: 1692-94.

Baumgartner, R., Heymsfield, S., Roche, A., and Bernardino, M., 1988, Abdominal fat depots measured with computed tomography: effect of degree of obesity, sex and age., Am. J. Clin. Nutr. 48: 936-45.

Bergstrom, R., Newell-Morris, L., Leonetti, D., Shuman, W., Wahl, P., and Fujimoto, W., 1990, Association of elevated fasting C-peptide level and increased intra-abdominal distribution with development of NIDDM in Japanese-American men, Diabetes 39: 104-11.

Bjorntorp, P., 1990, Portal adipose tissue as a generator of risk factors for cardiovascular disease and diabetes, Arterioscler. 10: 493-6.

Bjorntorp, P., 1991, Visceral fat accumulation: the missing link between psychosocial factors and cardiovascular disease?, J. Intern. Med. 230: 195-201.

Bjorntorp, P., 1993, Visceral obesity:A "civilization syndrome", Obes. Res. 1: 206-22.

Bouchard, C., andTremblay, A., 1990, Genetic effects in human energy expenditure components, Int. J. Obes. 14: 49-58.

Busetto, L., Baggio, M., Zurlo, F., Carraro, R., Digito, M., and Enzi, G., 1992, Assessment of abdominal fat distribution in obese patients: Anthropometry versus computerized tomography, Int J Obes. 16: 731-6.

Despres, J., Prud'homme, D., Pouliot, M., Tremblay, A., and Bouchard, C., 1991, Estimation of deep abdominal adipose-tissue accumulation from simple anthropometric measurements in men, Am. J. Clin. Nutr. 54: 471-7.

Donahue, R., Abbott, R., Bloom, E., Reed, D., and Yano, K., 1987, Central obesity and coronary heart disease in men, Lancet 1: 821-23.

Ducimetiere, P., Richard, J., and Cambien, F., 1986, The pattern of subcutaneous fat distribution in middle-aged men and the risk of coronary heart disease: the Paris Prospective Study, Int. J. Obes. 1986: 229-40.

Ferland, M., Despres, J., Tremblay, A., Pinault, S., Naddeau, S., Moorjani, S., and PJ, L., 1989, Assessment of adipose tissue distribution by computed axial tomography in obese women: association with body density and anthropometric measurements., Br. J. Nutr. 61: 139-48.

Gillum, R., 1987, The association of body fat distribution with hypertension, hypertensive heart disease, coronary hear disease, diabetes and cardiovascular risk factors in men and women aged 18-79 years, J Chronic Dis. 40: 421-28.

Haffner, S., Stern, M., Hazuda, H., Pugh, J., and Patterson, J., 1987, Do upper-body and centralized adiposity measure different aspects of regional body-fat distribution? Relationship to non-insulin-dependent diabetes mellitus, lipids, and lipoproteins, Diabetes 36: 43-51.

Hartz, A., Rupley, D., and Rimm, A., 1984, The association of girth measurements with disease in 32,856 women, Am. J. Epidemiol. 199: 71-80.

Jensen, M., Haymond, M., Rizza, R., Cryer, P., and Miles, J., 1989, Influence of body fat distribution on free fatty acid metabolism, J. Clin. Invest. 83: 1168-77.

Kalkhoff, R., Hartz, A., Rupley, D., Kissebah, A., and Selber, S., 1983, Relationship of body fat distribution to blood pressure, carbohydrate tolerance, and plasma lipids in healthy obese women, J. Lab. Clin. Med. 102: 621-27.

Kannel, W. B., andGordon, T., 1979, Physiological and medical concomitants of obesity: the Framingham Study in "Obesity in America", G.A. Bray, ed. US Department of Health, Education and Welfare, Public Health Services, National Institutes of Health Publication: 79-359, Bethesda.

Keil, J., Sutherland, S., Knapp, R., Lackland, D., Gazes, P., and Tyroler, H., 1993, Mortality rates and risk factors for coronary disease in black as compared with white men and women, New Eng. J. Med. 329: 73-8.

Kirschner, M., Samojlik, E., Drejka, M., Szmal, E., Schneider, G., and Ertel, N., 1990, Androgen-estrogen metabolism in women with upper versus lower body obesity, J. Clin. Endoc. Metab. 70: 473-79.

Kissebah, A., Freedman, D., and Peiris, A., 1989, Health risks of obesity, Med. Clin. North Am. 73: 111-38.

Kissebah, A., Vydelingum, N., Murray, R., Evans, D., Hartz, A., Kalkhoff, R., and Adams, P., 1982, Relation of body fat distribution to metabolic complications of obesity, J. Clin. Endocrinol Metab. 54: 254-60.

Koester, R., Hunter, G., Snyder, S., Khaled, M., and Berland, L., 1992, Estimation of computerized tomography derived abdominal fat distribution, Int. J. Obes. 16: 543-54.

Krotkiewski, M., andBjorntorp, P., 1986, Muscle tissue in obesity with different distribution of adipose tissue. effects of physical training, International J. of Obes. 10: 331-341.

Krotkiewski, M., Björntorp, P., Sjöström, L., and Smith, U., 1983, Impact of obesity on metabolism in men and women. Importance of regional adipose tissue distribution, J. Clin. Invest. 72: 1150-62.

Kvist, H., Chowdhury, B., Grangard, U., Tylen, U., and Sjostrom, L., 1988, Total and visceral adipose-tissue volumes derived from measurements with computed tomography in adult men and women: predictive equations. Am. J. Clin. Nutr. 48: 1351-61.

Lapidus, L., Bengtsson, C., Larsson, B., Pennert, K., Rybo, E., and Sjostrom, L., 1984, Distribution of adipose tissue and risk of cardiovascular disease and death: a 12 year follow up of participants in the population study of women in Gothenburg, Sweden, Br. Med. J. 289: 1257-61.

Larsson, B., Seidell, J., Svardsudd, K., Welin, L., Tibblin, G., Willhelmsen, L., and Bjorntorp, P., 1989, Obesity, adipose tissue distribution and health in men. The study of men born in 1913., Appetite 13: 37-44.

Larsson, B., Svardsudd, K., Welin, L., Wilhelmsen, L., Bjorntorp, P., and Tibblin, G., 1984, Abdominal adipose tissue distribution, obesity, and risk of cardiovascular disease and death: 13 year follow up of participants in the study of men born in 1913, Br. Med. J. 288: 1401-4.

Lundgren, H., Bengtsson, C., Blohme, G., Lapidus, L., and Sjostrom, L., 1989, Adiposity and adipose tissue distribution in relation to incidence of diabetes in women: results from a prospective population study in Gothenburg, Sweden, Int. J. Obes. 23: 413-423.

Marin, P., Holmang, S., Holm, G., Jonsson, L., and Bjorntorp, P., 1992, The effects of testosterone treatment on body composition and metabolism in middle-aged , obese men, Int. J. Obes. 16: 975-81.

Mueller, W., Deutsch, M., Malina, R., Bailey, D., and Mirwald, R., 1986, Subcutaneous fat topography: age changes and relationship to cardiovascular fitness in Canadians, Human Biol. 58: 955-73.

Ohlson, L., Larsson, B., Svardsudd, K., Eriksson, H., Wilhelmson, L., Bjorntorp, P., and Tibblin, G., 1985, The influence of body fat distribution on the incidence of diabetes mellitus: 13.5 years of follow-up of the participants in the Study of Men Born in 1913, Diabetes 34: 1055-58.

Ohlsson, L., Larsson, B., Svardsudd, K., Selin, L., Eriksson, H., Wilhelmsen, L., Bjorntorp, P., and Tibblin, G., 1985, The influence of body fat distribution on the incidence of diabetes mellitus, Diabetes 34: 1055-8.

Peiris, A., Aiman, E., Drucker, W., and Kissebah, A., 1989, The relative contributions of hepatic and peripheral tissues to insulin resistance in hyperandrogenic women, J. Clin. Endocrinol. Metab. 68: 715-20.

Rebuffé-Scrive, M., Andersson, B., Olbe, L., and Björntorp, P., 1989, Metabolism of adipose tissue in intraabdominal depots of nonobese men and women, Metabolism 38: 453-58.

Rebuffe-Scrive, M., Enk, L., Crona, N., Lonnroth, P., Abrahamson, L., Smith, U., and Bjorntorp, P., 1985, Fat cell metabolism in different regions in women. Effect of mentrual cycle, pregnancy and lactation J. Clin. Invest 75: 1973-6.

Reichley, K., Mueller, W., Hanis, C., Joos, S., Tulloch, B., Barton, S., and Schull, W., 1987, Centralized obesity and cardiovascular disease risk in Mexican Americans, Am J Epidemiol. 125: 373-86.

Seidell, J., 1992, Regional obesity and health, Int. J. Obesity. 16: S31-4.

Seidell, J., Bjorntorp, P., Sjostrom, L., Kvist, H., and Sannerstedt, R., 1990, Visceral fat accumulation in men is positively associated with insulin, glucose, and c-peptide levels, but negatively with testosterone, levels. Metabolism 39: 897-901.

Seidell, J., Cigolini, M., Deslypere, J., Charzewska, J., Ellsinger, B., and Cruz, A., 1991, Body fat distribution in relation to physical activity and smoking habits in 38-year-old European men, Am. J. Epidemiol. 133: 257-65.

Seidell, J., Oosterlee, A., Deurenberg, P., Hautvast, J., and Ruijs, J., 1988, Abdominal fat depots measured with computed tomography: effects of degree of obesity, sex, and age, Eur. J. Clin. Nutr. 42: 805-15.

Seidell, J., Oosterlee, A., Thijssen, M., Burema, J., Deurenberg, P., Hautvast, J., and Ruijs, J., 1987, Assessment of intra-abdominal and subcutaneous abdominal fat: relation between anthropometry and computed tomography., Am. J. Clin. Nutr. 45: 805-15.

Shimokata, H., Muller, D., and Andres, R., 1989, Studies in the distribution of body fat III. Effects of cigarette smoking, JAMA. 261: 1169-73.

Stern, M., Patterson, J., Mitchell, B., Haffner, S., and Hazuda, H., 1990, Overweight and mortality in Mexican Americans, Int. J. Obes. 14: 623-29.

Stevens, J., Keil, J., Rust, P., Tyroler, H., Davis, C., and Gazes, P., 1992a, BMI and body girths as predictors of mortality in black and white women, Arch. Intern. Med. 152: 1257-62.

Stevens, J., Keil, J., Rust, P., Verdugo, R., Davis, C., Tyroler, H., and Gazes, P., 1992b, Body mass index and body girths as predictors of mortality in black and white men, Amer. J. Epidemiol. 135: 1137-46.

Stevens, J., Knapp, R., Keil, J., and Verdugo, R., 1991, Changes in body weight and girths in black and white adults studied over a 25 year interval, Int. J. Obes. 15: 803-8.

Stevens, J., Plankey, M., Keil, J., Rust, P., Tyroler, H., and Davis, C., 1993, Black women have smaller abdominal girths than white women of the same relative weight., Am. J. Clin. Epidemiol. in press.

Stokes, J., and Garrison, R. K., 1985, The independent contribution of various indices of obesity to the 22-year incidence of coronary heart disease: The Framingham Heart Study, in "Metabolic Complications of Human Obesities", J.E.A. Vague, ed. Elsevier Science Publishers.New York. New York: Elsevier Science Publishers J.

Terry, R., Page, W., and Haskell, W., 1992. Waist/hip ratio, body mass index and premature cardiovascular disease mortality in US Army veterans during a twenty -three year follow-up study. Int. J. of Obes. 16: 417-23.

Tremblay, A., Despres, J., Leblanc, C., Craig, C., Ferris, B., Stephens, T., and Bouchard, C., 1990, Effect of intensity of physical activity on body fatnes and fat distribution, Am. J. Clin. Nutr. 51: 153-7.

Vague, J., 1956, The degree of masculine differentiation of obesities, Am. J. Clin. Nutr. 4: 20-33.

Weits, T., VanDerBeek, E., Wedel, M., and TerHaarRomeny, B., 1988, Compputed tomography measurement of abdominal fat deposition in relation to anthropometry, Int. J. Obes. 12: 217-25.

THE ROLE OF LIPOPROTEINS IN ATHEROGENESIS

John R. Guyton

Department of Medicine
Duke University Medical Center
Durham, NC 27710

INTRODUCTION

Lipoproteins unquestionably play a crucial role in atherogenesis, since populations around the world which have low plasma cholesterol levels do not experience much clinical atherosclerotic disease, regardless of the presence of other risk factors such as smoking and hypertension. Current knowledge about the role of lipoproteins in atherogenesis can be categorized into four general areas. The first concerns the transport and distribution of lipoproteins within the arterial wall. The most remarkable fact here is that the tissue concentration of low density lipoproteins (LDL) in the arterial intima is very high compared to all other connective tissues in the body.

The second general area covers the mechanisms of lipid accumulation in arterial cells and in the extracellular space from native and modified LDL in the tissue space. The mechanisms of lipid accumulation in arterial cells are well described in other reviews and will not be covered here (Brown and Goldstein, 1983; Schwartz et al., 1992; Fowler et al., 1991). This review will look in some detail at probable mechanisms of lipid droplet and vesicle formation from LDL occurring outside of cells in the arterial wall.

Knowledge in the third area derives primarily from epidemiologic studies showing that high density lipoprotein (HDL) levels correlate extremely well with protection from atherosclerosis. However, the mechanisms of this protection at the basic cellular level and in terms of reverse cholesterol transport are quite complex and incompletely understood. Other references should be consulted for the current state of knowledge (Johnson et al., 1991).

The fourth major area is that of lipoprotein oxidation, which has deleterious effects on both LDL and HDL. These effects have been demonstrated in vitro and in animal studies, but the clinical impact is largely unknown. In this volume Dr. Steinberg summarizes the current state of knowledge.

Nutrition and Biotechnology in Heart Disease and Cancer
Edited by J.B. Longenecker *et al.*, Plenum Press, New York, 1995

29

ARTERIAL INTIMAL LIPOPROTEIN CONCENTRATIONS

Immunostaining of human aortic sections for apolipoprotein B, the characteristic apoprotein of LDL, reveals intense reaction product in the intimal layer, but almost none in the medial layer (Bocan et al., 1988). Quantitative immunoassays have shown very high concentrations of LDL in the intima, but unmeasurable amounts in the media. In fact the concentration of LDL in the arterial intima is about equal to the plasma LDL concentration (Smith, 1974; Hoff et al., 1977). It is appropriate to ask how this compares to the usual LDL concentration found in the interstitial space of connective tissues. The best estimate of usual interstitial LDL concentration comes from measurements in peripheral lymph, which yield approximately one tenth of the plasma concentration. Therefore, we conclude that the connective tissue in the arterial intima sees an LDL concentration that is approximately ten times as high as that which is seen by any other connective tissue in the body.

How does this circumstance come about? **Figure 1** shows that the arterial intima is a

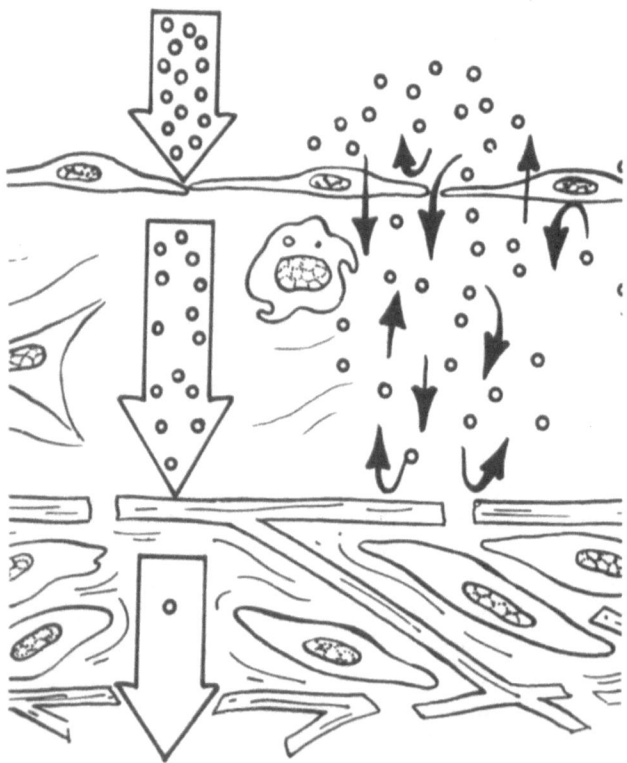

Figure 1. Lipoprotein transport in the arterial wall. Both endothelium (near top of figure) and internal elastic lamina are permeability barriers. Few LDL particles penetrate the tunica media. Both convection (left) and diffusion (right) of LDL are depicted. Reproduced with permission from Guyton, J.R. (1990).

loosely organized tissue, whereas the arterial media is tightly organized. Particles the size of LDL appear to be sterically excluded from the arterial media. The internal elastic lamina performs a barrier function in a manner somewhat analogous to the barrier function of the arterial endothelium. Since the internal elastic lamina is riddled with holes, or fenestrae, it must be the case that collagen fibers and proteoglycans are packed tightly in the fenestrae effectively excluding most lipoproteins. Therefore, the intima resides between two permeability barriers. This observation is important, since it suggests that LDL not only is present in the intima in high concentration, but also that the intima LDL is stagnant. The lifetime of LDL in the plasma is 2-3 days. Measurement of the lifetime of LDL in arterial intima is one goal of current research. Depending on the thickness of the intima, it may be as long as weeks or months.

The relationship of human aortic apolipoprotein concentrations to age, as described by Yla-Herttuala et al. (1987), is shown in **Figure 2**. Apolipoprotein A1 intimal concentrations do not rise very much with age, but apoB concentrations increase greatly as adult life is attained. By the age of 30-40 years, intimal apoB mass concentrations are 5-fold higher than apoA1 concentrations, despite the fact that apoA1 mass in plasma is usually greater than apoB mass. Since these apolipoproteins are found on LDL and HDL (high density lipoproteins), respectively, a corresponding increase in LDL relative to HDL in the arterial intima must occur.

To summarize these observations, LDL is found in a much higher concentration in arterial intima than in any other connective tissue in the body. This is almost certainly the reason that the cholesterol deposition regularly occurs with aging in the arterial intima, but very rarely occurs at all in other sites.

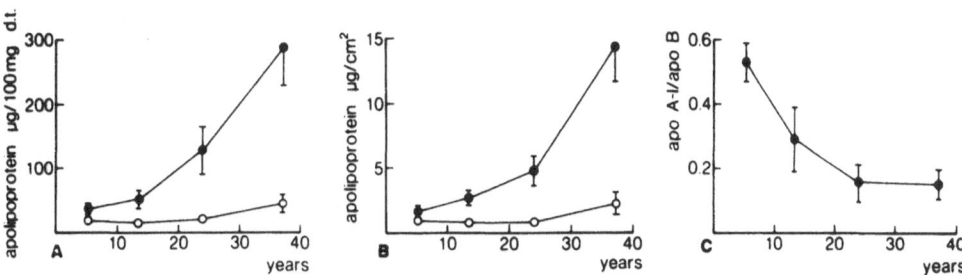

Figure 2. Age-related increases of apoB (closed circles) and apoA-1 (open circles) detected immunochemically in human aortic intima, expressed in two ways (Panels A and B). ApoB, associated with LDL, can be considered atherogenic, while apoA-1, associated with HDL, might be anti-atherogenic. The ratio of apoA-1 to apoB declines with age (Panel C). Reproduced with permission from Yla-Herttuala et al. (1987).

ATHEROSCLEROTIC CORE REGION

The development of the lipid-rich core region in atherosclerosis has been a focus of research interest in relatively few laboratories. The importance of the core is underscored by its role in setting the stage for plaque rupture and arterial thrombosis (Constantinides, 1966). Tracy (1979) showed that the core has a tendency to extend from a position initially deep in the intima toward the lumen of the artery with increasing age. This has the effect of eroding and weakening the fibrous cap which is providing support to the nonthrombogenic endothelial surface. When the cap is almost eroded an acute event may occur, perhaps arterial spasm or just a small rise in blood pressure, which causes the lesion finally to rupture, exposing flowing blood to the highly thrombogenic subendothelial material. In a pathological examination of coronary arteries from patients dying less than 2 weeks after myocardial infarction, Ridolfi and Hutchins (1977) found coronary thrombosis in 97% of cases, and rupture or ulceration of atherosclerotic plaques in 93%. Davies and co-workers (Davies and Thomas, 1985; Lendon et al., 1991) have studied extensively the features of the core and the fibrous cap which predispose to plaque rupture and thrombosis.

Early Core Development

Core development is not merely a feature of large, mature fibrous plaques, but has been seen in small raised lipid-containing lesions termed "fibrolipid lesions" (Bocan et al., 1986; Bocan and Guyton, 1985; Bocan et al., 1988; Guyton and Klemp, 1994). Fibrolipid lesions are thought to be derived from pre-existing fatty streaks, and this leads to a question: Does the core develop in the fatty streak, or does fibromuscular growth of the lesion come first and only afterward does the core develop? In a recent study (Guyton and Klemp, 1993), human aortic lesions with the surface appearance of fatty streaks were found to contain, in approximately 1/5 of cases, cholesterol clefts in the deep intima. Furthermore, in the lesions with clefts, the percent of total volume occupied by cells in the deep intima was 44% lower than it was in the deep intima of ordinary fatty streaks which lacked cholesterol clefts. Thus formation of cholesterol crystals and dropout of cells were evident, and these characteristics largely define an incipient core region. It seems evident that the core of the atherosclerotic plaque actually begins in the fatty streak stage of atherosclerosis, or at least in a stage of visually flat aortic lesions.

Figure 3 shows the cholesterol clefts in one of those lesions by electron microscopy, with the typical arrangement in which the cholesterol clefts were interdigitated and with the typical arrangement in which the cholesterol clefts were interdigitated and interspersed with collagen fibers. This appearance argues against that common notion that cholesterol crystals

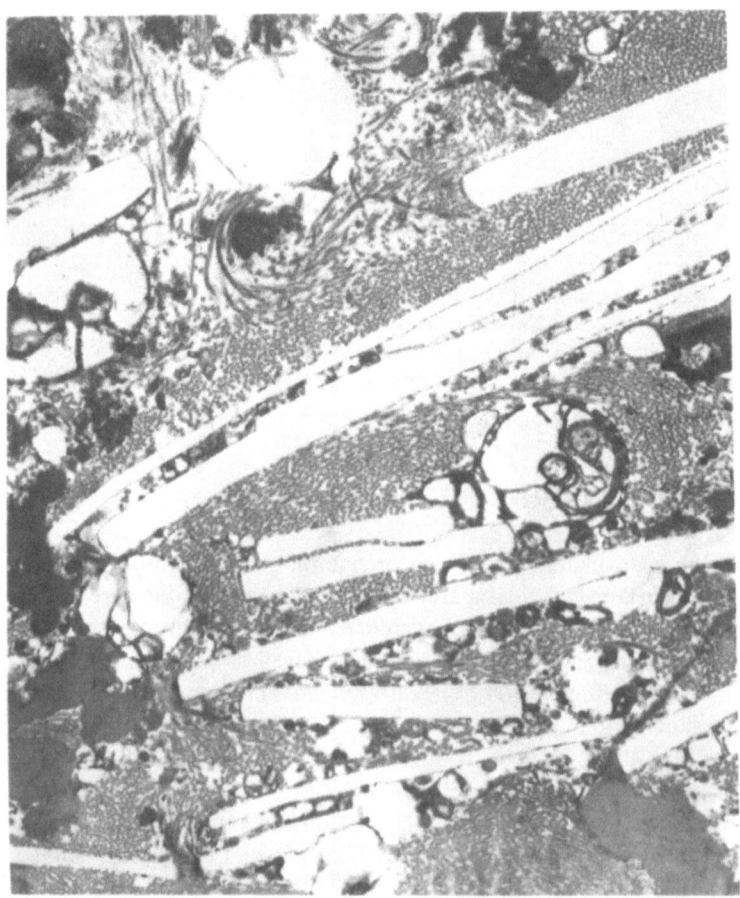

Figure 3. Cholesterol clefts inserted among collagen fibrils in the deep intima of a flat human aortic lesion resembling a fatty streak. Vesicular lipid structures are also present. Electron micrograph reproduced with permission from Guyton and Klemp (1993).

arise and grow within foam cells. Instead they are particularly associated with collagen fibers.

EXTRACELLULAR LIPID DEPOSITION

A strong case can be made that much of the lipid deposition in the core region of the fibrous plaque is due to an extracellular pathway. The alternative pathway, in which lipid is accumulated in cells with subsequent cell death, may be relatively less important. The argument for extracellular lipid deposition is based on 4 kinds of evidence: (1) Transitional morphologies, especially at the ultrastructural level, attest to the existence and frequency of extracellular modes of lipid accumulation. (2) The chemistry and structure of core lipid deposits favor an extracellular pathway. (3) Plausible mechanisms for extracellular deposition can be demonstrated. (4) Direct visualization of lipoprotein fusion in the extracellular space has been demonstrated in one situation.

Figure 4 is an electron micrograph showing extracellular lipid deposition which simply cannot be explained by foam cell death. The very dark structures are lipid droplets forming within elastic fibers, where foam cells do not reach (Guyton et al., 1985). Bundles of collagen fibers likewise have been found to contain interspersed deposits of lipid (Bocan et al., 1986). An impression from these studies is that elastin tends to accumulate lipid in oily droplets (cholesteryl ester), while collagen tends to accumulate vesicular lipid (phospholipid and free cholesterol). This impression needs to be confirmed, but the basic finding of lipid deposits intimately associated with connective tissue fibers is clear.

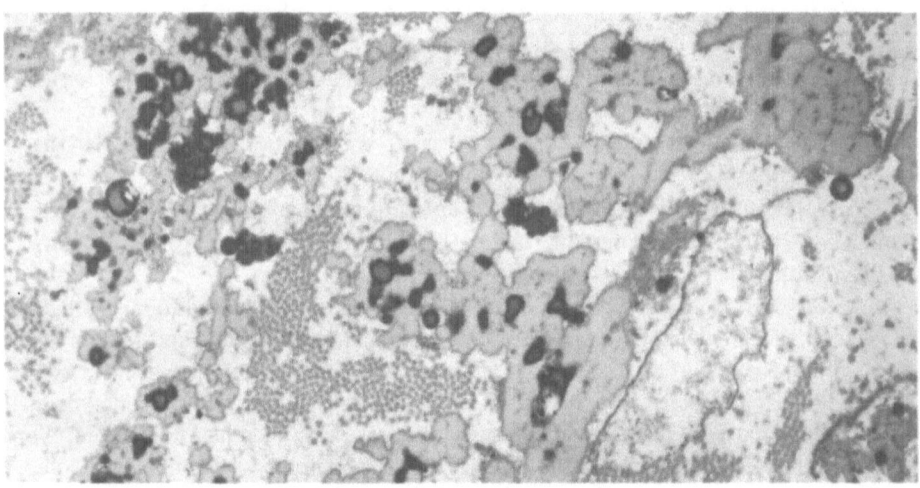

Figure 4. Deposits of oily droplets (presumed to be cholesteryl ester) found within elastic fibers in human aortic intima. Reproduced with permission from Guyton and Klemp (1993).

The vesicular type of lipid deposition was described about ten years ago by Kruth (1984, 1985) as filipin-staining deposits rich in free cholesterol, found both in human atherosclerosis and in a very early stage of atherosclerosis in cholesterol-fed rabbits. Subsequently Simionescu et al. (1985) and Guyton and Klemp (1992) elaborated the description of the rabbit subendothelial deposits with electron microscopic studies. The lipid, which is mostly vesicular, appeared just underneath the endothelium within two weeks after the onset of cholesterol feeding, a time just at the beginning of the migration of monocytes into the subendothelial space. Therefore, the lipid deposits could not be explained by the prior accumulation in foam cells and subsequent death of the cells.

Light microscopic studies also suggest that core lipids may not be derived from dead foam cells (Bocan and Guyton, 1985). By light microscopy with Oil Red O staining, the early lipid-rich core is found often to occur in the deep intima, while foam cells usually are seen in the superficial intima separated by some distance from the core. Moreover, the core lipids have a smudged, less distinct appearance compared to the cellular lipid droplets of the foam cells. This is due to the fact that much of the core lipid is vesicular, and lipid deposits within the core are much smaller than foam cell oily droplets. Thus the core lipids appear to be distinct from the foam cell lipids in terms of both location and light microscopic morphology.

The ultrastructure and chemistry of core lipid deposits give further evidence of extracellular derivation. Cytoplasmic lipid droplets in foam cells are almost always larger than 0.4 µm in profile diameter by electron microscopy. New techniques for preserving lipid structure during electron microscopic tissue preparation have only recently allowed adequate examination of extracellular lipid droplets, which were found to be almost always less than 0.4 µm in profile diameter. In a study of the core region in large mature fibrous plaques in human aorta, approximately 90% of the area occupied by neutral (oily) lipid in the core was found in extracellular droplets with profile diameters less than 0.4 µm (Guyton and Klemp, 1989). If this lipid is assumed to have originated in foam cells, then it is surprising not to find more oily droplets of the size found in foam cells. The simpler explanation favors direct extracellular formation of the droplets.

Recently we obtained new data on the lipid chemistry of the early core region in fibrolipid lesions and fibrous plaques, as compared to results obtained in fatty streaks and the cap regions of fibrolipid lesions and fibrous plaques. The fatty streak contains mostly cholesteryl ester, but the core of the fibrolipid lesion shows a substantial and very significant shift toward unesterified cholesterol. Interestingly, in the fibrous plaque core region the ratio of free to esterified cholesterol is much more variable. Indeed, it now appears that the earliest significant core lipids are comprised mostly of the free cholesterol/phospholipid vesicles described by Kruth and co-workers in variety of situations in atherosclerosis and demonstrated also by our electron microscopic studies (Guyton and Klemp, 1989). In the larger raised lesions which we designated as fibrous plaques, a secondary type of extracellular lipid deposition consisting mostly of cholesteryl ester-rich small droplets can occur (Guyton and Klemp, 1988).

About 30 years ago, Elspeth Smith (1974) described interesting differences between the cholesteryl esters accumulating in foam cells in human fatty streaks and the cholesteryl esters found in the core of fibrous plaques. The fatty streak cholesteryl esters show a shift in the pattern so that cholesteryl oleate becomes the major species, quite distinct from the pattern in plasma lipoproteins where cholesteryl linoleate is higher than the oleate (Smith, 1974). This shift in the cellular cholesteryl esters apparently occurs at the step of re-esterification of cholesterol in the cytoplasm, following initial lysosomal hydrolysis of the lipoprotein cholesteryl esters. In agreement with Smith, we have found that cholesteryl esters in fibrous plaque cores and in fibrolipid lesion cores more closely resemble the plasma lipoprotein pattern. This suggests that they may be derived directly from plasma lipoproteins without the intervening steps of uptake, hydrolysis, and re-esterification in cells.

LIPOPROTEIN AGGREGATION AND FUSION

If core lipids are not derived from cellular accumulation and cell death, then how are they deposited? Khoo, Steinberg and co-workers (1988) showed that vortexing LDL in saline buffer rapidly causes aggregation. The LDL aggregates were readily taken up by macrophages, which formed foam cells. However, since we were interested in an extracellular mechanism, it seemed appropriate to determine the ultrastructural properties of the aggregated LDL following vortexing, but in the absence of contact with cells. **Figure 5** shows the result of this inquiry. The LDL particles do not simply aggregate to form grape-like clusters, but they actually fuse with each other. The surface components of the LDL separate, forming multilammelar vesicles, similar to the early vesicles seen in human lesions. The oily core lipids also tend to separate, forming lipid droplets. Chemical analysis confirmed that aggregation by vortexing is associated with disruption of the usual stoichiometry of LDL lipids (Guyton et al., 1991).

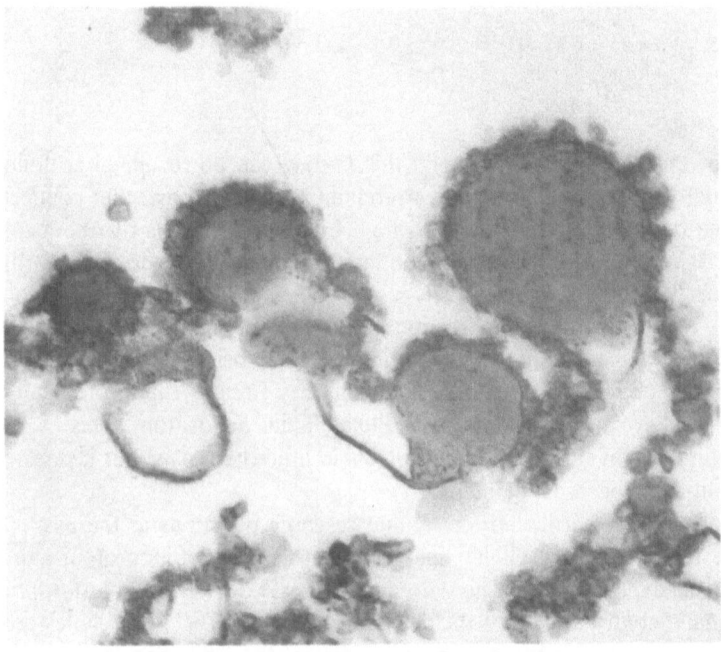

Figure 5. Electron micrograph of LDL which have self-aggregated after vortexing. The formation of lipid droplets and vesicles is demonstrated. Particles the size of LDL are also seen, but are among the smallest structures evident in the micrograph. Reproduced with permission from Guyton, Klemp, and Mims (1991).

Evidence from freeze etch electron microscopy performed on rabbit aortic subendothelial tissue strongly suggests that lipoprotein aggregation and fusion occur in vivo. This technique was first applied by Frank and Fogelman (1989) to tissue from rabbits developing atherosclerosis from genetic hypercholesterolemia or from cholesterol feeding. Clusters of round particles appeared in the subendothelial matrix. The particles had sizes appropriate for lipoproteins, lipid droplets, and lipid vesicles. A more dynamic study was subsequently performed by the same group. A large amount of human LDL was infused intravenously into a rabbit to produce a plasma cholesterol level of about 1000 mg/dl. After two hours of perfusion, the aorta was removed from the rabbit, and freeze-etch electron microscopy was performed. Round particles the size of plasma LDL and also larger particles, apparently the result of fusion of LDL, were found in the subendothelium of the experimental rabbits (Nievelstein et al., 1991).

The conditions that may allow lipoprotein fusion to occur and the possible differences in LDL that might exacerbate this process in some patients are largely unknown. Vortexing of LDL is not an acceptable model for the process that seems to occur in vivo, because of the air-fluid interface which tends to denature proteins during vortexing and probably is responsible for the demonstrated rapid aggregation and fusion of particles. Several groups have found that chemical modification of LDL - for example, treatment with phospholipase or with cholesteryl esterase or copper-catalyzed oxidation of LDL -have been shown to cause aggregation (Hoff et al., 1992; Suits et al., 1989; Musliner et al., 1987). Further progress in this area has been reported by Tertov and co-workers (1992). Incubation of LDL in tissue culture media for 48 hours sometimes leads to the formation of large particles demonstrated by elution in the void volume of a sizing column. LDL obtained from patients with coronary disease or diabetes mellitus were more likely to produce this early peak. Furthermore, LDL from the coronary patients tended to have loss of sialic acid residues from carbohydrate chains attached to apolipoprotein B; and treatment of normal LDL with neuraminidase, which removes sialic acid, appeared to cause aggregation. These findings bear further investigation.

SUMMARY

Some of the concepts presented in this review can be recapped as follows: LDL is found in a much higher concentration in arterial intima than in any other connective tissue in the body. One response of the intimal to high LDL levels appears to be a toxic response resulting in atherosclerotic core formation, with eventual breakdown and rupture of the intima causing arterial thrombosis. The core does not develop simply from foam cell necrosis, but from a complex interaction of tissue lipoproteins, cells, and extracellular matrix. Core development is an early event in atherosclerosis progression, since the features of early cores can be found in lesions resembling fatty streaks. Lipoprotein aggregation and fusion may be key processes in extracellular lipid deposition. This is obviously an incomplete summary of the role of lipoproteins in atherosclerosis, but it does point toward new significant areas of research interest.

There are several particularly intriguing research questions at the present time. How do the cholesterol-rich extracellular lipid deposits develop? Lipoprotein aggregation and fusion is a partial explanation, but how do deposits with 60% free cholesterol develop when the lipoproteins contributing to them have only 20-30% free cholesterol? Multiple hypotheses have been posed, but little evidence for any one pathway is available. Nevertheless, the extremely high levels of free cholesterol in the atherosclerotic core are likely to have effects on cellular membrane functions. Another intriguing question: How is core development related to the overall process of fatty streak to fibrous plaque conversion? The fibrous plaque has two hallmarks, one of which is a rather massive proliferation of cells and fibrous tissue, and the other is the development of the core. Our recent evidence suggests that core development may occur first (Guyton and Klemp, 1993). It is possible that the events surrounding core development - for example, cellular toxicity - or perhaps the lipids themselves in the core may trigger or at least contribute to the subsequent massive fibrocellular proliferation? Finally, are there clinical states that favor lipoprotein aggregation and fusion, and how may it be prevented? Khoo et al. (1990) found that the presence of HDL or apolipoprotein A1 could largely prevent the LDL aggregation which occurs with vortexing. As this process becomes better understood, we may find new ways to prevent it and thereby prevent extracellular lipid deposition in human atherosclerosis.

REFERENCES

Bocan, T.M. and Guyton, J.R., 1985, Human aortic fibrolipid lesions: progenitor lesions for fibrous plaques, exhibiting early formation of the cholesterol-rich core, Am. J. Pathol. 120:193.

Bocan, T.M., Brown, S.A. and Guyton, J.R., 1988, Human aortic fibrolipid lesions. Immunochemical localization of apolipoprotein B and apolipoprotein A, Arterioscler. 8:499.

Bocan, T.M., Schifani, T.A. and Guyton, J.R., 1986, Ultrastructure of the human aortic fibrolipid lesion. Formation of the atherosclerotic lipid-rich core, Am. J. Pathol. 123:413.

Brown, M.S. and Goldstein, J.L., 1983, Lipoprotein metabolism in the macrophage: implications for cholesterol deposition in atherosclerosis, Ann. Rev. Biochem. 52:223.

Constantinides, P., 1966, Plaque fissures in human coronary thrombosis, J. Atheroscler. Res. 6:1.

Davies, M.J. and Thomas, A.C., 1985, Plaque fissuring--the cause of acute myocardial infarction, sudden ischaemic death, and crescendo angina, Brit. Heart J. 53:363.

Fowler, S.D., Gasque-Carter, P.D., Pattillo-Adkisson, E., Sasiela, W.J. and Xenachis, D.N., 1991, Cellular models of atherosclerosis in the young, Ann. New York Acad. Sci. 623:60.

Frank, J.S. and Fogelman, A.M., 1989, Ultrastructure of the intima in WHHL and cholesterol-fed rabbit aortas prepared by ultra-rapid freezing and freeze-etching, J. Lipid Res. 30:967.

Guyton, J.R., Bocan, T.M. and Schifani, T.A., 1985, Quantitative ultrastructural analysis of perifibrous lipid and its association with elastin in nonatherosclerotic human aorta, Arterioscler. 5:644.

Guyton, J.R. and Klemp, K.F., 1988, Ultrastructural discrimination of lipid droplets and vesicles in atherosclerosis: value of osmium-thiocarbohydrazide-osmium and tannic acid-paraphenylenediamine techniques, J. Histochem. Cytochem. 36:1319.

Guyton, J.R. and Klemp, K.F., 1989, The lipid-rich core region of human atherosclerotic fibrous plaques: prevalence of small lipid droplets and vesicles by electron microscopy, Am. J. Pathol 134:705.

Guyton, J.R., 1990, Lipid metabolism and atherogenesis, in "The Science and Practice of Pediatric Cardiology," A. Garson, Jr., J.T. Bricker, and D.G. McNamara, eds., Lea and Febiger, Philadelphia.

Guyton, J.R., Klemp, K.F., and Mims, M.P., 1991, Altered ultrastructural morphology of self-aggregated low density lipoproteins: coalescence of lipid domains forming droplets and vesicles, J. Lipid Res. 32:953.

Guyton, J.R. and Klemp, K.F., 1992, Early extracellular and cellular lipid deposits in aorta of cholesterol-fed rabbits, Am. J. Pathol. 141:925.

Guyton, J.R. and Klemp, K.F., 1993, Transitional features in human atherosclerosis: intimal thickening, cholesterol clefts, and cell loss in human aortic fatty streaks, Am. J. Pathol. 143:1444.

Guyton, J.R. and Klemp, K.F., 1994, Development of the atherosclerotic core region: chemical and ultrastructural analysis of microdissected atherosclerotic lesions from human aorta, in press.

Hoff, H.F., Heideman, C.L., Gaubatz, J.W., Gotto, A.M., Jr., Erickson, E.E. and Jackson, R.L., 1977, Quantification of apolipoprotein B in grossly normal human aorta, Circ. Res. 40:56.

Hoff, H.F., Whitaker, T.E and O'Neil, J., 1992, Oxidation of low density lipoprotein leads to particle aggregation and altered macrophage recognition, J. Biol. Chem. 267:602.

Johnson, W.J., Mahlberg, F.H., Rothblat, G.H. and Phillips, M.C., 1991, Cholesterol transport between cells and high-density lipoproteins, Biochim. Biophys. Acta 1085:273.

Khoo, J.C., Miller, E., McLoughlin, P. and Steinberg, D., 1988, Enhanced macrophage uptake of low density lipoprotein after self-aggregation, Arterioscler. 8:348.

Khoo, J.C., Miller, E., McLoughlin, P. and Steinberg, D., 1990, Prevention of low density lipoprotein aggregation by high density lipoprotein or apolipoprotein A-I, J. Lipid Res. 31:645.

Kruth, H.S., 1984, Localization of unesterified cholesterol in human atherosclerotic lesions: demonstration of filipin-positive, oil-red-O-negative particles, Am. J. Pathol. 114:201.

Kruth, H.S., 1985, Subendothelial accumulation of unesterified cholesterol. An early event in atherosclerotic lesion development, Atheroscler. 57:337.

Lendon, C.L., Davies, M.J., Born, G.V. and Richardson, P.D., 1991, Atherosclerotic plaque caps are locally weakened when macrophages density is increased, Atheroscler. 87:87.

Musliner, T.A., McVicker, K.M., Iosefa, J.F. and Krauss, R.M., 1987, Lipolysis products promote the formation of complexes of very-low-density and low-density lipoproteins, Biochim. Biophys. Acta 919:97.

Nievelstein, P.F.E.M., Fogelman, A.M., Mottino, G. and Frank, J.S., 1991, Lipid accumulation in rabbit aortic intima 2 hours after bolus infusion of low density lipoprotein, Arterioscl. Thrombos. 11:1795.

Ridolfi, R.L. and Hutchins, G.M., 1977, The relationship between coronary artery lesions and myocardial infarction: ulceration of atherosclerotic plaques precipitating coronary thrombosis, Amer. Heart J. 93:468.

Schwartz, C.J., Valente, A.J., Sprague, E.A., Kelley, J.L., Cayatte, A.J. and Mowery, J., 1992, Atherosclerosis: potential targets for stabilization and regression, Circulation 86[Suppl III]:III-117.

Simionescu, N., Vasile, E., Lupu, F., Popescu, G. and Simionescu, M., 1985, Prelesional events in atherogenesis: accumulation of extracellular cholesterol-rich liposomes in the arterial intima and cardiac valves of the hyperlipidemic rabbit, Am. J. Pathol. 123:109.

Smith, E.B., 1974, The relationship between plasma and tissue lipids in human atherosclerosis, Adv. Lipid Res. 12:1.

Suits, A.G., Chait, A., Aviram, M. and Heinecke, J.W., 1989, Phagocytosis of aggregated lipoprotein by macrophages: low density lipoprotein receptor-dependent foam-cell formation, Proc. Natl. Acad. Sci. U. S. A. 86:2713.

Tertov, V.V., Orekhov, A.N., Sobenin, I.A., et al., 1992, Three types of naturally occurring modified lipoproteins induce intracellular lipid accumulation due to lipoprotein aggregation, Circ. Res. 71:218.

Tracy, R.E., Strong, J.P., Toca, V.T. and Lopez, C.R., 1979, Atheronecrosis and its fibroproliferative base and cap in the thoracic aorta, Lab. Invest. 41:546.

Yla-Herttuala, S., Solakivi, T., Hirvonen, J., et al., 1987, Glycosaminoglycans and apolipoproteins B and A-1 in human aortas, Arterioscler. 7:333.

ROLE OF OXIDIZED LDL AND ANTIOXIDANTS IN ATHEROSCLEROSIS

Daniel Steinberg

University of California, San Diego
Department of Medicine
Basic Science Building-Room 1080
9500 Gilman Drive, Dept. 0682
La Jolla, CA 92093-0682

INTRODUCTION

Atherosclerosis is responsible for more morbidity and mortality than any other single degenerative disease, including cancer. Its clinical expression in the form of myocardial infarction, stroke and peripheral vascular disease accounts for about 50% of all deaths in developed countries. It is a disease of aging, in the sense that the occurrence of clinically significant lesions increases sharply past the 5th decade, but it is equally clear that clinical expression is strongly linked to a number of environmental factors (e.g., hypercholesterolemia, hypertension, cigarette smoking, etc.). Correction of these "risk factors", especially hypercholesterolemia, has clearly been shown to reduce the number of clinical events. An important point that is often overlooked is that spontaneous atherosclerosis in animals, including non-human primates, is very rare. This stands in contrast to the very significant incidence of cancers in animals and their rising incidence with age. Animals rarely exhibit LDL cholesterol levels over 80 mg/dl; some have levels as low as 5-25 mg/dl. Furthermore, the apparent immunity of animals to atherosclerosis is almost certainly not due to an intrinsic resistance of their vessel wall to the atherogenic process -- lesions have been induced in virtually every mammalian species by simply raising the total plasma cholesterol level to a sufficiently high value. Thus there is good reason to indict hypercholesterolemia as a centrally important factor in human atherosclerosis.

Research over the last two decades has led to a new consensus on the sequence of events that initiate atherosclerotic lesions. Many of these events, including the accumulation of cholesterol in macrophages ("foam cells"), are accelerated by low density lipoprotein that has undergone oxidative modification. Presumably this oxidative modification is ultimately traceable to free radicals. In 7 of 8 published studies, involving rabbits or nonhuman primates, treatment with an antioxidant has significantly inhibited -- and strongly inhibited (35-80%) -- the rate of progression of experimental atherosclerosis.

Nutrition and Biotechnology in Heart Disease and Cancer
Edited by J.B. Longenecker *et al.*, Plenum Press, New York, 1995

39

Exactly how LDL is oxidized *in vivo* is not known. There are a number of "candidate" systems, identified by studies in cell culture, but there is no definitive information about the origin of the free radicals involved *in vivo*. Enzymes with the capacity to catalyze oxidative modification of LDL in isolated cells include 15-lipoxygenase, myeloperoxidase, other peroxidases, DPNH oxidase, and xanthine oxidase. Thus far, however, the relative importance of these *in vivo* remains unknown.

Whether or not the oxidative modification hypothesis is applicable to the human disease remains to be established. Epidemiologic studies and some biochemical studies have provided evidence consonant with that possibility. However, until we have clinical intervention trials we must suspend judgment regarding the importance of oxidative modification in human atherosclerosis. Intervention trials are now underway at several centers and a definitive answer may be forthcoming within the next 5 years.

The accumulation of lipoprotein cholesterol is clearly central to the initiation of the "fatty streak" -- the first anatomically defined lesion in atherosclerosis. However, even at the very earliest stages of lesion development a complex array of cytokines and growth factors are already beginning to influence the process. These become progressively more relevant as the lesions progress with cell growth and deposition of matrix materials that eventually cause stenosis. These factors are produced by endothelial cells, by macrophages, by T-lymphocytes and by smooth muscle cells. The development of the fibrous plaque lesion has many characteristics of a chronic inflammatory processes. Exactly which cytokines and growth factors are of the greatest importance remains to be determined.

Finally, it is important to note that death from coronary artery disease is almost always associated with an occlusive thrombosis at the site of an advanced atherosclerotic lesion. That thrombosis generally overlies a part of the lesion that has "ruptured" or ulcerated, exposing the blood to tissue in the subendothelial space thus triggering the clotting mechanisms. Lesions that contain a large quantity of extracellular lipid are more likely to "rupture" than lesions poor in lipid. There is some evidence that macrophages and oxidative processes may play a role in the erosion of endothelium leading to the fatal thrombosis, but this is mostly speculative and much remains to be learned.

CURRENT CONSENSUS REGARDING THE SEQUENCE OF EVENTS INITIATING ATHEROGENESIS (Ross and Glomset, 1976; Davies et al., 1976; Steinberg, 1983; Ross, 1986; Stary, 1987; Steinberg and Witztum, 1990) (see **Figure 1**)

It has been amply demonstrated that hypercholesterolemia is a dominant risk factor for atherosclerosis and that correction of hypercholesterolemia can sharply reduce risk. However, the cellular and molecular mechanisms by which low density lipoprotein -- the major carrier of plasma cholesterol -- initiates lesions has only recently been appreciated. For many years it was thought that loss of endothelial cells from the arterial lining as a result of some "injury" was a necessary first step in the development of atherosclerotic lesions. That concept was heavily influenced by the many experiments showing that if you do in fact denude arteries of their endothelium, you induce platelet adherence, release of platelet-derived growth factor (and probably other growth factors) leading to smooth muscle cell proliferation. Certainly smooth muscle cell proliferation is a characteristic accompaniment of atherosclerosis but it only becomes a dominant process later in lesion development. We now know that the earliest lesion -- the fatty streak -- develops under an unbroken layer of endothelial cells. The hallmark of this lesion is the so-called "foam cell" characterized by the accumulation of an enormous number of lipid droplets, mainly cholesterol ester. While the fatty streak lesion is itself benign, it is now recognized to be the precursor of the fibrous plaque, which in turn gives rise to the later lesions responsible for clinical expression of coronary heart disease. The later evolution of the lesion is extremely complex and poorly understood. However, if we could understand the mechanisms leading to the fatty streak, and if we could intervene to inhibit its formation, we might not ever have to deal with the later lesions and their clinical consequences.

Foam Cell Formation

The foam cells in a fatty streak arise mainly from circulating monocytes that penetrate the endothelium and take up residence beneath it. A smaller fraction of these foam cells arise also from smooth muscle cells that migrate from the media up into the subendothelial space. Soon after hypercholesterolemia is induced in experimental animals, there is an increase in the adhesion of circulating monocytes to the arterial endothelium. In part this adherence is induced by changes in the monocyte itself, and in part by induced expression of specific adhesion molecules on the endothelium. The adherent cells can then respond to chemotactic factors generated in the subendothelial space that induce monocyte migration

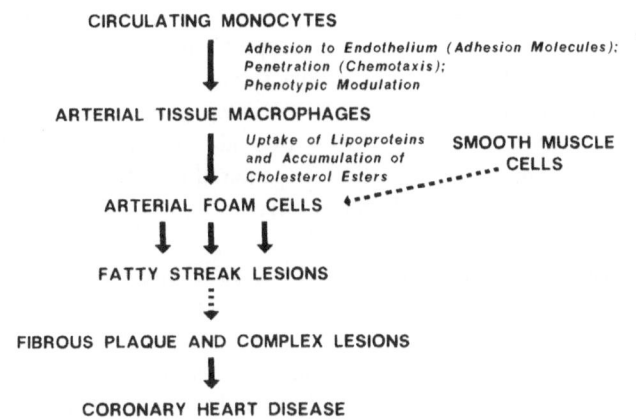

Figure 1. Current Consensus regarding the sequence of events initiating atherogenesis.

between endothelial cells into the intima. Several chemotactic factors that may play a role have been identified but their relative importance *in vivo* is not yet known. Once in the subendothelial space, monocytes undergo a phenomenal phenotypic modulation, taking on the properties of a tissue macrophage. Many different systems are modified in this process.

Role of Oxidative Modification of LDL (Steinberg et al., 1989)

Since the major source of cholesterol in the plasma is the low density lipoprotein (LDL) one would therefore anticipate that incubation of monocytes (or the macrophages derived from them) with a high concentration of LDL would convert them into foam cells. Yet, as first shown in the laboratories of Michael S. Brown and Joseph L. Goldstein (1983), this does not happen. The same is true with regard to smooth muscle cells: even though they can be precursors of foam cells, incubation of them *in vitro* with even high concentrations of LDL does not produce foam cells. It only increases the cell content of cholesterol by a small amount. These observations pointed up a paradox that could be best explained if it were not LDL itself (i.e., native LDL as it circulates in the plasma) but some modified form of LDL that was taken up by the macrophages and by the smooth muscle cells. The first "candidate" modified form of LDL was LDL treated *in vitro* with acetic anhydride -- acetyl LDL. An outstanding series of studies from the laboratories of Brown and Goldstein (1983) showed that this chemically modified form of LDL was taken up many times more rapidly than was native LDL and, unlike native LDL, could lead to accumulation of cholesterol in monocyte/macrophages. Moreover, they showed that acetyl LDL was taken up by a specific receptor distinct from the receptor for native LDL -- the acetyl LDL or "scavenger" receptor. Unlike the LDL receptor, the acetyl LDL receptor does not down-regulate as the cell cholesterol content increases. Recent studies show that rabbit arterial smooth muscle cells can express this same receptor. The acetyl LDL receptor was cloned in Krieger's laboratory in 1988 and has now been fully characterized (reviewed in Krieger et al., 1993). However, acetylation of LDL *in vivo* has never been demonstrated and is unlikely to occur. Studies by Henriksen et al. (1981) first suggested an alternative modification of LDL that *is* biologically plausible, namely, oxidative modification. Incubation of LDL with cultured endothelial cells, smooth muscle cells or monocyte/macrophages (i.e., any of the cell types found in atherosclerotic lesions) can modify LDL to a form recognized by the same receptor characterized by Brown and Goldstein i.e. the acetyl LDL or "scavenger" receptor. Indeed, most of the uptake and internalization of oxidized LDL by monocyte/macrophages occurs by way of the acetyl LDL receptor. A smaller fraction occurs via an independent receptor that recognizes oxidized LDL but not acetyl LDL. The discovery of oxidatively modified LDL generated by exposure to cultured cells suggested a solution to the paradox of foam cell formation -- they are induced by LDL, yes, but not by LDL in its *native* form.

Alternative Modifications Of LDL That May Promote Foam Cell Formation
(Steinberg and Witztum, 1990)

Before going on to discuss the evidence that oxidation of LDL does indeed occur and that it does play a role in atherogenesis, we want to emphasize that there are *other* modified forms of LDL that can be taken up avidly by macrophages and that could therefore also contribute to foam cell formation. While none of these has been as thoroughly studied as oxidative modification, neither have they been shown to be *un*important.

Self-aggregation of LDL was first shown by Khoo et al. (1988) to lead to enhanced uptake. Aggregation was induced in these early studies by surface denaturation (violent agitation of the solution), and later by treatment with phospholipase C, or by heat denaturation. The aggregates were taken up not by way of the acetyl LDL receptor but rather by way of the native LDL receptor and the uptake was phagocytotic rather than endocytotic. Why this uptake does not downregulate the LDL receptor and arrest further uptake is puzzling.

Immunoglobulin-LDL complexes can deliver LDL cholesterol to the macrophage by way of the Fc receptor. That uptake, independent of the LDL receptor, is not down-regulated as the cholesterol content of the cell builds up. It has been shown that massive amounts of cholesterol ester can accumulate in monocyte/macrophages exposed to immune complexes containing LDL.

As a final example, complexes of LDL with proteoglycans may increase foam cell formation either because the complex is more readily subjected to oxidative modification or because the complex is recognized by scavenger receptors.

Again, while there is some evidence that these modified forms occur *in vivo*, their role is much less well characterized than that of oxidatively modified LDL.

EVIDENCE FOR THE IMPORTANCE OF OXIDATIVE MODIFICATION OF LDL IN ANIMAL MODELS OF ATHEROSCLEROSIS

A number of studies have established that oxidative modification of LDL does in fact occur *in vivo*, both in experimental animals and in humans. However, we shall confine ourselves here to a brief summary of the most important line of evidence, namely, that inhibitors of LDL oxidation can slow the progression of atherosclerosis in animal models. This was first shown in LDL receptor-deficient rabbits by Carew et al. (1987) using probucol as the antioxidant. Probucol has been in use for many years as a cholesterol-lowering agent. However, its antioxidant activity, first brought to light by Parthasarathy et al. (1986), is probably not related to its cholesterol-lowering activity. Because probucol lowers LDL levels even in LDL receptor-deficient rabbits, the control group of animals in the Carew studies was treated with low doses of lovastatin, doses just sufficient such that the mean cholesterol level in this reference group was the same as that in the probucol-treated group. After seven months of treatment the extent of aortic lesions in the probucol-treated group was only about 50% that of the lovastatin-treated group. A similar study was independently reported by Kita et al. (1987), with even more dramatic effect on lesion development. Over the past four years, six additional studies have been carried out testing the antioxidant hypothesis in rabbits and in monkeys. With one exception (Stein et al., 1989) all of these studies have yielded positive results. Several different antioxidants have been used including probucol, butylated hydroxytoluene (Björkhem et al., 1991), N,N'-diphenylphenylene-diamine (Sparrow et al., 1992) and vitamin E (Verlangieri and Bush, 1992). The inhibition of lesion progression has ranged from 35 to 80%. In a recently completed study, probucol has been shown to be effective also in cholesterol-fed monkeys (R. Ross, personal communication) Moreover, this highly significant inhibition of atherogenesis has occurred in the face of extremely high plasma cholesterol levels, ranging from 600 to 800 mg/dl in LDL receptor-deficient rabbits to levels of over 1500 mg/dl in the cholesterol-fed animals. In some of the probucol studies, no attempt was made to adjust for the cholesterol-lowering effect of probucol but the differences in cholesterol level were relatively small, probably too small to account for the magnitude of the effects on lesion progression. In other studies, the cholesterol levels in the probucol-treated group and the control group were matched either by treating the control group with a different cholesterol-lowering agent or by adjusting the cholesterol intake of the diet as necessary to keep the cholesterol levels matched. Neither butylated hydroxytoluene nor diphenylphenylene-diamine have any cholesterol-lowering effect and so there was no problem about matching cholesterol levels in those studies.

Just as "specific" inhibitors sometimes turn out to have additional unsuspected effects, so might some of these antioxidant compounds have additional biological effects that could account for some or all of their antiatherogenic potency. This is especially the case with probucol, which has a number of known effects that could be relevant. Whether or not butylated hydroxytoluene shares these effects with probucol, to which it has a very close chemical homology, is not known. On the other hand, N,N'-diphenylphenylenediamine and vitamin E are quite unrelated in structure either to probucol or to each other. The fact that all of these antioxidants, representing overall a rather wide range of chemical structures, are effective as antiatherogenic agents supports the simplest interpretation -- that they work by virtue of their antioxidant activity. However, the possibility that several different modes of action are involved cannot be completely ruled out.

CLINICAL RELEVANCE OF THE OXIDATIVE MODIFICATION
HYPOTHESIS (Steinberg, 1991; Steinberg and Workshop Participants, 1992)

There are good reasons to predict that if atherogenesis in hypercholesterolemic animal models is inhibited by antioxidant compounds, then atherogenesis in hypercholesterolemic patients will also be inhibited. First of all, oxidation of LDL occurs also in humans, as shown by the presence of oxidized LDL in extracts of human atherosclerotic lesions, and, indirectly, by the presence of autoantibodies against oxidized LDL in human plasma. Second, antioxidants work in the LDL receptor-deficient rabbit, where the major lipoprotein elevated is LDL, and LDL is also the major lipoprotein elevated in most patients with severe hypercholesterolemia. Finally, the fact that antioxidants work also in non-human primates strongly supports the validity of the extrapolation to humans.

While, the extrapolation to humans is reasonable, it may nevertheless be invalid. For example, in all of the animal models studied thus far plasma lipoprotein levels have been extremely high and the lesions have been produced over a relatively short period of time. Under these conditions, antioxidant-sensitive phenomena may become of overriding importance and yet they may be much less important when lipoprotein levels are only modestly elevated. In this connection, when lipoprotein levels are very high, the rates of deposition of lipid in the developing lesions may be so great that changes in the rates of removal (which is presumed to depend upon HDL levels) may become insignificant. In humans, with limited "pressure" from elevated LDL and a limited rate of incorporation of lipid into the artery, the rate of lesion development may depend much more strongly on the rate of efflux (and therefore perhaps the concentration of plasma HDL). This may be relevant to the interpretation of the preliminary negative report from the PQRST study in Sweden on the effects of probucol on femoral atherosclerosis (see below).

A number of epidemiologic studies have demonstrated correlations between either blood levels of antioxidant vitamins or estimated dietary intakes of antioxidant vitamins and risk of coronary heart disease (Gey et al, 1989; Stampfer et al., 1993; Rimm et al., 1993). Like all epidemiologic observational studies however, these only demonstrate a correlation and do not establish a cause-and-effect relationship. Furthermore, the epidemiologic studies have sometimes yielded conflicting results, particularly with regard to the importance of vitamin C intake. Another intriguing line of evidence is the finding that cigarette smoking, which has been documented to increase CHD risk strikingly, is associated with an increased susceptibility of plasma LDL to oxidative modification *ex vivo* (Harats et al., 1990). This has been shown by comparing plasma LDL from smokers with that from non-smokers and also by demonstrating an acute increase in susceptibility of plasma LDL to oxidative modification within 90 minutes of the onset of cigarette smoking (6 or 7 cigarettes over the span of 90 minutes).

While all of these observations taken together support the proposition that oxidative modification is important in the human disease, that proposition will only be definitively established by appropriately large, double-blind clinical intervention trials using as end-point either CHD events or objectively measured changes in progression of lesions (by angiography or by ultrasound measurements). There are at present no published trials of this kind. One preliminary report by Gaziano et al. (1990) suggests that ß-carotene may decrease CHD risk. That result was based on results in a subset of men entering the Harvard Physicians Health Survey, begun almost 8 or 9 years ago to test the hypothesis that aspirin might prevent CHD and that β-carotene might prevent cancer. The β-carotene part of the study has been extended and final results will not be available for several more years. In the interim, the investigators decided to analyze results in a subset of 333 men who had had manifestations of coronary heart disease when the study began (even though the intention was not to include men with established coronary heart disease). Within that subset of 333 men there were about 40% fewer major coronary events in those taking ß-carotene. No results on the larger cohort were reported nor will they be until the study extension has been completed several years from now.

The only other interventional result is the still unpublished report of the results of the Swedish PQRST trial, which was presented orally by Drs. G. Walldius and A.G. Olsson at the 1993 meeting of the European Atherosclerosis Society in Jerusalem. This was a double-blind, 3-year intervention trial based on quantitative femoral angiography. Patients received either cholestyramine alone or cholestyramine plus probucol. The only datum reported in Jerusalem was the change in femoral volume, which was the primary end-point of the study. The patients receiving both cholestyramine and probucol did no better than the patients getting cholestyramine alone, in whom there was a modest increase in femoral volume over the 3-year period of observation. This failure of probucol to improve things was noted despite the fact that the total cholesterol in the group getting both drugs was distinctly lower than that in the group getting only cholestyramine. However, a large part of the drop in cholesterol level was probably due to a drop in HDL cholesterol. In fact this may account for the negative result. It is conceivable that the rather consistent positive results in experimental animals (even though there is some drop in HDL with probucol treatment of animals), is due to the extremely high levels of plasma cholesterol in the experimental models, ranging from 600-800 mg/dl in the receptor-deficient rabbit to well over 1500 mg/dl in the cholesterol-fed rabbits or monkeys. It is possible that when the rates of deposition of lipid are extremely high, changes in the rate of removal, possibly depending upon HDL, make little difference. On the other hand, when the elevation of cholesterol levels is more modest, as in the patients studied in most clinical trials, changes in HDL may be much more significant and could mask any beneficial effect of probucol acting as an antioxidant. We must await the complete report of the PQRST study in order to evaluate it properly and more studies will be needed.

What conclusions can we come to, then, with regard to the use of antioxidants in patients? In the absence of even one completely reported clinical intervention trial, we certainly cannot make any recommendations regarding therapy. It has been argued that because the antioxidant vitamins are relatively non toxic and unlikely to do any harm, there is no real reason *not* to recommend their use based on the experimental and animal model studies. That would overlook the *indirect* harm that can, and probably would, occur. Some patients will assume that by taking vitamin E they solve their problems and will then neglect to adhere to a prescribed diet, slack off on their exercising program and give up on trying to stop smoking. In any case, we must not abandon our criteria for proof of efficacy and must not make recommendations until we are sure that there is some measurable benefit to be conferred. That we will only know after the appropriate clinical intervention trials are reported.

COMPLICATED LESIONS AND CLINICAL EXPRESSION
OF THE DISEASE (Ross, 1986; Davies and Woolf, 1990)

The fatty streak is a clinically silent lesion. It is only slightly elevated and does not significantly interfere with blood flow. Only after cell proliferation and deposition of connective tissue matrix does the lesion begin to intrude on the lumen and interfere with blood flow. Thus, the further development of the fatty streak lesion is linked to a complex pattern of cytokines and growth factors that regulate cell proliferation, matrix deposition and remodeling of the vessel. The first of these growth factors to be described was the platelet-derived growth factor (PDGF). It is now recognized that PDGF can be synthesized and secreted not only by platelets but also by endothelial cells, smooth muscle cells and macrophages. Over the past decade at least 15 different cytokines and growth factors have been identified as products of cells within atherosclerotic lesions and the presence of these has in most instances been documented by immunohistochemistry and/or in situ hybridization. Most, but not all, of the factors potentially available in the lesion are proatherogenic. For example, interleukin-1 stimulates smooth muscle cell growth. On the other hand, interferon gamma, produced by T cells, *inhibits* smooth muscle cell growth and collagen synthesis. Whatever the factors contributing may be, the naturally history of the lesion is one of expansion with cell replication and matrix deposition until eventually it can even completely occlude the lumen. Much more common, however, is the development of a thinning of the cap of a partially stenotic lipid-rich lesion with ulceration and plaque rupture. It is this that precipitates the fatal thrombosis in most cases.

Are oxidative processes relevant to the development of the stenotic lesion and the cataclysmic perforation and thrombosis? Very possibly. For example, oxidized LDL can stimulate the release of interleukin 1, a growth factor for smooth muscle cells. Also, oxidized LDL itself is directly cytotoxic for endothelial cells in culture (and for many other cells as well) and could conceivably play a role in the ultimate erosion of the endothelial surface that precipitates thrombosis. The work of Dr. Michael Davies in London has shown that the terminal rupture of the lesion generally occurs at the shoulder region and that this region is relatively rich in macrophages. Thus it appears that there is continuing inflammatory activity even in the very old lesions, at least at the margins, which is where plaque rupture occurs. Whether or not antioxidant interventions can influence plaque rupture remains to be tested. Unfortunately, there are no good animal models for studying this crucially important event.

IMPLICATIONS AND CONCLUSIONS

If oxidative modification of LDL is critically important in accelerating the atherogenic process, as strongly suggested by the experimental evidence, then we would have a completely new mode of intervention available to us. We believe that the rate of progression of lesions is proportional to plasma LDL levels; it may also be proportional to the rate at which LDL undergoes oxidative modification. Interventions directed at lowering cholesterol levels and interventions directed at inhibiting LDL oxidation should then be additive. One reason this possibility is so attractive is that there are many antioxidants that are relatively non-toxic, especially the naturally occurring antioxidants, such as vitamin E. Exactly what level of antioxidative protection will be necessary remains to be determined. There are better antioxidants than vitamin E, at least when tested in an *in vitro* system. Probucol is in fact more potent than vitamin E. However, we are really not certain that the antiatherogenic potential of a compound can be assessed exclusively on the basis of the extent to which it protects circulating LDL against oxidative modification in one of our *in vitro* testing systems. We know that antioxidants *do* have effects at the intracellular level. Those could even be more important than the direct protection afforded by the antioxidants residing within the LDL particle. Only additional animal experiments will help us resolve that question. Another caveat should be noted with regard to mechanisms of action. As already mentioned, probucol has a number of biological effects over and above its antioxidant effect and its cholesterol-lowering effect, some of which could be very relevant to its antiatherogenic potential. The fact that other antioxidants of quite different structure are also effective makes it seem more likely that the operative mechanism is an antioxidant one, but that is not certain.

In summary, the oxidative modification hypothesis of atherogenesis has been strongly supported by a number of lines of evidence from *in vitro* and animal model studies. The hypothesis is robust but it is not yet fully established in respect to the human disease. Which enzyme systems *in vivo* are responsible for oxidative modification remains to be established. It also remains to be established to what extent antioxidants exert their protective effects by acting on cells and to what extent by simply residing in the LDL particle along with the naturally occurring LDL antioxidants. Thus, much additional experimental and clinical investigation is needed, but the hypothesis remains viable.

REFERENCES

Björkhem, I., Henrikkson-Freyschuss, A., Breuer, O., Diczfalusy, U., Berglund, L., and Henriksson, P., 1991, The antioxidant butylated hydroxytoluene protects against atherosclerosis, Arterioscler. Thromb. 11:15-22.

Brown, M.S. and Goldstein, J.L., 1983, Lipoprotein metabolism in the macrophage: Implications for cholesterol deposition in atherosclerosis, Ann. Rev. Biochem. 52:223-261.

Carew, T., Schwenke, D.C., and Steinberg, D., 1987, Antiatherogenic effect of probucol unrelated to it hypocholesterolemic effect: evidence that antioxidants in vivo can selectively inhibit low-density lipoprotein degradation in macrophage-rich fatty streaks and slow the progression of atherosclerosis in the Watanabe heritable hyperlipidemic rabbit, Proc. Natl. Acad. Sci. USA. 84:7725-7729.

Daugherty, A., Zweifel, B.S., and Schonfeld, G., 1989, Probucol attenuates the development of aortic atherosclerosis in cholesterol-fed rabbit, Br. J. Pharmacol. 98:612-618.

Davies, P.F., Reidy, M.A., Goode, T.B., and Bowyer, D.E., 1971, Scanning electron microscopy in the evaluation of endothelial integrity of the fatty lesion in atherosclerosis, Atheroscler. 25:125-130.

Davies, M.J. and Woolf, M., 1990, Atheroma: Atherosclerosis in ischaemic heart disease, in: "The Mechanisms", Volume I,. Science Press, London.

Gaziano, J.M., Manson, J.E., Ridker, P.M., Buring, J.E., and Hennekens, C.H., 1990, Beta-carotene therapy for chronic stable angina (Abstract), Circulation 82(Suppl. III):201.

Gey, K.F. and Puska, P., 1989, Plasma vitamins E and A inversely correlated to mortality from ischemic heart disease in cross-cultural epidemiology, Ann. N.Y. Acad. Sci. 570:268-82.

Harats, D., Ben-Naim, M., Dabach, Y., et al., 1990, Effect of vitamin C and E supplementation on susceptibility of plasma lipoproteins to peroxidation induced by acute smoking, Atheroscler. 85:47-54.

Henriksen, T., Mahoney, E.M., and Steinberg, D., 1981, Enhanced macrophage degradation of low-density lipoprotein previously incubated with cultured endothelial cells: recognition by receptors for acetylated low-density lipoprotein, Proc. Natl. Acad. Sci. USA. 78:6499-6503.

Khoo, J.C., Miller, E., McLoughlin, P., and Steinberg, D., 1988, Enhanced macrophage uptake of low-density lipoprotein after self-aggregation, Arterioscler. 8:348-358.

Kita, T., Nagano, Y., Kokode, M., et al., 1987, Probucol prevents the progression of atherosclerosis in Watanabe heritable hyperlipidemic rabbit, an animal model for familial hypercholesterolemia, Proc. Natl. Acad. Sci. USA. 84:5928-5931.

Krieger, M., Acton, S., Ashkenas, J., Pearson, A., Penman, M., and Resnick, D., 1993, Molecular flypaper, host defense, and atherosclerosis, J. Biol. Chem. 268:4569-4572.

Parthasarathy, S., Young, S.G., Witztum, J.L., Pittman, R.C., and Steinberg, D., 1986, Probucol inhibits oxidative modification of low-density lipoprotein, J. Clin. Invest. 77:641-644.

Rimm, E.B., Stampfer, M.J., Ascherio, A., Giovannucci,, E., Colditz, G.A., and Willett, W.C., 1993, Vitamin E consumption and the risk of coronary disease in men, New Engl. J. Med. 328:1450-1456.

Ross, R. and Glomset, J.A., 1976, The pathogenesis of atherosclerosis, New Engl. J. Med. 295:369-377; 420-425.

Ross, R. 1986, The pathogenesis of atherosclerosis -- an update, New Engl. J. Med. 314:488-500.

Sparrow, C.P., Doebber, T.W., Olszewski, J., et al., 1992, Low density lipoprotein is protected from oxidation and the progression of atherosclerosis is slowed in cholesterol-fed rabbits by the antioxidant N.N'-diphenyl-phenylenediamine, J. Clin. Invest. 89:1885-1891.

Stampfer, M.J., Hennekens, C.H., Manson, J.E., Colditz, G.A., Rosner, B., and Willett, W.C., 1993, Vitamin E consumption and the risk of coronary disease in women, New Engl. J. Med. 328:1444-1449.

Stary, H.C., 1987, Macrophages, macrophage foam cells, and eccentric intimal thickening in the coronary intima of young children, Atheroscler. 64:91-108.

Stein, Y., Stein, O., Delplanque, B., Fesmire, J.D., Lee, D.M., and Alaupovic, P., 1989, Lack of effect of probucol on atheroma formation in cholesterol-fed rabbits kept at comparable plasma cholesterol levels, Atheroscler. 75:145-155.

Steinberg, D., 1983, Lipoproteins and atherosclerosis -- a look back and a look ahead, Arterioscler. 3:283-301.

Steinberg, D., 1991, Antioxidants and atherosclerosis: A current assessment, Circulation 84(3):1420-1425.

Steinberg, D., 1993, Antioxidant vitamins and coronary heart disease, New Engl. J. Med. 328:1487-1489.

Steinberg, D., Parthasarathy, S., Carew, T.E., Khoo, J.C., and Witztum, J.L., 1989, Beyond cholesterol, New Engl. J. Med. 320:915-924.

Steinberg, D. and Witztum, J.L., 1990, Lipoproteins and atherogenesis, Current concepts, JAMA 264:3047-3052.

Steinberg, D. and Workshop Participants, 1992, Antioxidants in the Prevention of Human Atherosclerosis, Summary of the Proceedings of a National Heart, Lung, and Blood Institute Workshop: September 5-6, 1991, Bethesda.. Circulation 85(6):2338-2344.

Verlangieri, A.J. and Bush, M.J., 1992, Effects of d-α-tocopherol supplementation on experimentally induced primate atherosclerosis, J. Amer. Coll. Nutr. 11(2):131-138.

SIGNAL TRANSDUCTION IN ATHEROSCLEROSIS: SECOND MESSENGERS AND REGULATION OF CELLULAR CHOLESTEROL TRAFFICKING

Kenneth B. Pomerantz[1], Andrew C. Nicholson[2], and David P. Hajjar[3]

[1]Department of Medicine
[2]Department of Pathology
[3]Department of Biochemistry
Cornell University Medical College
New York, NY 10021

INTRODUCTION

While the anatomic pathology of atherosclerotic plaque development is well understood, the cellular mechanisms which contribute to lesion progression remain undefined. Prior to the 1970's, the prevailing view of atherogenesis was the concept that lipid infiltration from plasma-derived LDL was necessary and sufficient to elicit an atherogenic response. However, a fresh and intriguing view was proposed by Ross and Glomset in 1976. They suggested that atherogenesis was a consequence of the cellular response to injury (Ross and Glomset, 1976), and that humoral and cell-derived mediators, such as platelet-derived growth factor (PDGF) played a role in atherogenesis by promoting arterial smooth muscle cell hyperplasia (Ross et al., 1974). These seminal studies formed the basis for a novel approach to the study of atherosclerosis which was based on the concept that atherosclerosis may be viewed as a chronic inflammatory response to high concentrations of native or modified low density lipoprotein (LDL), which in turn elicits a series of cellular and biochemical events which result in alterations in endothelial cell phenotype, macrophage infiltration, smooth muscle cell migration, proliferation, foam cell development, and matrix deposition (Ross, 1986). Implicit in the Ross hypothesis is the concept that mediators released by cells comprising the lesion contribute to its development. Endothelial cells, macrophages, and arterial smooth muscle cells are all sources of, and targets for, a variety of humoral- and cell-derived mediators which are known to modulate cholesterol metabolism. In this review, we address the hypothesis that soluble mediators (growth factors and cytokines), derived from vascular and inflammatory cells, directly influence processes controlling cholesterol metabolism through the generation of second messengers including eicosanoids and cyclic nucleotides. We also discuss the possibility that derangements in intracellular cholesterol trafficking within smooth muscle cell and/or macrophage-derived foam cells, may be a result of disruption of the cytokine/eicosanoid network.

Nutrition and Biotechnology in Heart Disease and Cancer
Edited by J.B. Longenecker *et al.*, Plenum Press, New York, 1995

SIGNAL TRANSDUCTION MECHANISMS IN ARTERIAL CELLS

Receptor Signal Transduction Systems

The signal transduction pathways leading to activation of intracellular mediator synthesis is beyond the scope of this review, and is described elsewhere (Koch et al., 1991). Briefly, two major receptor systems mediate signal transduction pathways leading to second messenger synthesis.

First, ligand-association with receptors possessing tyrosine kinase activity (such as PDGF and EGF receptors) promotes autophosphorylation, and phosphorylation of other proteins, including phospholipase Cγ (**PLC-1γ**), **p21ras** (through adaptor proteins **SOS** and **GRB**), GTPase activating protein (**GAP**), and phosphatidylinositol 3'-kinase (**PI3K**) (Williams, 1989). Activated p21ras elicits phosphorylation of major downstream kinases, including **Raf**, and subsequently, Mitogen-Activated Protein Kinase Kinase (**MAPKK**), and MAP kinase (**MAPK**). The intrinsic activity of p21ras is negatively regulated by **GAP**, which facilitates the intrinsic GTPase activity of p21ras (**Figure 1**). Activation of PI3K

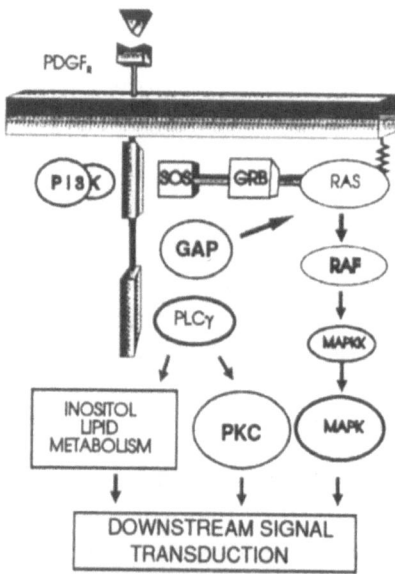

Figure 1. Receptor tyrosine kinase-mediated signal transduction: Ligands, such as PDGF signals through its receptor to phosphorylate itself and four principal substrates (PLCy, GAP, PI3 kinase, and p21ras) to initiate signals that mediate response to receptor occupancy.

promotes the phosphorylation of phosphatidyl inositol, and important prerequisite of inositol polyphosphate formation. Activation of PLCγ and subsequent generation of diacyl glycerol (DAG) and inositol phosphates leads to protein kinase C (**PKC**) activation and calcium flux, as will be discussed below. The overall effect of kinase activation is downstream activation of elements responsible for increasing the expression of numerous gene products (see below).

Second, a family of seven-transmembrane domain-containing receptors (which bind thrombin, acetylcholine, angiotensin II, bradykinin, and adenine nucleotides) interact with a family of guanine nucleotide-binding proteins, collectively known as heterotrimeric G-proteins, so named for their ability to bind guanine nucleotides, which is also requisite for their activation. G-proteins exist as a heterotrimeric complex of a catalytic Gα subunit (of which over 16 are known), and a ßγ regulatory subunit. There are at least 3-4 ß and γ isoforms. Gα subunits are linked to numerous effectors, including adenylyl cyclase (**A/C**, resulting in either adenylyl cyclase activation or inhibition), phospholipase C-1ß (**PLC1ß**), phosphodiesterase (**PDE**), and potentially, phospholipase A_2 (**PLA$_2$**), and others. Downstream effectors of G-protein activation thus include protein kinase A (**PKA**) and protein kinase C (**PKC**). A detailed description of G-protein dependent signalling is discussed elsewhere (Simon, Strathmann and Gautam, 1991). After ligand interaction with its receptor, the Gα subunit dissociates from the ßγ subunit, and promotes the downstream activation of its effectors. The receptor itself undergoes desensitization by phosphorylation by the ß-adrenergic receptor kinase (**ßARK**), or related enzymes. Interestingly, the ßγ subunit may also stimulate PLCß, suggesting that heterotrimeric G-proteins activation may activate more than one signalling pathway simultaneously. An important consequence of PLC activation is generation of diacylglycerol and subsequent activation of PKC, and adenylyl cyclase. PKC activation through this pathway also leads to MAPKK stimulation, and subsequent downstream effects (**Figure 2**).

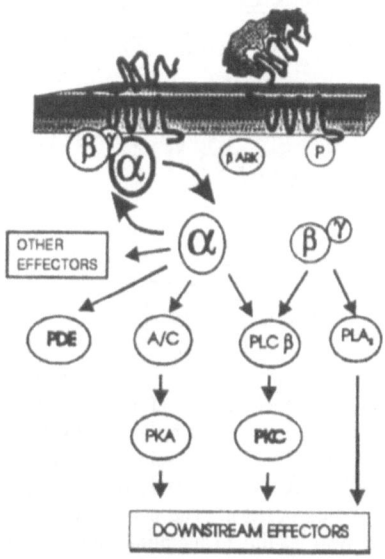

Figure 2. G-protein-dependent signal transductional. Ligands, such as thrombin, bradykinin, and angiotensin-II initiate activation of membrane-associated G-proteins, whose a-submit activates (or inhibits) downstream effectors, including PDE, A/C, PLCB-isozymes, and potentially PLA$_2$.

It is important to note that these distinct signalling pathways have a number of conversion points; **MAP kinase** stimulates the activity of transcription factors such as **jun** and **fos**, resulting in enhanced gene transcription; MAP kinase may also stimulate cytosolic **PLA$_2$**, thus promoting arachidonate release and subsequent eicosanoid synthesis. **PLC** isozyme activation results in liberation of DAG, resulting in diglyceride-lipase (**DGL**)-mediated arachidonate release and subsequent eicosanoid production. The influence of augmented eicosanoid production is diverse, and will be developed below. DAG also stimulates **PKC** activity, whose down-stream signals include adenylyl cyclase (**A/C**), which, by generating cyclic AMP (**cAMP**), serves to antagonize the mitogenic effects of growth factors, and a family of proteins which regulate cytoskeletal elements, known as Myristoylated Alanine-Rich C Kinase Substrates (**MARCKS**). MARCKS are responsible for cytoskeletal events following cytokine challenge. Finally, IP$_3$ generated from PLC promotes calcium release from the smooth endoplasmic reticulum, (**SER**) which regulates calcium-dependent control of gene transcription through the activation of calmodulin-dependent protein kinase

(**CLM-KINASE**), which serves to mediate numerous calcium-dependent processes (**Figure 3**). Thus, these parallel systems interact to mediate and regulate the proliferative responses to mitogenic cytokines and growth factors. Such proliferative responses may include, but are not limited to generation of other cytokines, as well as early response genes, which mediate responses to cytokine activation.

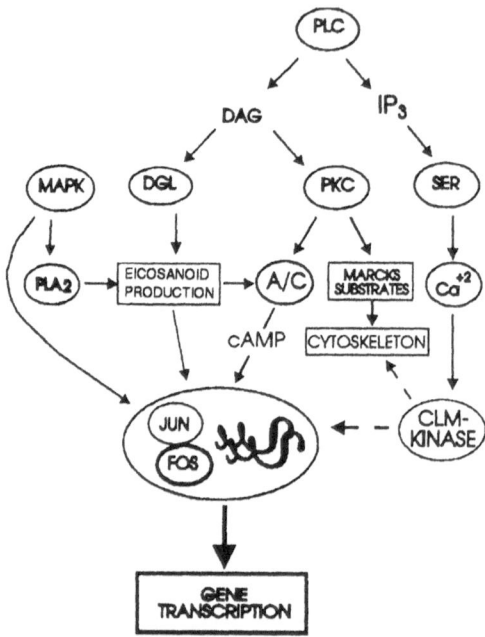

Figure 3. Intracellular signal transduction pathways common to receptor tyrosine kinases and G-proteins. Downstream signals from receptor tyrosine kinases and G-protein-dependent activation converge at numerous sites which serve to mediate and/or modulate receptor activation. For illustration, pro-mitogenic effects of PDGF, or angiotensin-II may be mediated through activation of MAPK, and calcium-release mediated by IP$_3$. However, these effects may be antagonized though signals generated simultaneously, including synthesis of eicosanoids and cyclic AMP. The net effect of receptor occupancy will thus be a reflection of the activities of each pathway leading to alterations in gene transcription.

Growth Factor/Cytokine Synthesis

Growth factors, including platelet-derived growth factor (PDGF), basic fibroblast growth factor (bFGF), and transforming growth factor-ß (TGF-ß) are synthesized by vascular cells. PDGF is secreted in biologically active form by endothelial cells, and is inducible by endotoxin (LPS), thrombin, tumor necrosis factor (TNF), and TGF-ß. PDGF stimulates smooth muscle cell proliferation, migration, chemotaxis, and chemokinesis. However, bFGF is probably not secreted, but is localized either intracellularly or bound to extracellular matrix. Basic FGF release may require either cell injury or exposure of the extracellular matrix to heparin, heparinases or other proteases. Basic FGF also directly stimulates PDGF release from cultured endothelial cells. TGF-ß is secreted into the extracellular milieu in a latent form, and must be activated proteolytic cleavage (**Figure 4**). Cytokines, including interleukin-1 (IL-1), TNF, and granulocyte-macrophage colony stimulating factor (GM-CSF), are synthesized by inflammatory cells and activated arterial endothelial and smooth muscle cells. The mechanisms which promote cytokine secretion are complex, and involve multiple, and overlapping autocrine and paracrine loops. TNF and LPS increase PDGF and IL-1 production in numerous cell types. LPS increases TNF,

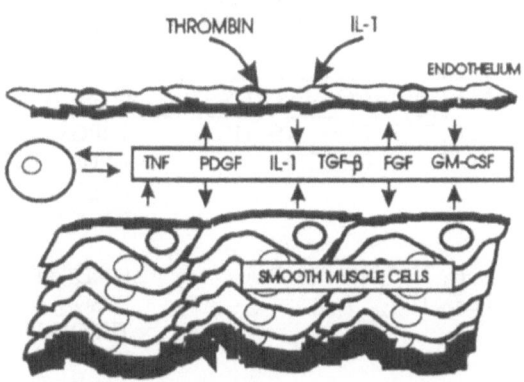

Figure 4. Cell-cell communication and the growth factor/cytokine network. Cytokines and growth factors are synthesized by endothelial cells, smooth muscle cells, and macrophages, and may elicit the generation of the same and/or other cytokines from neighboring cells. This effect serves to mediate and amplify responses to cytokine activation, which may include endothelial cell adhesion molecule expression, monocyte binding and diapedesis, and smooth muscle cell chemotaxix and proliferation.

IL-1, and IL-6 and interferon-γ (INF-γ) synthesis and secretion from vascular tissue. IL-1 can stimulate IL-6 production, and IL-6 can induce PDGF expression. These observations demonstrate that cytokines can rapidly amplify initial stimuli to stimulate cell proliferation and other consequences of cytokine activation.

Eicosanoid Biosynthesis

As introduced above, a major consequence of PLA_2 and PLC activation is eicosanoid synthesis. The biological activity of the principal eicosanoids, particularly PGI_2, appears to be in the regulation of other second messenger systems, most importantly, cyclic AMP. Endothelial cells synthesize mainly cyclooxygenase (COX) products ($PGI_2 > PGF_{2\alpha} > PGE_2$), hydroxyeicosatetrenoic acids (15-, 12- and 5-HETE's), and epoxyeicosatetraenoic acids (EETS). Endothelium can also convert neutrophil-derived leukotriene (LT) A_4 to LTC_4 during transcellular metabolism. Smooth muscle cells synthesize PGI_2 and PGE_2, but at levels that are about 6-10-fold less than endothelial cells, due to the lesser content of COX. Eicosanoid synthesis by macrophages depend on the type of macrophage studied, phenotype, and calcium concentration. Briefly, eicosanoid synthesis by macrophages is dependent upon the activity of PKC, which mediates eicosanoid synthesis following a variety of stimuli, including zymosan, LPS, and TNF. Unactivated macrophages synthesize mainly PGE_2; however, leukotrienes are also synthesized following acute stimulation by f-Met-Leu-Phe (**Table 1**).

Cytokines/Growth Factors And The Eicosanoid Network

Receptor ligands which utilize tyrosine kinase receptors or G-protein linked receptors are all potentially capable of eliciting eicosanoid generation following activation of PLC/PLA_2. Interestingly, growth factors and cytokines have a multiplicity of effects which modulate eicosanoid synthesis. Eicosanoids derived from the COX pathway are derived from two distinct enzymes. The first is a constitutively expressed enzyme, known as COX-1, which is responsible for generating eicosanoids following acute activation of PLA_2 by agonists, such as PDGF and IL-1 (Habenicht et al., 1985; Breviairio et al., 1990). The second COX enzyme, known as COX-2 is not normally expressed in unstimulated cells, but it is expressed following stimulation by cytokines or growth factors (O'Banion et al., 1991; Kujubu and Herschman, 1992; Hla and Neilson, 1992). Originally, the fact that soluble

Table 1. Eicosanoid synthesis by vascular cells

CELL TYPE	MAJOR EICOSANOIDS	PRINCIPAL AGONISTS
ENDOTHELIUM, SMOOTH MUSCLE CELLS	PGI_2, PGE_2, $PGF_{2\alpha}$, 15-, 11-HETE, 5-HETE, EET'S.	THROMBIN, BRADYKININ, ANGIOTENSIN II, ADP ACETYLCHOLINE PDGF, IL-1ß, TNF
MONOCYTES, MACROPHAGES	LTB_4, LTC_4, HETES, TxA_2, PGE_2, $PGF_{2\alpha}$, PGI_2.	F-MET-LEU-PHE OPSONIZED ZYMOSAN THROMBIN LPS, TNF
PLATELETS	TxA_2, PGE_2, $PGF_{2\alpha}$, 12-HETES.	THROMBIN, ADP, COLLAGEN PDGF

mediators could modify eicosanoid metabolism was first inferred from by the observation that exposure of fibroblasts to macrophage products produced a persistent increase in PGE_2 production (Korn, 1983). Further studies then proved that PDGF stimulated eicosanoid production by activating PLA_2 (Habenicht et al., 1981), and by up-regulating transcription of COX-2 and PGI_2 synthase genes (Habenicht et al., 1985). Epidermal growth factor (EGF) also stimulates smooth muscle cell PGI_2 generation (Blay and Hollenberg, 1989) by promoting the translocation of PLC-γ1 from the cytosol to the plasma membrane (Kim, Sim and Rhee, 1990). LPS is capable of eliciting eicosanoid production by endothelial cells (Bull et al., 1990) and macrophages (Glaser, Asmis and Dennis, 1990), and these effects could be mimicked by IL-1 (Rustin, Bull and Dowd, 1989), TNF (Atkinson et al., 1990), and INF-γ by direct phospholipase activation (Rossi et al., 1985) and increased transcription of COX mRNA (Kerr et al., 1989). In addition, TNF and IL-1 are synergistic (Floege et al., 1990) in stimulating eicosanoid synthesis, suggesting that multiple, and overlapping signal transduction pathways mediate TNF/IL-1-induced eicosanoid production.

Interactions Between Eicosanoids And Growth Factors/Cytokines

The significance of cytokine/growth factor-mediated eicosanoid metabolism lies in their ability to either mediate or modulate those pathways activated by those cytokines. For example, in macrophages, LPS stimulated IL-1 production, whose synthesis was augmented by indomethacin, but inhibited by exogenous PGE_2 (Libby, Warner and Friedman, 1988). In addition, the upregulation of macrophage TNF gene expression by phorbol ester is blunted by PGE_2 (Horiguchi et al., 1989). PGE_2 also inhibits CSF-1 gene expression in TNF-stimulated macrophages (Sherman et al., 1990). In contrast, LTB_4 mediates phorbol ester-induced TNF gene expression (Horiguchi et al., 1989). This may represent part of a positive feed-back loop since LTB_4 can stimulate IL-1 and TNF release (Gagnon et al., 1989). In addition, TNF-induced c-fos expression and subsequent proliferation in macrophages is mediated by lipoxygenase products, since activation was blocked by nordihydroguarietic acid (a non-selective lipoxygenase inhibitor). Add-back experiments demonstrated that 15-, 12- and 5-HPETE's alone induce c-fos expression (Haliday, Ramesha and Ringold, 1991). These data demonstrate that COX products down-regulate gene expression and effects, while lipoxygenase products, and possibly leukotrienes mediate and amplify the effects of cytokine activation.

The mechanism(s) by which eicosanoids mediate signal transduction are equally diverse. The most potent COX products such as PGI_2 and PGE_1 increase cyclic AMP (Dembinska-kiec, Rucker and Schonhofer, 1980). Cyclic AMP inhibits smooth muscle cell proliferation by decreasing DNA synthesis. In fibroblasts, TNF and IL-1 increase cyclic AMP, which is followed by a reduction in the expression of c-myc mRNA (Heldin et al., 1989). In addition, ß-adrenergic agents stimulate cyclic AMP and inhibit induction of c-sis and PDGF-B chain synthesis (Kavanaugh et al., 1988). In addition, cyclic AMP inhibits both PDGF and EGF-activated MAP kinase signalling pathways (Graves et al., 1993). Finally, cyclic AMP may inhibit cellular proliferation by inhibiting EGF-dependent diacylglycerol production and EGF receptor phosphorylation (Iwashita et al., 1990). Thus, elevations in cyclic AMP serve to antagonize or modulate growth factor-induced DNA synthesis and cell proliferation. However, lipoxygenase products, including leukotrienes, have a less clearly defined role in the regulation of cell function following activation by cytokines.

Regulation Of Cholesterol Metabolism And The Influence Of Specific Signal Transducing Mechanisms

Cellular cholesterol content is a reflection of the activities associated with cholesterol delivery, metabolism, and efflux. Cholesterol delivery is mediated by the LDL receptor (LDL-R), which recognizes apo-B-containing lipoproteins, or by a scavenger receptor which recognizes modified forms of LDL. LDL-derived cholesteryl esters (CE) are hydrolyzed by lysosomal (acid) CE hydrolase (ACEH); liberated free cholesterol then enters a pool of cellular free cholesterol also derived from *de novo* synthesis by HMG-CoA reductase. Free cholesterol may enter a "futile" cycle of transesterification and hydrolysis by the combined activities of Acyl Co-Cholesterol Acyltransferase (ACAT) and a cytoplasmic (neutral) CE hydrolase (NCEH). Free cholesterol is either used metabolically by the cell or is translocated to the cell membrane for net efflux, mediated principally by high density lipoproteins (HDL), or plasma albumin. Ultimately, HDL-associated CE are removed by hepatic extraction.

It is now well-established that the control of the enzymes that stimulate cholesterol availability are negatively regulated by cholesterol. Perhaps the most intensively studied in this regard is the LDL-receptor (LDL-R) and the HMG-CoA reductase genes. Mutational analysis of the 5' sequence of the promoter region of the LDL-R gene reveals a 45-base pair sequence that contains 16 base-pair repeats (Mehta et al., 1991). Repeat 3 possesses constitutive positive regulatory elements that act independently of sterols, and binds a nuclear binding protein known as Sp-1. Binding of Sp-1 to its binding site allows transcription of the LDL-R to proceed. Importantly, Repeat 2 inhibits the activity of Repeat 3, but only in the presence of sterols (Dawson et al., 1988). The actual mechanism by which this occurs is unclear. It has been postulated that either cholesterol directly binds to Repeat 3, or, cholesterol acts as an allosteric effector on a nuclear sterol binding protein that interferes with Sp-1 activation of LDL-R gene transcription (Dawson et al., 1988). More recently, a cluster of proteins of 59-68 kDa have been identified as a sterol-responsive element binding protein (SREBP), which binds specifically to a 10-base pair SRE-1 embedded within the 16-base pair Repeat 2 sequence (Briggs et al., 1993; Wang et al., 1993). It has been concluded that SREBP and Sp-1 cooperate to effectively mediate repression of the LDL-R gene by sterols.

Similarly, the expression of the HMG-CoA reductase gene is also subject to sterol-mediated repression, in addition to allosteric inhibition of the enzyme. The promoter region of the HMG-CoA reductase gene contains an octanucleotide segment of DNA that is 7/8 similar to repeat 2 of the LDL-R gene promoter, which bind specific nuclear proteins

(NF-1). It is believed that the binding of cholesterol to the NF-1/promoter complex results in repression of transcription.

However, the scavenger receptor, which internalizes modified forms of LDL is not down-regulated by cholesterol, and may thus be important in the accumulation of intracellular cholesterol when extracellular LDL levels are high. Similarly, there is no evidence that other enzymes important in cholesterol trafficking, including ACEH, NCEH, and ACAT, are subject to regulation by cholesterol by means other than by acting either as substrate (ACAT) or end-product inhibitor. However, each of these proteins/receptors are regulated by cytokines/growth factors, as detailed below (**Table 2**).

Table 2. Regulation of cholesterol trafficking by growth factors and cytokines

CELLS	CYTOKINE / GROWTH FACTOR	EFFECTOR	RESPONSE
SMOOTH MUSCLE CELLS	PDGF	LDL-RECEPTOR, SCAVENGER R, ACEH, NCEH, HMG -CoA REDUCTASE	INCREASE
		ACAT	NO CHANGE
		HDL RECEPTOR	DECREASE
SMOOTH MUSCLE CELLS	bFGF	LDL- RECEPTOR	INCREASE
	TGF-ß	LDL-RECEPTOR	INCREASE
	TNF	SCAVENGER - R	INCREASE
MONOCYTE/ MACROPHAGES	TGF-ß, TNF, LPS, IFN-γ	SCAVENGER -R	DECREASE
	GM-CSF, M-CSF	SCAVENGER R	INCREASE
	GM-CSF	LDL-RECEPTOR	INCREASE
	M-CSF	CE HYDROLASES	INCREASE
HEP-G2 CELLS	IL-1ß, TNF	LDL -RECEPTOR	INCREASE

The premise that regulation of cholesterol metabolism could be influenced by mediators *other than cholesterol* was derived from early studies demonstrating that growth factors in general increase in the availability of free cholesterol, which is necessary for cell growth and anabolism. The most studied growth factor in this regard is PDGF. PDGF promotes LDL binding and uptake (Cuthbert and Lipsky, 1990), and occurs by LDL-R dependent and independent means (Chait, Ross and Bierman, 1988), principally by releasing the LDL-R gene from cholesterol-induced suppression (Mazzone et al., 1990), and by increasing scavenger receptor activity in smooth muscle cells (Inaba et al., 1992). PDGF also increases *de novo* cholesterol synthesis from acetate by increasing HMG-CoA reductase expression (Habernicht, Glomset and Ross, 1980). PDGF also stimulates CE hydrolase activity, commensurate with the need to degrade lysosomal and cytoplasmic CE. Importantly, PDGF also down-regulates HDL receptor activity (Oppenheimer, Oram and Bierman, 1987). Basic FGF also promotes net free cholesterol synthesis from LDL by activating LDL-R activity (Chen, Hoshi and McKeehan, 1988). TGF-ß inhibits scavenger receptor activity in phorbol-ester-activated THP-1 cells (Bottalico et al., 1991), but stimulates LDL-R in arterial smooth muscle cells (Nicholson and Hajjar, 1992). Thus, the stimulation of LDL-R activity by growth factors may be due to the requisite need for cholesterol to provide membrane biogenesis accompanying the proliferative response to these agents.

The influence of cytokines on cellular cholesterol trafficking is diverse and cell-specific. In macrophages, LPS down-regulates the scavenger receptor (Van Lentin et al., 1985), and thus inhibits acetyl LDL degradation (Geng and Hansson, 1991); this effect is mediated by TNF (Van Lenten and Fogelman, 1992); INF-γ also decreases scavenger receptor activity in macrophages (Schectman et al., 1992; Fong, Fong and Cooper, 1990; Geng and Hansson, 1991). This is in contrast, however, to the effects of GM-CSF (Fibbe et al., 1986), which *stimulates* LDL and acetyl LDL uptake and degradation in macrophages (Ishibashi et al., 1990). In addition, M-CSF increases NCEH and ACEH activities in macrophages (Inaba et al., 1993). TNF increases scavenger receptor activity in arterial smooth muscle cells (Morisaki et al., 1993), and IL-1ß and TNF stimulates LDL-R expression in HEP-G2 cells by increasing transcription without altering mRNA stability (Stopeck et al., 1993), which could not override transcriptional inhibition by LDL-derived cholesterol. If these *in vitro* findings could be extrapolated to *in vivo*, these data suggest that the effects of growth factors and cytokines will depend on their local concentration and site of synthesis and action. Since the liver is the major site of LDL-R expression, levels of circulating cytokines which increase LDL expression may promote clearance of LDL cholesterol from the plasma. However, local production of cytokines and growth factors by intimal and subintimal cells could promote (M-CSF, GM-CSF), or inhibit (TNF, INF-γ) foam cell development by up- or down-regulating expression of the scavenger receptor, respectively. In fact, cytokines, including IL-2 (Wilson et al., 1989), TNF (Feinberg et al., 1988), and INF-γ (Wilson et al., 1990) have been clearly shown to promote hypocholesterolemia in humans and in experimental animals.

Second Messengers And Regulation of Cholesterol Metabolism

LDL-R expression in THP-1 cells is increased by agents which stimulate PKC (Auwerx et al., 1989), by mechanisms which may include increased mRNA transcription of the LDL-receptor gene (Auwerx, Chait and Deeb, 1989). However, expression of the LDL-receptor gene in response to PKC activation is not uniform. Exposure of HL-60 cells to phorbol esters promotes their differentiation, with commensurate loss of LDL-R expression (Frasier-Scott et al., 1988). Since activation of PKC leads eventually to PGI_2 production, and subsequently, increased levels of cyclic AMP, the effects of PGI_2 and cyclic AMP on LDL-R function has been studied. In monocytes, PGI_2 reduces LDL-R activity (Krone et al., 1988). However, in human smooth muscle cells, the effects of cyclic AMP on LDL-R activity are biphasic. Dibutryl cyclic AMP transiently reduces LDL-receptor activity, then dramatically increases LDL-receptor activity (B_{max}, but not K_d) over 24-hours; the latter effect occurs independently of mRNA or protein synthesis (Middleton and Middleton, 1990). However, if cyclic AMP is increased by exposure of smooth muscle cells to phosphodiesterase inhibitors, LDL-receptor activity remains inhibited (Middleton and Middleton, 1990). These activities occur even though a cyclic AMP-responsive element in the LDL-R promoter could not be identified (Takagi and Strauss, 1989). These observations support the premise that stimulation of PKC leads to increased LDL-R expression. Expression of the LDL-receptor is then in turn modulated by the actions of PGI_2 and cyclic AMP.

Since the expression of the scavenger receptor is not modulated by cholesterol, regulation by other cellular mediators may have significant impact on the ability of cells to control the amount of cholesterol incorporated by this pathway. In HL-60 cells, the activity of the scavenger receptor is resistant to PKC activation (Jouni and McNamara, 1991). Again, the response of immortalized macrophage cell lines to phorbol esters is not uniform, since phorbol ester, which promote differentiation in U-937 and THP-1 cells, increased scavenger receptor activity in these cell types (Hayashi et al., 1991). Interestingly, PMA or serum transiently induced the expression of the scavenger receptor in rabbit arterial smooth muscle cells (Pitas, 1990). The promoter region of the scavenger receptor has been cloned. Consensus or near-consensus AP-1 binding sites exist in the upstream regulatory regions of the human and bovine scavenger receptor. Consequently, treatment of cells with phorbol esters increases AP-1 activity of these elements. Unfortunately, there is a paucity of data on the role of other second messengers, such as cyclic AMP and eicosanoids in modulating the activity or expression of the scavenger receptor. In one study, lipoxygenase inhibitors prevent cholesterol accumulation *in macrophages* exposed to acetylated LDL (Schroeff et al., 1985), suggesting that macrophage HETE or leukotriene production may promote cholesterol deposition.

Recent evidence supports the contention that activation of the protein kinase cascade can modulate HMG-CoA reductase activity. The regulation of the expression of hepatic HMG-CoA reductase activity by the processes of phosphorylation and dephosphorylation has been intensively studied (Beg, Stonik and Brewer, 1987). The activity of HMG-CoA reductase is abolished when the enzyme is phosphorylated (either by PKC, AMP-activated protein kinase, or calmodulin-dependent kinase), and reactivated by dephosphorylation (by HMG-CoA reductase phosphatase). Unfortunately, little is known about distal regulatory mechanisms following PKC activation of HMG-CoA reductase activities in vascular cells. In one paper, it was demonstrated that PGI_2, a PGI_2 analog (iloprost) and PGE_1 reduce HMG-CoA reductase in monocytes (Krone et al., 1988), presumably through their ability to stimulate cyclic AMP production.

CE-hydrolytic activities have been demonstrated to be subject to regulation by second messengers. ACEH activity in arterial smooth muscle cells is dependent upon elevations in cyclic AMP production following stimulation by PGI_2, some of its metabolites, 12-hydroxyeicosatetraenoic acid, and 12-20-dihydroxyeicosatetraenoic acid (Hajjar et al., 1982). NCEH activity is stimulated by PGI_2 through activation of the cyclic AMP-dependent protein kinase (Hajjar, 1986; Khoo et al., 1993). Similarly, we and others have shown that NCEH activation is PKC-dependent, and is regulated by phosphorylation/dephosphorylation mechanisms. NCEH activity can be stimulated by eicosanoids, such as PGI_2 and PGE_1, both of which increase cyclic AMP formation. On the other hand, the overwhelming body of evidence suggests that CE synthesis through ACAT is regulated essentially by levels of free cellular cholesterol. However, there is at least one study to suggest that ACAT activity could be regulated by a PKC-dependent mechanism, since phorbol esters increased ACAT in U937 macrophages (Hayashi et al., 1991).

Vascular cells, macrophages, and hepatocytes possess an HDL receptor, which mediates cholesterol desorption from the cell membrane to HDL (Phillips et al., 1980). Recent evidence suggests that HDL may also stimulate intracellular cholesterol flux from the cytoplasm and endoplasmic reticulum to the cell membrane in a PKC-dependent (Mendez, Oram and Bierman, 1991) and cyclic AMP-dependent (Bernard et al., 1991) manner. The mechanism by which PKC activation promotes translocation of cholesterol from intracellular sites to the plasma membrane is as yet unclear, but may be due to its ability to stimulate the synthesis of other second messengers crucial to cholesterol trafficking. HDL stimulates PGI_2 production from smooth muscle cells (Pomerantz et al., 1984). However, the role of eicosanoids in mediating HDL-induced cholesterol efflux is somewhat controversial, since this process may (Shaknov et al., 1989) or may not (Pomerantz and Hajjar, 1990) be eicosanoid-dependent .

In summary, growth factors and cytokines are important cell-derived mediators that can, through signal transduction pathways, alter the expression of receptors and activities of enzymes that regulate cholesterol trafficking. In addition, these functions may be either mediated, or modulated by eicosanoids. These mediators may function in an autocrine and paracrine manner to amplify and/or modulate cytokine-growth factor effects. COX-generated eicosanoids appear to promote net CE hydrolysis in response to growth factor/cytokine activation, and antagonize the proliferative influence of these mediators. However, lipoxygenase-derived eicosanoids may serve to mediate or amplify the effects of cytokine activation. Thus, the net influence of cytokine effects in vascular cells may ultimately depend in part on eicosanoid-modulation of their biological activities.

ARE THESE SYSTEMS ALTERED IN ATHEROSCLEROSIS?

Atherosclerosis is characterized by intimal smooth muscle cell hyperplasia, with CE deposition in the extracellular matrix and in macrophage- and smooth muscle cell-derived foam cells. Histologic evidence of an inflammatory infiltrate in areas of cholesterol accumulation suggests that products of inflammatory cells could predispose to or exacerbate neutral lipid accumulation. This implies that cell-derived mediators, including cytokines, growth factors, and eicosanoids may regulate processes affecting cholesterol trafficking and accumulation. There is now convincing evidence that alterations in growth factor metabolism are associated with atherosclerosis. PDGF mRNA and protein are increased in atheromatous but not normal regions of aorta; these observations assume significance because antibodies to PDGF inhibit neointimal smooth muscle cell accumulation in a vascular injury model (Golden et al., 1991). Interestingly, antibodies to basic FGF also inhibits smooth muscle cell proliferation in response to intimal injury (Lindner and Reidy, 1991). This may be due to either increased bFGF synthesis, as has been described in in vitro studies demonstrating increased bFGF release in smooth muscle cell- derived foam cells (Kraemer et al., 1993), or due to neutralization of bFGF associated with cell matrix or from dead or injured cells. In fact, extracellular matrix may be a physiologically important reservoir for growth factors, as has been demonstrated in an elegant series of studies where macrophage-derived foam cells promoted release of bFGF and TGF-ß from smooth muscle-cell-derived matrices to a greater extent than normal cells (Falcone and Ferenc, 1988). This mechanism of increased growth factor release in atheromatous tissue is attractive since Ross has shown that PDGF production by macrophage-derived foam cells is actually reduced (Ross et al., 1990). IL-1 expression is increased in atherosclerotic areas of aortae from cholesterol-fed rabbits (Moyer et al., 1991). Presence of elevated levels of this cytokine in atheroma is consistent with endothelial and smooth muscle cell activation and insulation of macrophages and lymphocytes. While the influence of purified IL-1 on cholesterol trafficking in vivo has not been elucidated, other cytokines, including TNF, GM-CSF, and IL-2 lower serum cholesterol in humans (Wilson et al., 1989; Feinberg et al., 1988; Motoyoshi and Takaku, 1989), and INF-γ inhibits atherogenesis in rabbits (Wilson et al., 1990). The mechanism rendering cytokines with anti-atherogenic effects is unclear, but may be due to increased hepatic cholesterol removal and clearance, thus depleting the plasma of LDL-derived CE, thereby reducing net cholesterol load to the arterial vasculature. Thus, effects of growth factors and cytokines on net cholesterol and CE content may ultimately depend on their local concentration and site of synthesis and action. Hepatic extraction of LDL via the apo-B receptor is can be activated by cytokines, and may represent a significant mechanism by which plasma LDL is regulated. What is not well understood is the possibility that cytokines may also alter hepatic LDL synthesis and release. This aspect of lipoprotein metabolism needs to be explored. Since the vascular wall may also be a site of local cytokine/growth factor production, cholesterol flux through the arterial wall can also be modulated by these endogenous factors. Since local production of cytokines and growth factors by intimal and subintimal cells could promote (M-CSF, GM-CSF), or inhibit (TNF, INF-γ) foam cell development, the net influence of these cytokines on cholesterol flux may ultimately depend on which cytokine effect predominates. This aspect of cytokine pathobiology still has to be adequately addressed.

Finally, the interaction of the cytokine/growth factor network with second messengers in the context of atherosclerosis has been explored. The proposition that eicosanoids may be involved in atherogenesis is based on pathophysiologic data demonstrating that atherosclerosis is accompanied by reductions in PGI_2 synthetic capacity, with either no change or increase in thromboxane and hydroxyeicosatetraenoic acid production (Pomerantz and Hajjar, 1989; Myers et al., 1991; Wang and Powell, 1991). Furthermore, treatment of cholesterol-fed rabbits with stable PGI_2 analogs ameliorates disease progression (Braun et al., 1993). We have demonstrated that expression of the constitutive and inducible forms of COX are reduced in smooth muscle cell-derived foam cells (Pomerantz, Summers and Hajjar, 1993), in addition to reduced arachidonic acid availability (Pomerantz and Hajjar, 1989). The most obvious aspect of the significance of reduced eicosanoid-synthetic capacity in atherosclerosis is the fact that PGI_2 is a major vasodilator, inhibitor of platelet aggregation, and inhibits smooth muscle cell proliferation. Thus, the loss of PGI_2-dependent processes related to CE hydrolysis and HDL-induced cholesterol efflux could potentially lead to unrestricted cholesterol deposition. In addition, loss of COX-dependent inhibition of cytokine/growth factor-induced smooth muscle cell proliferation and net cholesterol retention could exacerbate the atherogenic process.

SUMMARY

The data summarized in this review demonstrate that the regulation of intracellular cholesterol trafficking is mediated not only by extracellular lipoprotein concentrations and transcriptional responses to alterations in intracellular free cholesterol content. Rather, the modulation of cholesterol trafficking is also regulated by the products synthesized following activation of signal transduction pathways originating at the cell surface. Furthermore, we have identified those cell-derived factors which utilize these signal transduction pathways to elicit alterations in cholesterol trafficking, and demonstrated the importance of the generation of second messengers, most notably eicosanoids, and cyclic AMP in promoting a modulatory influence on specific pro-atherogenic effects of mitogens.

REFERENCES

Atkinson, Y., Murray, A., Krilis, S., Vaas, M., and Lopez, A., 1990, Human tumour necrosis factor alpha (TNF-a) directly stimulates arachidonic acid release in human neutrophils, Immunol. 70:82.

Auwerx, J., Chait, A., and Deeb, S., 1989, Regulation of the low density lipoprotein receptor and hydroxymethylglutaryl coenzyme A reductase genese by protein kinase C and a putative negative regulatory protein, Proc. Natl. Acad. Sci. USA 86:1133.

Auwerx, J., Chait, A., Wolfbauer, G., and Deeb, S., 1989, Involvement of second messengers in regulation of the low-density lipoprotein receptor gene, Mol. Cell. Biol. 9:2298.

Beg, Z., Stonik, J., and Brewer, H., 1987, Modulation of the enzymic activity of 3-hydroxy-3-methylglutarryl coenzyme A reductase by multiple kinase systems involving reversible phosphorylation: a review, Metabolism 36:900.

Bernard, D., Rodriguez, A., Rothblat, G., and Glick, J., 1991, cAMP stimulates cholesteryl ester clearance to high density lipoproteins in J774 macrophages, J. Biol. Chem. 266:710.

Blay, J., and Hollenberg, M., 1989, Epidermal growth factor stimulation of prostacyclin produciton by cultured aortic smooth muscle cells: requirement for increased cellular calcium levels. J. Cell. Physiol. 139:524.

Bottalico, L., Wagner, R., Agellon, L., Assoian, R., and Tabas, I., 1991, Transforming growth factor-ß1 inhibits scavenger receptor activity in THP-1 human macrophages, J. Biol. Chem. 266:22866.

Braun, M., Hohlfeld, T., Kienbaum, P., Weber, A.A., Sarbia, M., and Schror, K., 1993, Antiatherosclerotic effects of oral cicaprost in experimental hypercholesterolemia in rabbits, Athero. 103:93.

Breviairio, F., Proserpio, P., Bertocchi, F., Lampugnani, M., Mantovani, A., and Dejana, E., 1990, Interleukin-1 stimulates prostacyclin production by cultured human endothelial cells by increasing mobilization and conversion, Arterioscler. 10:129.

Briggs, M., Yokoyama, C., Wang, X., Brown, M., and Goldstein, J., 1993, Nuclear protein that binds sterol regulatory element of low density lipoprotein receptor promoter. I. Identification of the protein and delineation of its target nucleotide sequence, J. Biol. Chem. 268:14490.

Bull, H., Rustin, M., Spaull, J., Cohen, J., Wilson-Jones, E., and Dowd, P., 1990, Pro-inflammatory mediators induce sustained release of prostaglandin E_2 from human dermal microvascular endothelial cells, Br. J. Dermatol. 122:153.

Chait, A., Ross, R., and Bierman, E., 1988, Stimulation of receptor-dependent and receptor-independent pathways of low density lipoprotein degradation in arterial smooth muscle cells by platelet-derived growth factor, Biochim. Biophys. Acta 960:183.

Chen, J., Hoshi, H., and McKeehan, W., 1988, Heparin-binding growth factor type one and platelet-derived growth factor are required for the optimal expression of cell surface low density lipoprotein receptor binding activity in human adult arterial smooth muscle cells, In Vitro 24:199.

Cuthbert, J., and Lipsky, P., 1990, Mitogenic stimulation alters the regulation of the LDL receptor gene expression in human lymphocytes, J. Lipid. Res. 31:2067.

Dawson, P., Hofmann, S., Westhuyzen van de, D., Sudhof, T., Brown, M., and Goldstein, J., 1988, Sterol-dependent repression of low density lipoprotein receptor promoter mediated by 16-base pair sequence adjacent to binding site for transcription factor Sp1, J. Biol. Chem. 263:3372.

Dembinska-kiec, A., Rucker, W., and Schonhofer, P., 1980, Effects of PGI_2 and PGI_2 analogs on cAMP levels in cultured endothelial and smooth muscle cells derived from bovine arteries, Naunyn-Schmied. Arch. Pharm. 311:67.

Falcone, D., and Ferenc, M., 1988, Acetyl-LDL stimulates macrophage-dependent plasminogen activation and degradation of extracellular matrix. J. Cell. Physiol. 135:387.

Feinberg, B., Kurzrock, R., Talpaz, M., Blick, M., and Saks, S., 1988, A phase-I trial of intravenously-administered recombinant tumor necrosis factor in cancer patients, J. Clin. Oncol. 6:1328.

Fibbe, W., Vandamme, J., Biliau, A., Voogt, P., Duinkerken, N., Kluck, P., and Falkenburg, F., 1986, Interleukin-1 (22-K Factor) induces the release of granulocyte-macrophage colony stimulating factor from human mononuclear phagocytes, Blood 68:1316.

Floege, J., Topley, N., Wessel, K., Kaever, V., Radeke, H., Hoppe, J., Kishimoto, T., and Resch, K., 1990, Monokines and platelet-derived growth factor modulate prostanoid production in growth arrested, human mesangial cells, Kidney Int. 37:859.

Fong, L., Fong, A., and Cooper, A., 1990, Inhibition of mouse macrophage degradation of acetyl-low density lipoprotein by interferon-gamma, J. Biol. Chem. 265:11751.

Frasier-Scott, K., Hatzakis, H., Seong, D., Jones, C., and Wu, K., 1988, Influence of natural and recombinant interleukin 2 on endothelial cell arachidonate metabolism: induction of de novo synthesis of prostaglandin H synthase, J. Clin. Invest. 82:1877.

Gagnon, L., Filion, L., Dubois, C., and Rola-Pleszczynski, M., 1989, Leukotrienes and macrophage activation: augmented cytotoxic activity and enhanced interleukin 1, tumor necrosis factor, and hydrogen peroxide production, Agents and Actions 26:142.

Geng, Y., and Hansson, G., 1991, Interferon-gamma inhibits scavenger receptor expression and foam cell formation in human macrophages, FASEB J. 5:A1246.

Glaser, K., Asmis, R., and Dennis, E., 1990, Bacterial lipopolysacchride priming of the P388D1 macrophage-like cells for enhanced arachidonic acid metabolism. Platelet-activating factor activation of phospholipase A_2, J. Biol. Chem. 265:8658.

Golden, M., Au, Y., Kirkman, T., Wilcox, J., Raines, E., Ross, R., and Clowes, A., 1991, Platelet-derived growth factor activity and mRNA expression in healing vascular grafts in baboons. Association in vivo of platelet-derived growth factor mRNA and protein with cellular proliferation, J. Clin. Invest. 87:406.

Graves, L.M., Bornfeldt, K.E., Raines, E.W., Potts, B.C., Macdonald, S.G., Ross, R., and Krebs, E.G., 1993, Protein kinase-A antagonizes platelet-derived growth factor- induced signalling by mitogen-activated protein kinase in human arterial smooth muscle cells, Proc. Natl. Acad. Sci. USA 90:10300.

Habenicht, A., Glomset, J., King, W., Nist, C., and Mitchel, C., 1981, Early changes in phosphatidylinositol and arachidonic acid metabolism in quiescent Swiss 3T3 cells stimulated to divide by platelet-derived growth factor, J. Biol. Chem. 256:12329.

Habenicht, A., Goerig, M., Grulich, J., Rothe, D., Gronwald, R., Loth, U., Schettler, G., Kommerell, B., and Ross, R., 1985, Human platelet-derived growth factor stimulates prostaglandin synthesis by activation and by rapid de novo synthesis of cyclooxygenase, J. Clin. Invest. 75:1381.

Habernicht, A., Glomset, J., and Ross, R., 1980, Relation of cholesterol and mevalonic acid to the cell cycle in smooth muscle and swiss 3T3 cells stimulated to divide by platelet-derived growth factor, J. Biol. Chem. 255:5134.

Hajjar, D., 1986, Regulation of neutral cholesteryl esterase in arterial smooth muscle cells: stimulation by agonists of adenylate cyclase and cyclic AMP-dependent protein kinase, Arch. Biochem. Biophys. 247:49.

Hajjar, D., Weksler, B., Falcone, D., Hefton, J., Tack-Goldman, K., and Minick, C., 1982, Prostacyclin modulates cholesteryl ester hydrolytic activity by its effect on cyclic adenosine monophosphate in rabbit aortic smooth muscle cells. J. Clin. Invest. 70:479.

Haliday, E., Ramesha, C., and Ringold, G., 1991, TNF induces c-fos via a novel pathway requiring conversion of arachidonic acid to a lipoxygenase metabolite, Embo J. 10:109.

Hayashi, K., Dojo, S., Hirata, Y., Ohtani, K., Nakashima, E., Nishio, H., Kurushima, H., Saeki, M., and Kajiyama, G., 1991, Metabolic changes in LDL receptors and an appearance of scavenger receptors after phorbol ester-induced differentiation of U937 cells, Biochim. Biophys. Acta 1082:152.

Heldin, N., Paulsson, Y., Forsberg, K., Heldin, C., and Westermark, B., 1989, Induction of cyclic AMP synthesis is followed by a reduction in the expression of c-myc messenger RNA and inhibition of ^{3}H-thymidine incorporation in human fibroblasts, J. Cell. Physiol. 138:17.

Hla, T., and Neilson, K., 1992, Human cyclooxygenase-2 cDNA, Proc. Natl. Acad. Sci. USA 89:7384.

Horiguchi, J., Spriggs, D., Imamura, K., Stone, R., Luebbers, R., and Kufe, D., 1989, Role of arachidonic acid metabolism in transcriptional induction of tumor necrosis factor gene expression by phorbol ester, Mol. Cell. Biol. 9:252.

Inaba, T., Gotoda, T., Shimano, H., Shimada, M., Harada, K., Kozaki, K., Watanabe, Y., Eitetsu, H., Motoyoshi, K., Yazaki, Y., and Yamada, N., 1992, Platelet-derived growth factor induces c-fms and scavenger receptor genes in vascular smooth muscle cells, J. Biol. Chem. 267:13107.

Inaba, T., Shimano, H., Gotoda, T., Harada, K., Shimada, M., Kawamura, M., Yazaki, Y., and Yamada, N., 1993, Macrophage colony-stimulating factor regulates both activities of neutral and acidic cholesteryl ester hydrolases in human monocyte-derived macrophages, J. Clin. Invest. 92:750.

Ishibashi, S., Inaba, T., Shimano, H., Herada, K., Inoue, I., Mokuno, H., Mori, N., Gotoda, T., Takaku, F., and Yamada, N., 1990, Monocyte colony-stimulating factor enhances uptake and degradation of acetylated low density lipoproteins and cholesterol esterification in human monocyte-derived macrophages, J. Biol. Chem. 265:14109.

Iwashita, S., Mitsui, M., Shoji-Kasai, Y., and Senshu-Miyaike, M., 1990, cAMP-mediated modulation of signal transduction of epidermal growth factor (EGF) receptor systems n huan epidermoid carcinoma A431 cells. Depression of EGF-dependent diacylglycerol production and EGF receptor phosphorylation, J. Biol. Chem. 265:10702.

Jouni, Z., and McNamara, D., 1991, Lipoprotein receptors of HL-60 macrophages. Effect of differentiation with tetramyristic phorbol acetate and 1,25-dihydroxyvitamin D3, Arterio. and Thromb. 11:995.

Kavanaugh, W., Harsh, G., Starksen, N., Rocco, C., and Williams, L., 1988, Transcriptional regulation of the A and B chain genes of platelet-derived growth factor in microvascular endothelial cells, J. Biol. Chem. 263:8470.

Kerr, J., Stevens, T., Davis, G., McLaughlin, J., and Harris, R., 1989, Effects of recombinant interleukin-1 beta on phospholipase A_2 activity, phospholipase A_2 mRNA levels, and eicosanoid formation in rabbit chondrocytes, Biochem. Biophys. Res. Comm. 165:1079.

Khoo, J.C., Reue, K., Steinberg, D., and Schotz, M.C., 1993, Expression of hormone-sensitive lipase messenger RNA in macrophages, J. Lipid Res 34:1969.

Kim, U., Sim, H. and Rhee, S., 1990, Epidermal growth factor and platelet-derived growth factor promote translocation of phospholipase C-_ from cytosol to membrane, FEBS Lett. 270:33.

Koch, C., Anderson, D., Moran, M., Ellis, C., and Pawson, T., 1991, SH2 and SH3 domains: elements that control interactions of cytoplasmic signalling proteins, Science 252:668.

Korn, J., 1983, Fibroblast prostaglandin E_2 synthesis. Persistance of an abnormal phenotype after short-term exposure to mononuclear cell products, J. Clin. Invest. 71:1240.

Kraemer, R., Pomerantz, K., Joseph-Silverstein, J., and Hajjar, D., 1993, Induction of basic FGF mRNA and protein synthesis in smooth muscle cells by cholesterol-enrichment and 25-hydroxy-cholesterol, J. Biol. Chem. 268:8040.

Krone, W., Klass, A., Nagele, H., Behnke, B., and Greten, H., 1988, Effect of prostaglandins on LDL receptor activity and cholesterol synthesis in freshly isolated human mononuclear leukocytes, J. Lipid. Res. 29:1663.

Kujubu, D., and Herschman, H., 1992, Dexamethasone inhibits mitogen induction of the TIS10 prostaglandin synthase cyclooxygenase gene, J. Biol. Chem. 267:7991.

Libby, P., Warner, S., and Friedman, G., 1988, Interleukin 1: a mitogen for human vascular smooth muscle cells that induces the release of growth-inhibitory prostanoids, J. Clin. Invest. 81:487.

Lindner, V., and Reidy, M., 1991, Proliferation of smooth muscle cells after vascular injury is inhibited by an antibody against basic fibroblast growth factor, Proc. Natl. Acad. Sci. USA 88:3739.

Mazzone, T., Basheeruddin, K., Ping, L., and Schick, C., 1990, Relation of growth- and sterol-regulatory pathways for low density lipoprotein receptor gene expression, J. Biol. Chem. 265:5145.

Mehta, K., Brown, M., Bilheimer, D., and Goldstein, J., 1991, The low density receptor in Xenopus laevis. II. Feedback repression mediated by conserved sterol regulatory element, J. Biol. Chem. 266:10415.

Mendez, A., Oram, J., and Bierman, E., 1991, Protein kinase C as a mediator of high density lipoprotein receptor-dependent efflux of intracellular cholesterol, J. Biol. Chem. 266:10104.

Middleton, A., and Middleton, B., 1990, Agents which increase cyclic AMP have diverse effects on low-density-lipoprotein-receptor function in human vascular smooth muscle cells and skin fibroblasts, Biochem. J. 267:607.

Morisaki, N., Xu, Q.P., Koshikawa, T., Saito, Y., Yoshida, S., and Ueda, S., 1993, Tumour necrosis factor-alpha can modulate the phenotype of aortic smooth muscle cells, Scand. J. Clin. Lab. Invest. 53:347.

Motoyoshi, K., and Takaku, F., 1989, Serum cholesterol-lowering activity of human monocytic colony-stimulating factor, Lancet 326.

Moyer, C., Sajuthi, D., Tulli, H., and Williams, J., 1991, Synthesis of IL-1 alpha and IL-1 beta by arterial cells in atherosclerosis, Amer. J. Pathol. 138:951.

Myers, S., Russell, D., Parks, L., and Reed, M., 1991, Triphasic response of prostacyclin production in rabbit thoracic aorta in early atherosclerosis, Prosta. Leuko. Essent. Fatty Acids 44:31.

Nicholson, A., and Hajjar, D., 1992, Transforming growth factor-beta up-regulates low density lipoprotein receptor-mediated cholesterol metabolism in vascular smooth muscle cells, J. Biol. Chem. 267:25982.

O'Banion, M., Sadowski, H., Winn, V., and Young, D., 1991, A serum- and glucocorticoid-regulated 4-kilobase mRNA encodes a cyclooxygenase-related protein, J. Biol. Chem. 266:23261.

Oppenheimer, M., Oram, J., and Bierman, E., 1987, Downregulation of high density lipoprotein receptor activity of cultured fibroblasts by platelet-derived growth factor, Arterio. 7:325.

Phillips, M., McClean, L., Stoudt, G., and Rothblat, G., 1980, Mechanism of cholesterol efflux from cells. Atheroscler. 36:409.

Pitas, R., 1990, Expression of the acetyl low density lipoprotein receptor by rabbit fibroblasts and smooth muscle cells, J. Biol. Chem. 265:12722.

Pomerantz, K., and Hajjar, D., 1989, Eicosanoid metabolism in cholesterol-enriched arterial smooth muscle cells: reduced arachidonate release with concommitant decrease in cyclooxygenase products, J. Lipid. Res. 30:1219.

Pomerantz, K., and Hajjar, D., 1989, Eicosanoids in regulation of arterial smooth muscle cell phenotype, proliferative capacity, and cholesterol metabolism, Arterioscler. 9:413.

Pomerantz, K., and Hajjar, D., 1990, High density lipoprotein-induced cholesterol efflux from arterial smooth muscle cell-derived foam cells: functional relationship of the cholesteryl ester cycle and eicosanoid biosynthesis, Biochem. 29:1892.

Pomerantz, K., Summers, B., and Hajjar, D., 1993, Eicosanoid metabolism in cholesterol-enriched smooth muscle cells - evidence for reduced post-transcriptional processing of cyclooxygenase I and reduced cyclooxygenase II gene expression, Biochem. 32:13624.

Pomerantz, K., Tall, A., Feinmark, S., and Cannon, P., 1984, Stimulation of vascular smooth muscle cell prostacyclin and prostaglandin E_2 synthesis by plasma high and low density lipoproteins, Circ. Res. 54:554.

Ross, R., 1986, The pathogenesis of atherosclerosis - an update, N. Engl. J. Med. 314:488.

Ross, R., and Glomset, J., 1976, The pathogenesis of atherosclerosis, N. Engl. J. Med. 295:369.

Ross, R., Glomset, J., Kariya, B., and Harker, L., 1974, A platelet-dependent serum factor that stimulates the proliferation of arterial smooth muscle cells in vitro, Proc. Natl. Acad. Sci. USA 71:1207.

Ross, R., Masuda, J., Raines, E., Gown, A., Kutsuda, S., Sasahara, M., Malden, L., Masuko, H., and Sato, H., 1990, Localization of PDGF-B protein in macrophages in all phases of atherogenesis, Science 248:1009.

Rossi, V., Breviario, F., Ghezzi, P., Dejana, E., and Mantovani, A., 1985, Prostacyclin synthesis induced in vascular cells by interleukin-1, Science 229:174.

Rustin, M., Bull, H., and Dowd, P., 1989, Effect of human recombinant interleukin -1alpha on release of prostacyclin from human endothelial cells, Br. J. Dermatol. 120:153.

Schectman, G., Kaul, S., Mueller, R., Borden, E., and Kissebah, A., 1992, The effect of interferon on the metabolism of LDLs, Arterioscler. and Thromb. 12:1053.

Schroeff, J., Havekes, L., Weerheim, A., Emeis, J,. and Vermeer, B., 1985, Suppression of cholesteryl ester accumulation in cultured human monocyte-derived macrophages by lipoxygenase inhibitors, Biochem. Biophys. Res. Comm. 127:366.

Shaknov, Y., Larrue, J., Perova, N., Dorian, B., Daret, D., Shcherbakova, I., Bricaud, H., and Oganov, R., 1989, Prostacyclin-mediated efflux from high density lipoproteins as cellular cholesterol acceptors on aortic smooth muscle cells, J. Mol. Cell. Cardiol. 21:461.

Sherman, M., Weber, B., Datta, R., and Kufe, D., 1990, Transcriptional and posttranscriptional regulation of macrophage-specific colony stimulating factor gene expression by tumor necrosis factor. Involvement of arachidonic acid metabolites, J. Clin. Invest. 85:442.

Simon, M., Strathmann, M., and Gautam, N., 1991, Diversity of G-proteins in signal transduction, Science 252:802.

Stopeck, A., Nicholson, A., Mancini, F., and Hajjar, D., 1993, Cytokine regulation of LDL receptor gene transcription in HEP G2 cells, J. Biol. Chem. 268:17489.

Takagi, K. and Strauss, F., 1989, Control of low density lipoprotein receptor gene expression in steroidogenic cells, Can. J. Physiol. Pharmacol. 67:968.

Van Lenten, B., and Fogelman, A., 1992, Lipopoly-sacchride-induced inhibition of scavenger receptor expression in human monocyte-macrophages is mediated through tumor necrosis factor-alpha, J. Immunol. 148:112.

Van Lentin, B., Fogelman, A., Seager, J., Ribi, E., Haberland M., and Edwards, P., 1985, Bacterial endotoxin selectively prevents the expression of scavenger-receptor activity on human monocyte-macrophages, J. Immunol. 134:3718.

Wang, T. and Powell, W., 1991, Increased levels of monohydroxy metabolites of arachidonic acid and linoleic acid in LDL and aorta from atherosclerotic rabbits, Biochim. Biophys. Acta 1084:129.

Wang, X., Briggs, M., Hua, X., Yokoyama, C., Goldstein J., and Brown, M., 1993, Nuclear protein that binds sterol regulatory element of low density lipoprotein receptor promoter. II. Purification and characterization, J. Biol. Chem. 268:14497.

Williams, L., 1989, Signal transduction by the platelet-derived growth factor, Science 243:1564.

Wilson, A., Schaub, R., Goldstein, R., and Kuo, P., 1990, Suppression of aortic atherosclerossis in cholesterol-fed rabbits by purified rabbit interferon, Arterioscler. 10:208.

Wilson, E., Birchfield, G., Hejazi, J., Ward, J., and Samlowski, W., 1989, Hypocholesterolemia in patients treated with recombinant interleukin-2: appearance of renmant-like lipoproteins, J. Clin. Oncol. 7:1573.

GENETIC DETERMINANTS OF MYOCARDIAL INFARCTION

Jan L. Breslow and Marilyn Dammerman

Laboratory of Biochemical
    Genetics and Metabolism
The Rockefellow University
New York, NY 10021

INTRODUCTION

A family history of myocardial infarction (MI), especially MI at an early age, is a potent risk factor for coronary artery disease (CAD). The risk increases with the number of first-degree relatives affected and is inversely related to the age at which they became affected (Roncaglioni et al., 1992). While some of this increased risk is due to shared environment, genetic factors appear to predominate (Nora et al., 1980). Monozygotic (identical) twins are significantly more likely to be concordant for CAD than are dizygotic (fraternal) twins (Goldbourt and Neufeld, 1986). Dyslipidemia, diabetes mellitus, hypertension and obesity, the major metabolic risk factors for CAD, are in large measure genetically determined. In addition, a family history of MI confers increased risk in both genders independent of other known risk factors (Colditz et al., 1991; Roncaglioni et al., 1992).

CAD and the conditions predisposing to it are, in their common forms, complex disorders with multiple genetic and environmental components. Such disorders are difficult to study genetically due to late age of onset, the likelihood that not all individuals with a genetic predisposition develop the disease, and the probability that multiple genes--and different genes in different families--are involved. However, clues to the genetic factors underlying CAD have been obtained from studies of single-gene disorders. Considerable progress has been made in elucidating the genetic basis of lipoprotein abnormalities. The genes for many proteins involved in lipid metabolism have been isolated, and relatively common as well as rare mutations have been found (Breslow, 1991; Zannis et al., 1993). Single genes causing rare forms of diabetes have been identified (Steiner et al., 1990; Froguel et al., 1993; van den Ouweland et al., 1992; Taylor et al., 1992), as well as common susceptibility genes for hypertension and MI (Jeunemaitre et al., 1992; Cambien et al., 1992).

In the vast majority of individuals, the genetic basis of CAD is likely to lie in multiple subtly abnormal genes, and it is clear that adverse lifestyle choices can lead to expression of this genetic potential. It is probable that the development of appropriate genetic markers will facilitate identification of those members of CAD-prone families, and perhaps members of the general public as well, who are likely to develop CAD or its predisposing disorders.

Nutrition and Biotechnology in Heart Disease and Cancer
Edited by J.B. Longenecker *et al.*, Plenum Press, New York, 1995

This information would presumably take the form of a genetic profile specifying the alleles carried at genes known to strongly influence CAD risk. In the future, intensive preventive efforts with respect to modifiable CAD risk factors may focus on carriers of high-risk alleles. At present, all patients with a family history of CAD or related disorders should receive special attention with respect to modifiable risk factors.

GENETIC FACTORS PREDISPOSING TO ATHEROSCLEROSIS: LIPOPROTEIN DISORDERS

Lipoprotein disorders result from abnormal synthesis, processing or catabolism of plasma lipoprotein particles. These particles consists of a core of cholesterol ester and triglyceride enclosed in a coat of phospholipids and apolipoproteins. More than half of patients with angiographically confirmed CAD before age 60 have a familial lipoprotein disorder (Genest et al., 1992). The association is most striking among younger patients and declines with increasing age at first MI. This suggests the presence of genetic factors that accelerate age-associated cardiovascular changes seen in the general population (Bierman, 1991). Severe hyperlipidemia (total cholesterol above 300 mg/dL or triglycerides above 500 mg/dL) usually indicates a genetic disorder, and xanthomas almost always signal an underlying genetic defect. These findings warrant examination of the patient's first-degree relatives (Bierman, 1991). Four types of lipoprotein abnormalities are observed: elevated LDL cholesterol; reduced high density lipoprotein (HDL) cholesterol, usually with increased triglycerides and very low density lipoprotein (VLDL) cholesterol; elevated levels of chylomicron remnants and intermediate density lipoproteins (IDL); and elevated levels of Lp(a) particles (Breslow, 1991).

Elevated LDL Cholesterol

Cholesterol levels over 240 mg/dL are associated with a three-fold increased risk of death from ischemic heart disease in men relative to cholesterol levels below 200 mg/dL, and there is a continuous risk gradient as cholesterol rises (Bierman, 1991). Elevated total cholesterol primarily reflects elevated LDL cholesterol, which constitutes 70% of plasma cholesterol. Disorders characterized by elevation of cholesterol alone are classified as Fredrickson Type IIa hyperlipoproteinemia.

Familial hypercholesterolemia is a relatively common cause of elevated LDL cholesterol and is present in approximately 5% of patients with MI (Brown and Goldstein, 1991). An autosomal dominant disorder, FH is due to a defective LDL receptor gene on chromosome 19. As in other dominant disorders, 50% of first-degree relatives are affected. Since there is a gene dosage effect, patients inheriting a defective gene from both parents are severely affected. Reduction in or absence of functional cell surface LDL receptor molecules impairs LDL catabolism, and failure to carry out receptor-mediated IDL uptake results in enhanced IDL conversion to LDL.

FH homozygotes typically have six-fold elevations in LDL cholesterol, with total cholesterol levels of 650 to 1000 mg/dL, and they can be identified at birth by markedly elevated cholesterol in umbilical cord blood (Goldstein and Brown, 1989). CAD is often clinically apparent before age 10, with MI as early as 18 months of age (Brown and Goldstein, 1991), and most homozygotes suffer fatal MI by age 30. A unique type of planar cutaneous xanthoma is present at birth or develops in childhood, often between the thumb and index finger, and many patients are first identified by dermatologists. However, diagnosis may be delayed until the appearance of angina pectoris or until an episode of syncope due to xanthomatous aortic stenosis (Brown and Goldstein, 1991). A diagnosis of pediatric FH often serves as the impetus for further case-finding in the family.

FH heterozygotes have LDL cholesterol levels twice normal or approximately 140 mg/dL higher than family members with two normal genes. Five percent of males have had an MI by age 30, 25% have died of MI by age 50, and 50% have died by age 60 (Goldstein and Brown, 1989). Onset of MI is typically delayed by 10 years in women. Heterozygous FH may be distinguishable from most other hypercholesterolemias by the presence of nodular xanthomas of the Achilles and other tendons, seen in up to 75% of heterozygotes (Brown and Goldstein, 1991). Diabetes and obesity are not associated with FH, and a slender physique is typical (Brown and Goldstein, 1991). Clinical presentation may be influenced by other genetic and lifestyle factors, however.

At least 150 different LDL receptor mutations have been described. In families in which the specific mutation is known, genetic testing can be used to identify additional affected individuals, including prenatally. In certain ethnic groups, FH is unusually common and one or a few mutations predominate due to a founder effect, making it feasible to genetically screen members of these groups. These include French Canadians, Lebanese Christians, Jews of Lithuanian origin, Finns and South African Afrikaners.

A second relatively common single-gene disorder causing elevated LDL cholesterol, familial defective apo B100 (FDB), is due to a mutation in the apo B gene on chromosome 2 (Breslow, 1991; Zannis et al., 1993). The resulting amino acid substitution disrupts apo B binding to the LDL receptor, impairing LDL uptake. Heterozygosity for this disorder increases LDL cholesterol levels by at least 50% (60 to 80 mg/dL) relative to unaffected family members. In general FDB may be clinically milder than FH, but many patients have tendon xanthomas and cholesterol levels may fall within the FH range. In some cases the two disorders are distinguishable only by genetic tests, and the approach to treatment is the same.

In the general population, genetic variation in apo E accounts for 3%-5% of the variance in LDL cholesterol levels. Apo E mediates hepatic uptake of chylomicron remnants as well as IDL particles. Apo E is encoded by a gene cluster on chromosome 19 that also encodes apolipoproteins CI and CII. Apo E genotyping has revealed three common alleles in the population: E3 (Caucasian frequency 77%), E4 (15%) and E2 (8%). Individuals with the E4/3 genotype (the E4 protein is synthesized from one parental allele and E3 from the other) have mean LDL cholesterol levels 5-10 mg/dL higher than subjects with the most common genotype, E3/3, while individuals with E3/2 have LDL cholesterol levels 10-20 mg/dL lower than E3/3 subjects. Apo E genetic variation also influences atherosclerosis progression in the general population. In an autopsy study of more than 500 young male trauma victims, E3/2 was associated with reduced atherosclerosis relative to E3/3 (Hixon, 1991). This was observed in both whites and blacks and was only partially explained by apo E-associated differences in cholesterol levels. Apo E genotypes accounted for 6%-7% of the variance in aortic lesion extent.

Familial combined hyperlipidemia (FCHL), a complex disorder of unknown etiology, is the most common genetic hypercholesterolemia and one of the most frequent causes of premature CAD. FCHL (Type IIb hyperlipoproteinemia) is characterized by elevations of LDL cholesterol and VLDL triglyceride (>95th percentile for age and gender) within the same family. Affected subjects may have one or both abnormalities, and the lipid profile may vary over time. Xanthomas are uncommon. VLDL and apo B are overproduced in FCHL, and this may lead to elevated plasma VLDL and hypertriglyceridemia in some family members, while in others with more efficient lipolysis the consequence is elevated LDL. Hyperlipidemia appears in 10%-20% of patients in childhood, usually as hypertriglyceridemia (Kane and Havel, 1989). Patients usually have a strong family history of premature CAD, and hypertriglyceridemic family members seem to be at the same increased risk as those with hypercholesterolemia (Bierman, 1989). FCHL accounts for 10% of individuals with LDL cholesterol >95th percentile and is present in approximately 10% of patients with MI (Brown and Goldstein, 1991). The disorder is exacerbated by obesity, diabetes, hypothyroidism, exogenous estrogen and alcohol. Patients are often responsive to low-fat diet and exercise, but drug therapy may be necessary.

Family and twin studies indicate that half the population variance in LDL cholesterol is genetic. Approximately 7% of the variance is explained by known LDL receptor, apo B and apo E mutations, with the preponderance unexplained (Breslow, 1991). Individuals differ in their responses to dietary fat and cholesterol, and genes controlling diet response may explain much of the variation in LDL cholesterol. These genes may influence such processes as cholesterol absorption and synthesis, bile acid synthesis, and LDL receptor synthesis and catabolism (Brown and Goldstein, 1991).

Reduced HDL Cholesterol and Elevated Triglycerides

Reduced HDL cholesterol (below 35 mg/dL) has been found, in a large number of studies, to be a potent independent risk factor for CAD. In a study of patients with confirmed CAD prior to age 60, reduced HDL cholesterol was the most common lipoprotein abnormality (39% of cases) (Genest et al., 1992). Approximately half the population variance in HDL cholesterol is likely to be attributable to genetic factors. HDL cholesterol levels are strongly correlated with apo AI levels. Inter-individual differences in HDL cholesterol and apo AI levels correlate best with the apo AI fractional catabolic rate (FCR) rather than with the synthetic rate. FCR increases with decreasing particle size, which in turn correlates with increased triglycerides (Breslow, 1991). Thus, in many individuals elevated triglycerides may be the driving force lowering HDL cholesterol levels. This may result from increased exchange of HDL cholesterol ester for VLDL triglycerides and from decreasing HDL size due to subsequent hepatic lipase action. In individuals with normal triglycerides, other mechanisms may result in reduced HDL cholesterol levels. For example, a low-fat diet, which is associated with protection against CAD, lowers HDL cholesterol by decreasing apo AI production (Brinton et al., 1990).

A small number of patients with defective HDL production due to mutations in the apo AI gene on chromosome 11 have been described. These mutations preclude synthesis of the protein and, in the homozygous state, are characterized by very low HDL cholesterol (Breslow, 1989). Most of these patients develop planar xanthomas and CAD between the ages of 25 and 50. Deficiency of the LCAT enzyme results in abnormal lipoprotein particles of all types due to inability to esterify free cholesterol, large quantities of which accumulate in plasma and tissues (Zannis et al., 1993). This rare disorder is caused by homozygosity for mutations in the LCAT gene on chromosome 16. HDL cholesterol may be markedly reduced, and patients may have premature CAD as well as renal damage due to the lipoprotein disorder. Several other disorders have been described that result in low HDL cholesterol but not in premature CAD. These include Tangier disease, characterized by abnormally rapid HDL clearance; dyslipidemia due to certain apo AI amino acid substitutions; and homozygous LPL deficiency. While it is clear that low HDL cholesterol is an important risk factor for CAD in the general population, whether this is a direct cause-and-effect relationship or whether low HDL cholesterol exacerbates the risk of an otherwise atherogenic lipoprotein profile remains to be resolved.

Two deficiencies of proteins involved in HDL processing have been found to raise HDL cholesterol levels (Breslow, 1991; Zannis et al., 1993). In hepatic lipase deficiency, there is premature CAD despite elevated HDL cholesterol. This rare disorder is characterized by a failure to remodel HDL_2 to HDL_3 due to a mutation in the hepatic lipase gene on chromosome 15. The second disorder, CETP deficiency, is due to mutations in the CETP gene on chromosome 16. These result in elevated apo AI and HDL cholesterol levels and reduced apo B and LDL cholesterol levels, a profile associated with protection against CAD and with longevity. CETP mutations occur frequently among Japanese and in this population are a common cause of high HDL cholesterol (Tall, 1993). Apart from cases of

CETP deficiency, families with unusually high HDL cholesterol and apo AI levels have been reported. Simple Mendelian inheritance is generally not observed, and the genetic factors responsible have yet to be identified (Breslow, 1989).

Single-gene syndromes probably account for a very small fraction (perhaps 1%) of the population variance in HDL cholesterol levels. A large number of studies have failed to demonstrate a major single-gene effect on HDL cholesterol in the general population (Breslow, 1989). Based on family studies, however, a very significant proportion of the variance (40%-60%) appears attributable to polygenic inheritance. This could result from common sequence variations in a small number of genes, and it seems likely that genes regulating HDL particle size, including those influencing the activity of HDL-processing proteins, may play an important role. The LPL gene was shown to influence HDL_3 cholesterol levels in a recent study of families of CAD victims (Coresh et al., 1993).

Elevated triglycerides are generally found together with low HDL cholesterol and these appear to be causally linked. In relatives of CAD patients with familial lipoprotein disorders, an elevated triglyceride level with reduced HDL cholesterol was nearly four-fold more common than reduced HDL cholesterol alone (Genest et al., 1992). In some but not all studies, elevated triglycerides have been shown to be an independent CAD risk factor (Austin, 1991). Nearly one fourth of patients with CAD prior to age 60 had hypertriglyceridemia in a recent study (Genest et al., 1992). These cases were more often familial than sporadic and were almost always associated with low HDL cholesterol. Among diabetics, hypertriglyceridemia constitutes a major CAD risk factor (Chait and Brunzell, 1990).

Elevated triglycerides are common in the population. Ten percent of males aged 35-39, for example, have triglycerides >250 mg/dL. Genetic hypertriglyceridemia may present as elevated triglycerides in the absence of secondary causes or as unusual triglyceride sensitivity to alcohol, exogenous estrogen, weight gain, hyperglycemia or hypothyroidism. While moderate triglyceride elevations may be secondary to other metabolic abnormalities or to exogenous agents, severe hypertriglyceridemia (>1000 mg/dL) generally reflects a genetic propensity.

Familial hypertriglyceridemia is genetically heterogeneous, and in most cases the genetic basis is unknown. In some families the disorder appears to be inherited in an autosomal dominant manner, while in others recessive or non-Mendelian patterns are observed. Hypertriglyceridemia is strongly associated with hyperinsulinemia, hyperglycemia and hypertension, a constellation of features termed syndrome X (Reaven, 1988). Severe fasting chylomicronemia (Type I hyperlipoproteinemia) is due to homozygosity for mutations in the gene for LPL (chromosome 8) or for the LPL cofactor apo CII (chromosome 19). Type I disease is characterized by very low or undetectable LPL activity and very high triglycerides but not by premature CAD. This suggests that elevated triglycerides are not inherently atherogenic but are a feature of several different metabolic disorders, only some of which are atherogenic.

In Caucasians hypertriglyceridemia is strongly associated with the minor (less common) allele of a two-allele polymorphism near the coding sequence for apolipoprotein CIII (apo CIII) on chromosome 11 (Rees et al., 1983). In one study, this marker was present in 44% of patients with severe hypertriglyceridemia and in 17% of controls (Dammerman et al., 1993). Association of this and other polymorphisms in the apo AI/CIII/AIV gene cluster with CAD among subjects with a family history of CAD has been reported (Price et al., 1989). Genetic factors conferring protection against hypertriglyceridemia also reside in this region (Dammerman et al., 1993). Apo CIII is a major protein constituent of chylomicrons and VLDL particles. Normolipidemic subjects carrying the minor allele of the apo CIII polymorphism have higher apo CIII levels than noncarriers (Shoulders et al., 1991), and apo CIII and triglyceride levels are strongly correlated. Apo CIII inhibits LPL *in vitro* and may also inhibit hepatic uptake of triglyceride-rich particles and their remnants. Transgenic mice overexpressing the human apo CIII gene have severe hypertriglyceridemia (Breslow, 1993), making this the strongest candidate gene in this region with respect to regulation of triglycerides. Apo E4 carriers and heterozygotes for LPL mutations are also at increased risk of hypertriglyceridemia (Ghiselli et al., 1982; Babirak et al., 1989; Wilson et al., 1990).

Elevated Chylomicron Remnants and IDL Cholesterol

Type III hyperlipoproteinemia is characterized by plasma accumulation of chylomicron remnants and IDL particles due to impaired catabolism. Both cholesterol and triglycerides are elevated, with mean levels of 450 mg/dL and 700 mg/dL, respectively (Mahley and Rall, 1989). LDL cholesterol is generally low due to reduced conversion of VLDL to LDL and/or to upregulation of LDL receptors. Patients are susceptible to severe premature CAD, strokes and peripheral vascular disease.

Chylomicron remnants and IDL particles are normally cleared by hepatic remnant and LDL receptors, which recognize apo E on the surface of these particles, as described previously. A common impairment of this pathway is due to the apo E2 allele, which encodes a protein with only 1%-2% of normal receptor binding activity. One percent of the population is homozygous for E2 (E2/2 genotype). These individuals do not generally exhibit fasting hyperlipidemia but have difficulty clearing chylomicron remnants from plasma postprandially (Brown and Goldstein, 1991). Approximately 1 in 50 E2/2 individuals is unable to compensate for the defective apo E protein and develops the fasting lipid elevations characteristic of Type III hyperlipoproteinemia. Xanthoma striata palmaris, orange or yellow discolorations of the palmar and digital creases, are pathognomonic of Type III disease (Mahley and Rall, 1989). The difference between E2/2 individuals with and without fasting hyperlipidemia is presumably due to factors that affect IDL metabolism. These may include aging, exogenous estrogen, obesity, glucose intolerance and hypothyroidism, as well as heterozygosity for another genetic defect, such as FH (Mahley and Rall, 1989; Brown and Goldstein, 1991). Type III hyperlipoproteinemia has been reported in several patients with the apo E2/2 genotype who are also heterozygous for an LPL mutation (Ma et al., 1993), and there are almost certainly other mutant genes that can serve as a "second hit" resulting in Type III disease. Type III patients are highly diet- and weight-responsive, but drug therapy is often required.

Homozygous apo E mutations resulting in very low to undetectable levels of plasma apo E have recently been described (Zannis et al., 1993). Apo E deficiency is associated with very high plasma levels of VLDL- plus IDL-cholesterol and with atherosclerosis. In mice, germ-line ablation of both copies of the apo E gene results in advanced atherosclerotic lesions similar to those observed in human CAD (Breslow, 1993).

Elevated Lipoprotein(a)

In case-control studies, elevated plasma levels of Lp(a) have been found to be an independent risk factor for CAD (Berg, 1992). Levels of Lp(a), an LDL-like particle, vary greatly in the population (from <0.1 to >200 mg/dL) and are almost entirely genetically determined. Plasma Lp(a) concentrations >20 mg/dL have been reported to confer increased risk of CAD, as well as cerebrovascular and peripheral vascular disease. Lp(a) >39 mg/dL was the most common familial dyslipidemia in a study of patients with confirmed CAD before age 60, accounting for 19% of patients with a familial dyslipidemia (Genest et al., 1992). Among Hawaiian men of Japanese ancestry, 14% of all cases of MI and 28% of cases prior to age 60 were attributed to elevated Lp(a) (Rhoads et al., 1986). In contrast to case-control studies, prospective studies of Lp(a) and CAD have produced inconsistent results. While two small Swedish studies detected an association (Berg, 1992), a prospective analysis of Helsinki Heart Study participants found no relationship between Lp(a) levels and future coronary events (Jauhiainen et al., 1991). Similarly, no association between Lp(a) levels and future MI in men was detected in the Physicians' Health Study (296 cases, 296 controls) (Ridker et al., 1993a). Blacks have higher Lp(a) levels than Caucasians, but this does not appear to translate into increased CAD risk (Sorrentino et al., 1992).

The protein component of the Lp(a) particle consists of one molecule of apo B disulfide-bonded to one molecule of apolipoprotein(a) [apo (a)], a very large protein of unknown function. Apo(a) bears a striking resemblance to the fibrinolytic enzyme precursor plasminogen, and the genes for these two proteins are closely linked on chromosome 6 (Berg, 1992). Plasminogen consists of five pretzel-shaped domains, termed kringles, and a protease domain, while apo(a) consists of multiple tandem copies of a sequence resembling the plasminogen kringle IV domain, plus single copies of plasminogen-like kringle V and protease domains. The protease domain of apo(a) is unable to degrade fibrin. Both the apo(a) gene and protein vary widely in size within the population due to differing numbers of copies of the kringle IV sequence (Berg, 1992).

Approximately 90% of the variability in Lp(a) levels in the population is attributable to the apo(a) gene (Boerwinkle et al., 1992). Lp(a) levels are inversely correlated with apo(a) size, and individuals with smaller apo(a) isoforms are therefore presumed to be at greater risk of CAD (Berg, 1992). Size polymorphisms account for nearly 70% of the variance in Lp(a) levels (Boerwinkle et al., 1992). Apo(a) alleles of the same size are heterogeneous at the DNA sequence level, and it has been estimated that there may be more than 100 alleles in total (Cohen et al., 1993). Alleles of the same size but differing sequence are coinherited with different Lp(a) levels in families (Cohen et al., 1993), and variation in regions of the apo(a) gene that regulate messenger RNA levels are likely to make an independent contribution to variability in Lp(a) levels. Sequence heterogeneity in the apo(a) gene may contribute to genetic variation in CAD risk not only via effects on Lp(a) levels but also via qualitative differences in the apo(a) molecule.

DIABETES MELLITUS

Diabetes is an independent risk factor for CAD, conferring a relative risk of 2 in men and nearly 5 in women for death from CAD (Kannel and McGee, 1979). Non-insulin-dependent diabetes mellitus (NIDDM) is strongly associated with a constellation of CAD risk factors, notably obesity, dyslipidemia and hypertension. There is a strong correlation between diabetic nephropathy and CAD in both NIDDM and IDDM (Mattock et al., 1988). A third of IDDM patients develop nephropathy, and genetic factors, notably familial predisposition to hypertension, appear to play a role (Krolewski et al., 1988). The familial nature of diabetes is well-known. Hyperglycemia is a feature of at least 60 rare genetic syndromes, demonstrating that mutations in many different genes can produce diabetes (Shohat et al., 1992). While genetic factors play a role in IDDM, they appear to be a very strong determinant of NIDDM susceptibility.

Insulin-Dependent Diabetes Mellitus

A widely held view of IDDM posits the coincidence of a genetic predisposition with a viral infection or other environmental insult, leading to T cell-mediated destruction of pancreatic beta cells (Shohat et al., 1992; Thorsby and Ronningen, 1993). IDDM is a potent risk factor for premature CAD. Cardiovascular mortality among IDDM patients aged 30-34 has been estimated at 32 times the US population rate for this age group (Shohat et al., 1992). Having a first-degree relative with IDDM was an independent CAD risk factor in one study (Nora et al., 1980), suggesting common underlying genetic factors. The inheritance pattern of IDDM is complex, apparently involving multiple genes. Concordance in monozygotic twins is only 35%-50% (Thorsby and Ronningen, 1993), confirming the importance of environmental factors.

The human leukocyte antigen (HLA) gene complex on chromosome 6, which encodes highly polymorphic molecules of the immune system, accounts for 60%-70% of the genetic susceptibility to IDDM (Shohat et al., 1992). The HLA D (or Class II) region contains the genes for the cell-surface molecules DP, DQ and DR, which are expressed on B lymphocytes, monocytes, macrophages and endothelial cells. The function of these proteins is to present foreign antigen fragments to helper T cells (Thorsby and Ronningen, 1993). Approximately 95% of Caucasians with IDDM carry the DR3 or DR4 alleles or both, compared with 50% of nondiabetic controls, while DR2 is associated with resistance to IDDM (Shohat et al., 1992). Attention has recently focused on the DQ genes, which are adjacent and tightly linked to the DR genes (Thorsby and Ronningen, 1993). The relative importance of the DR and DQ genes in IDDM is controversial, and the precise molecular bases of HLA-mediated susceptibility and resistance to IDDM are unknown.

Polymorphisms near the insulin gene on chromosome 11 are associated with IDDM in population studies, and some but not all family studies have shown linkage of IDDM to this gene (Julier et al., 1991). The insulin amino acid sequence does not differ between IDDM-associated and other alleles, suggesting that genetic variation may instead affect regulation of insulin synthesis. There is some evidence for HLA and insulin gene interaction in IDDM (Julier et al., 1991).

Non-Insulin-Dependent Diabetes Mellitus

Normal glucose homeostasis is a balance between glucose production in the liver and glucose clearance into muscle and other peripheral tissues, and pancreatic beta cells constantly adjust insulin release to maintain this balance (DeFronzo et al., 1992; Leahy and Boyd, 1993). NIDDM is characterized by multiple defects in these processes, notably defective insulin secretion and insulin resistance. Both mutations that impair insulin release and mutations that reduce glucose clearance have been identified.

Insulin gene mutations are a rare cause of NIDDM (Steiner et al., 1990). Certain insulin amino acid substitutions have been shown to cause greatly reduced affinity for the insulin receptor, resulting in hyperinsulinemia and glucose intolerance or mild diabetes, with autosomal dominant inheritance. Other insulin mutations resulting in hyperproinsulinemia and mild diabetes have also been described.

Maturity-onset diabetes of the young (MODY) is an autosomal dominant form of NIDDM with an early onset, usually before age 25. MODY is not associated with obesity and generally has a mild course. One form of MODY is caused by a defective glucokinase gene on chromosome 7 (Froguel et al., 1993), while a second form has been mapped via family studies to chromosome 20 (Bell et al., 1991). More than 25 different glucokinase mutations have been identified. Such mutations have been found in the majority of MODY families characterized and account for approximately 6% of familial NIDDM in France (Froguel et al., 1993).

Heritable mutations of mitochondrial DNA have been shown to cause diabetes in a small number of families (Ballinger et al., 1992; van den Ouweland et al., 1992). The circular mitochondrial genome, present in thousands of copies per cell, is inherited only from the mother, and maternal inheritance of diabetes is observed in these families. In one family diabetes and deafness is due to a mitochondrial DNA deletion affecting 16 of the 22 mitochondrial transfer RNA (tRNA) genes and the genes for 12 of the 13 respiratory chain enzyme subunits (Ballinger et al., 1992). In the other families this syndrome is due to a single nucleotide substitution in the gene encoding the tRNA for leucine, which lies outside the region of the large deletion (van den Ouweland et al., 1992). This mutation causes mitochondrial myopathy, encephalopathy, lactic acidosis and stroke-like episodes (MELAS) in some families and diabetes with deafness in others; the former appears to be associated with a higher proportion of mutant mitochondrial DNA molecules in muscle (van den Ouweland et al., 1992). Diabetes caused by mitochondrial mutations may present as IDDM (early onset, low to undetectable insulin levels, ketoacidosis) or as NIDDM (late onset, normal to high insulin levels, without ketoacidosis).

Mutations in the insulin receptor gene on chromosome 19 result in several rare syndromes characterized by insulin resistance, hyperinsulinemia, glucose intolerance or frank diabetes, and acanthosis nigricans; in females, hyperandrogenism may also be present (Taylor, 1992). More than 40 insulin receptor mutations are known (Leahy and Boyd, 1993). Some result in decreased receptor number, while others reduce receptor affinity for insulin or impair tyrosine kinase activity. While patients with two mutant alleles generally have extreme insulin resistance, heterozygotes may have a clinical presentation similar to common NIDDM.

Single-gene disorders characterized to date are likely to account for only a small fraction of the cases of NIDDM. In a Finnish study, an association between common NIDDM and a polymorphism in the glycogen synthase gene on chromosome 19 has been reported (Groop et al., 1993). The minor allele for this polymorphism, which does not affect the amino acid sequence of the protein, was found in 30% of NIDDM patients, compared with 8% of controls (Groop et al., 1993). This allele is associated with marked insulin resistance, hypertension and a strong family history of NIDDM.

OBESITY

Obesity is an important contributor to all of the major metabolic disorders associated with CAD: dyslipidemia, notably elevated triglycerides and reduced HDL cholesterol; insulin resistance and NIDDM; and hypertension. Apart from these correlations, longitudinal studies with long follow-up periods (≥ 20 years) have generally supported obesity as an independent CAD risk factor (Pi-Sunyer, 1993). Abdominal adiposity shows an especially strong association with dyslipidemia, insulin resistance and CAD (Despres et al., 1992). The common form of obesity is a consequence of a genetic predisposition together with environmental factors. Studies of twins reared apart indicate that genetic factors predominate, accounting for up to 70% of the variance in body mass index (BMI) (Stunkard et al., 1990). The results of several studies indicate that BMI is strongly influenced by a major recessive gene. Genetic factors are also thought to play a role in abdominal adiposity, which is associated with increased glucocorticoid levels and, in women, with increased androgens (Despres et al., 1992).

Efforts to find human genes predisposing to obesity have not so far been successful. It has been proposed that the Pima and other ethnic groups with a high prevalence of obesity harbor a "thrifty" gene. This gene (or set of genes) is thought to confer a selective advantage in times of famine but to predispose to obesity and diabetes when calorie intake is high and energy expenditure is low (Neel, 1962). Such a gene may confer a low relative metabolic rate and/or favor fat storage, with reduced fat oxidation and depletion of carbohydrate stores. The expression of the "thrifty" trait appears most pronounced in societies undergoing a rapid shift from an agrarian way of life to a more sedentary Western lifestyle. This shift is likely to account for the sharp rise in the prevalence of obesity and NIDDM observed worldwide, most notably in rapidly developing countries (Diamond, 1992). Similar factors may explain the rising prevalence of obesity in the US, especially among the less educated, who consume more calories and fat than more educated individuals (Stamler, 1993). These trends illustrate the importance of the interaction between lifestyle and common genetic factors.

HYPERTENSION

Hypertension is an independent risk factor for both myocardial infarction and stroke. While overt renal, endocrine or neurologic dysfunction may disrupt blood pressure homeostasis, essential hypertension of unknown etiology accounts for more than 90% of cases. Untreated hypertension leads to atherosclerotic complications in nearly 30% of patients and to organ damage involving the heart, brain, eyes or kidney in more than half of patients over a period of 7 to 10 years (Williams, 1991). Hypertension and hypertension-associated morbidity and mortality are more common among blacks than among Caucasians. Hypertension may enhance atherogenesis via mechanical injury to endothelial cells, changes in membrane permeability to atherogenic lipoproteins, increased release of lysosomal enzymes, and increased thickness of the intimal smooth-muscle layer (Bierman, 1991). The major effect on the heart is left ventricular hypertrophy (LVH) in response to the excessive work load, and sustained hypertension ultimately leads to deterioration of this chamber, resulting in heart failure (Williams, 1991).

Blood pressure is known to be influenced by genetic factors, although the observed strength of this influence has differed widely among various studies, accounting for between 20% and 60% of the variance (Kurtz, 1993). Individuals with essential hypertension are twice as likely to have a hypertensive parent than are normotensives (Burke and Motulsky, 1992). The blood pressure correlations observed between monozygotic twins (0.6 for diastolic and 0.55-0.7 for systolic) are at least two-fold greater than those observed between dizygotic twins (Burke and Motulsky, 1992). Several lines of evidence indicate that non-modulating hypertension has a strong genetic component (Williams and Hollenberg, 1993). A family history of hypertension can be elicited from 85% of non-modulators, compared with 25%-30% of other normal- or high-renin hypertensives. In addition, a high degree of sibling concordance has been observed for the non-modulating trait. Multiple genes are presumed to interact with obesity, sedentary lifestyle, high salt intake, noise, crowding and stress to influence blood pressure (Burke and Motulsky, 1992).

Single-gene disorders leading to neurogenic, renal or endocrine hypertension have been described (Williams, 1991). The most common is autosomal dominant polycystic kidney disease, a genetically heterogeneous disorder that results in hypertension in 75% of cases. Glucocorticoid-remediable aldosteronism (GRA), a rare autosomal dominant disorder causing hypertension, has been mapped to the adjacent 11 ß-hydroxylase and aldosterone synthase genes on chromosome 8 (Lifton et al., 1992).

Variation in the angiotensinogen gene on chromosome 1 appears to influence susceptibility to essential hypertension and preeclampsia in the general population. An association has been demonstrated between hypertension and the minor allele of an angiotensinogen polymorphism, and hypertension has been shown to be linked to this gene in studies of affected sibling pairs (Jeunemaitre et al., 1992). The polymorphism results in a methionine to threonine amino acid substitution in angiotensinogen, with a Caucasian allele frequency of 35%. The threonine allele is associated with elevated plasma angiotensinogen, especially in women (Jeunemaitre et al., 1992), and confers a relative risk of severe hypertension of 1.85. Plasma angiotensinogen levels are strongly correlated with blood pressure and are elevated in hypertensive families. Blood pressure in rodents is increased by injection of angiotensinogen and decreased by injection of antibodies to angiotensinogen (Jeunemaitre et al., 1992). It has been proposed that overexpression of angiotensinogen in carriers of the threonine allele may result in a small increase in angiotensin II levels, causing hyper-reactivity of the renin-angiotensin system in response to salt or other environmental factors. This may lead to hyperaldosteronism with sodium retention, as well as vascular hypertrophy and increased peripheral vascular resistance (Jeunemaitre et al., 1992). The threonine allele has also been implicated in preeclampsia in both Caucasian and Japanese women (Ward et al., 1993).

The renin-angiotensin system appears to play a role in MI apart from its role in hypertension. The ACE gene on chromosome 17 has been implicated as a common contributor to MI in men. An ACE gene polymorphism is associated, in the homozygous state, with MI in both Caucasian and Japanese men (Cambien et al., 1992; Ohishi et al., 1993). Subjects carrying the MI-associated allele have higher plasma ACE levels than non-carriers. ACE cleaves angiotensin I to generate angiotensin II and also degrades bradykinin. Angiotensin II and bradykinin have opposing effects on smooth muscle cells and on vascular tone. Angiotensin II promotes growth of myocardial and vascular smooth muscle cells and promotes neointimal hyperplasia after arterial wall injury, while bradykinin appears to be a growth inhibitor. It has been proposed that the ACE polymorphism may influence MI risk by modulating levels of angiotensin II and bradykinin in coronary arteries (Cambien et al., 1992). Angiotensin II may also reduce fibrinolysis by increasing plasma levels of plasminogen activator inhibitor-1 (PAI-1), which inhibits endogenous t-PA (Ridker et al., 1993b). While ACE genetic variation affects blood pressure in rodents, no such association has been observed in humans.

SUMMARY

There is a strong familial and genetic nature of CAD and predisposing metabolic disorders. This should encourage health care workers to focus additional attention on the younger members of affected families, particularly the families of patients with MI prior to age 55. This should take the form of genetic counseling as well as patient education and follow-up with respect to hygienic measures of proven efficacy, and aggressive treatment of metabolic disorders that prove resistant to changes in lifestyle. In certain cases, as discussed in this review, genetic testing may prove helpful.

ACKNOWLEDGMENTS

The authors thank the following colleagues for helpful comments on the manuscript: Drs. Noah Robbins, Mark Rabinovitch, Neal Azrolan, Markus Stoffel, Wendy Chung, Rudolph Leibel, Peter Weinstock and Chithranjan Nath. We also thank Andrew Arsham and Cristina Alfaro, who provided expert research assistance, and Marjan Kamali, who assisted in manuscript preparation.

REFERENCES

Austin, M.A., 1991, Plasma triglyceride and coronary heart disease, Arterioscl. Thromb. 11:2-14.

Babirak, S.P., Iverius, P.-H., Fujimoto, W.Y., and Brunzell, J.D., 1989, Detection and characterization of the heterozygote state for lipoprotein lipase deficiency, Arterioscl. 9:326-34.

Ballinger, S.W., Shoffner, J.M., Hedaya, E.V. et al., 1992, Maternally transmitted diabetes and deafness associated with a 10.4kb mitochondrial DNA deletion, Nature Genet.. 1:11-15.

Bell, G.I., Xiang, K.-S., and Newman, M.V. et al., 1991, Gene for non-insulin-dependent diabetes mellitus (maturity-onset diabetes of the young subtype) is linked to DNA polymorphism on human chromosome 20q., Proc. Natl. Acad. Sci. USA 88:1484-8.

Berg, K., 1992, Lp(a) lipoprotein: An important genetic risk factor for atherosclerosis. in: "Molecular Genetics of Coronary Artery Disease, Candidate Genes and Processes in Atherosclerosis," Monogr. Hum. Genet. Vol. 14, A.J. Lusis, J.I. Rotter, and R.S. Sparkes, eds., Karger, Basel, 189-207.

Bierman, E.L., 1991, Atherosclerosis and other forms of arteriosclerosis, in: "Harrison's Principles of Internal Medicine," 12th ed., J.D. Wilson, E. Braunwald, and K.J. Isselbacher et al., eds., McGraw-Hill, New York, 992-1001.

Boerwinkle, E., Leffert, C.C., Lin, J., Lackner, C., Chiesa, G., and Hobbs, H.H., 1992, Apolipoprotein(a) gene accounts for greater than 90% of the variation in plasma lipoprotein(a) concentrations, J. Clin. Invest. 90:52-60.

Breslow, J.L., 1989, Familial disorders of high density lipoprotein metabolism, in: "The Metabolic Basis of Inherited Disease," Vol. I. 6th ed., C.R. Scrivner, A.L. Beaudet, W.S. Sly, and D. Valle, eds., McGraw-Hill, New York, 1251-66.

Breslow, J.L., 1991, Lipoprotein transport gene abnormalities underlying coronary heart disease susceptibility. Annu. Rev. Med. 42:357-71.

Breslow, J.L., 1993, Transgenic mouse models of lipoprotein metabolism and atherosclerosis, Proc. Natl. Acad. Sci. USA 90:8314-8.

Brinton, E.A., Eisenberg, S., and Breslow, J.L., 1990, A low-fat diet decreases high density lipoprotein (HDL) cholesterol levels by decreasing HDL apolipoprotein transport rates, J. Clin. Invest. 85:144-51.

Brown, M.S. and Goldstein, J.L., 1992, The hyperlipoproteinemias and other disorders of lipid metabolism, in: "Molecular Genetics of Coronary Artery Disease. Candidate Genes and Processes in Atherosclerosis," Monogr. Hum. Genet. Vol. 14, A.J. Lusis, J.I. Rotter, and R.S. Sparkes, eds., Karger, Basel.

Cambien, F., Poirier, O., and Lecerf, L. et al., 1992, Deletion polymorphism in the gene for angiotensin-converting enzyme is a potent risk factor for myocardial infarction, Nature 359:641-4.

Chait, A. and Brunzell, J.D., 1990, Acquired hyperlipidemia (secondary dyslipoproteinemias), Endocrinol. Metab. Clinics N. Amer. 19:259-78.

Cohen, J.C., Chiesa, G., and Hobbs, H.H., 1993, Sequence polymorphisms in the apolipoprotein(a) gene. Evidence for dissociation between apolipoprotein(a) size and plasma lipoprotein(a) levels, J. Clin. Invest. 91:1630-6.

Colditz, G.A., Rimm, E.B., Giovannucci, E., Stampfer, M.J., Rosner, B., and Willett, W.C., 1991, A prospective study of parental history of myocardial infarction and coronary artery disease in men, Am. J. Cardiol. 67:933-8.

Coresh, J., Svenson, K.L., Beaty, T.H., Kwiterovich, P.O., and Lusis, A.J., 1993, Sib-pair linkage analysis of the lipoprotein lipase gene and lipoprotein levels: The Johns Hopkins Coronary Artery Disease Family Study, Am. J. Hum. Genet. 53:Suppl:Abstract 788.

Dammerman, M., Sandkuijl, L.A., Halaas, J.L., Chung, W., and Breslow, J.L., 1993, An apolipoprotein CIII haplotype protective against hyper-triglyceridemia is specified by promoter and 3' untranslated region polymorphisms, Proc. Nat'l. Acad. Sci. USA 90:4562-6.

DeFronzo, R.A., Bonadonna, R.C., and Ferrannini E., 1992, Pathogenesis of NIDDM: A balanced overview, Diabetes Care 15:318-68.

Despres, J.-P., Moorjani, S., Lupien, P.J., Tremblay, A., Nadeau, A., and Bouchard, C., 1992, Genetic aspects of susceptibility to obesity and related dyslipidemias, Mol. Cell Biochem. 113:151-69.

Diamond, J.M., 1992, Diabetes running wild, Nature 357:362-3.

Froguel, P., Zouali, H., and Vionnet, N. et al., 1993, Familial hyperglycemia due to mutations in glucokinase. Definition of a subtype of diabetes mellitus, N. Eng. J. Med. 328:697-702.

Genest, J.J. Jr., Martin-Munley, S.S., and McNamara, J.R. et al., 1992, Familial lipoprotein disorders in patients with premature coronary artery disease, Circulation 85:2025-33.

Ghiselli, G., Schaefer, E.J., Zech, L.A., Gregg, R.E., and Brewer, H.B. Jr., 1982, Increased prevalence of apolipoprotein E4 in Type V hyperlipoproteinemia, J. Clin. Invest. 70:474-7.

Goldbourt, U. and Neufeld, H.N., 1986, Genetic aspects of arteriosclerosis, Arteriosc. 6:357-77.

Goldstein, J.L. and Brown, M.S., 1989, Familial hypercholesterolemia. in: "The Metabolic Basis of Inherited Disease," Vol. I. 6th ed., C.R. Scrivner, A.L. Beaudet, W.S. Sly, and D. Valle, eds., McGraw-Hill, New York, 1215-50.

Groop, L.C., Kankuri, M., and Schalin-Jantti, C. et al., 1993, Association between polymorphism of the glycogen synthase gene and non-insulin-dependent diabetes mellitus, N. Eng. J. Med. 328:10-14.

Hixson, J.E., 1991, Pathobiological Determinants of Atherosclerosis in Youth (PDAY) Research Group. Apolipoprotein E polymorphisms affect atherosclerosis in young males, Arteriosc. Thromb. 11:1237-44.

Jauhiainen, M., Koskinen, P., and Ehnholm, C. et al., 1991, Lipoprotein (a) and coronary heart disease risk: a nested case-control study of the Helsinki Heart Study participants, Atherosc. 89:59-67.

Jeunemaitre, X., Soubrier, F., and Kotelevtsev, Y.V. et al., 1992, Molecular basis of human hypertension: Role of angiotensinogen, Cell 71:169-80.

Julier, C., Hyer, R.N., and Davies, J. et al., 1991, Insulin-IGF2 region on chromosome 11p encodes a gene implicated in HLA-DR4-dependent diabetes susceptibility, Nature 354:155-9.

Kane, J.P. and Havel, R.J., 1989, Disorders of the biogenesis and secretion of lipoproteins containing the B apolipoproteins, in: "The Metabolic Basis of Inherited Disease," Vol. I. 6th ed., C.R. Scrivner, A.L. Beaudet, W.S. Sly, D. Valle. eds., McGraw-Hill, New York, 1139-64.

Kannel, W.B. and McGee, D.L., 1979, Diabetes and cardiovascular disease. The Framingham study, J.A.M.A. 241:2035-8.

Krolewski, A.S., Canessa, M., and Warram, J.H. et al., 1988, Predisposition to hypertension and susceptibility to renal disease in insulin-dependent diabetes mellitus, N. Engl. J. Med 318:140-5.

Kurtz, T.W., 1993, Genetics of essential hypertension, Am. J. Med. 94:77-84.

Leahy, J.L. and Boyd, A.E., III, 1993, Diabetes genes in non-insulin-dependent diabetes mellitus, N. Engl. J. Med 328:56-7.

Lifton, R.P., Dluhy, R.G., and Powers, M. et al., 1992, A chimaeric 11 ß-hydroxylase/aldosterone synthase gene causes glucocorticoid-remediable aldosteronism and human hypertension, Nature 355:262-5.

Ma, Y., Zhang, H., Liu, M.-S., Frohlich, J., Brunzell, J.D., and Hayden, M.R., 1993, Type III hyperlipoproteinemia in apo E2/2 homozygotes: Possible role of mutations in the lipoprotein lipase gene, Circulation 88:Suppl:I-179.

Mahley, R.W. and Rall, S.C. Jr., 1989, Type III hyperlipoproteinemia (Dysbetalipoproteinemia): The role of apolipoprotein E in normal and abnormal lipoprotein metabolism, in: "The Metabolic Basis of Inherited Disease," Vol. I. 6th ed., C.R. Scrivner, A.L. Beaudet, W.S. Sly, and D. Valle, eds., McGraw-Hill, New York, 1195-1213.

Mark, A.L., 1992, `Syndrome X': Is it a significant cause of hypertension? Negative, Hosp. Pract. 27:Suppl 1:41-4.

Mattock, M.B., Keen, H, and Viberti, G.C. et al., 1988, Coronary heart disease and urinary albumin excretion rate in Type 2 (non-insulin-dependent) diabetic patients, Diabetologia 31:82-7.

Neel, J.V., 1962, Diabetes mellitus: A "thrifty" genotype rendered detrimental by "progress"? Am. J. Hum. Genet. 14:353-62.

Nora, J.J., Lortscher, R.H., Spangler, R.D., Nora, A.H., and Kimberling, W.J., 1980, Genetic-epidemiologic study of early-onset ischemic heart disease, Circulation 61:503-8.

Ohishi, M., Fujii, K., and Minamino, T. et al., 1993, A potent genetic risk factor for restenosis, (Letter.) Nature Genet. 5:324-5.

Pi-Sunyer, F.X., 1993, Medical hazards of obesity, Ann. Intern. Med. 119:655-60.

Price, W.H., Kitchin, A.H., Burgon, P.R.S., Morris, S.W., Wenham, P.R., and Donald, P.M., 1989, DNA restriction fragment length polymorphisms as markers of familial coronary heart disease, Lancet i:1407-11.

Reaven, G.M., 1988, Role of insulin resistance in human disease, Diabetes 37:1595-607.

Rees, A., Shoulders, C.C., Stocks, J., Galton, D.J., and Baralle, F.E., 1983, DNA polymorphism adjacent to the human apoprotein AI gene: Relation to hypertriglyceridemia, Lancet i:444-6.

Rhoads, G.G., Dahlen, G., Berg, K., Morton, N.E., and Dannenberg, A.L., 1986, Lp(a) lipoprotein as a risk factor for myocardial infarction, J.A.M.A. 256:2540-4.

Ridker, P.M., Hennekens, C.H., and Stampfer, M.J., 1993a, A prospective study of lipoprotein(a) and the risk of myocardial infarction, J.A.M.A. 270:2195-9.

Ridker, P.M., Gaboury, C.L., Conlin, P.R., Seely, E.W., Williams, G.H., and Vaughan, D.E., 1993b, Stimulation of plasminogen activator inhibitor in vivo by infusion of angiotensin II. Evidence of a potential interaction between the renin-angiotensin system and fibrinolytic function, Circulation 87:1969-73.

Roncaglioni, M.C., Santoro, L., and D'Avanzo, B. et al., 1992, Role of family history in patients with myocardial infarction. An Italian case-control study, Circulation 85:2065-72.

Shohat, T., Raffel, L.F., Vadheim, C.M., and Rotter, J.I., 1992, Diabetes mellitus and coronary heart disease genetics, in: "Molecular Genetics of Coronary Artery Disease. Candidate Genes and Processes in Atherosclerosis," Monogr. Hum. Genet. Vol. 14, A.J. Lusis, J.I. Rotter, and R.S. Sparkes, eds., Karger, Basel.

Shohat, T., Raffel, L.F., Vadheim, C.M., and Rotter, J.I., 1992, Diabetes mellitus and coronary heart disease genetics, in: "Molecular Genetics of Hypertension. Candidate Genes and Processes in Atherosclerosis," Monogr. Hum. Genet. Vol. 14, A.J. Lusis, J.I. Rotter, and R.S. Sparkes, eds., Karger, Basel.

Shoulders, C.C., Harry, P. J., and Lagrost, L. et al., 1991, Variation at the apo AI/CIII/AIV gene complex is associated with elevated plasma levels of apo CIII, Atheroscl. 87:239-47.

Sorrentino, M.J., Vielhauer, C., Eisenbart, J.D., Fless, G.M., Scanu, A.M., and Feldman T. 1992, Plasma lipoprotein(a) protein concentration and coronary artery disease in black patients compared with white patients, Am. J. Med. 93:658-62.

Stamler, J., 1993, Epidemic obesity in the United States, Arch Intern Med 153:1040-3.

Steiner, D.F., Tager, H.S., Chan, S.J., Nanjo, K., Sanke, T., and Rubenstein, A.H., 1990, Lessons learned from molecular biology of insulin-gene mutations, Diabetes Care 13:600-9.

Stunkard, A.J., Harris, J.R., Pedersen, N.L., and McClearn, G.E., 1990, The body-mass index of twins who have been reared apart, N. Engl. J. Med. 322:1483-7.

Tall, A.R., 1993, Plasma cholesteryl ester transfer protein, J. Lipid Res. 34:1255-74.

Taylor, S.I., 1992, Lilly Lecture: Molecular mechanisms of insulin resistance. Lessons from patients with mutations in the insulin-receptor gene, Diabetes 41:1473-90.

Thorsby, E. and Ronningen, K.S., 1993, Particular HLA-DQ molecules play a dominant role in determining susceptibility or resistance to Type 1 (insulin-dependent) diabetes mellitus, Diabetologia 36:371-7.

Van den Ouweland, J.M.W., Lemkes, H.H.P.J., and Ruitenbeek, W. et al., 1992, Muation in mitochondrial tRNA$^{Leu(UUR)}$ gene in a large pedigree with maternally transmitted type II diabetes mellitus and deafness, Nature Genet. 1:368-71.

Ward, K., Hata, A., and Jeunemaitre, X. et al., 1993, A molecular variant of angiotensinogen associated with preeclampsia, Nature Genet. 4:59-61.

Williams, G.H., 1991, Hypertensive vascular disease, in: "Harrison's Principles of Internal Medicine," 12th ed., J.D. Wilson, E. Braunwald, and K.J. Isselbacher et al., eds., McGraw-Hill, New York, 1001-15.

Williams, G.H. and Hollenberg, N.K., 1993, Derangements in renin-angiotensin regulation in the pathogenesis of hypertension, in: "Cellular and Molecular Biology of The Renin-Angiotensin System," M.K. Raizada, M.I. Phillips, and C. Sumners, eds., CRC Press, Boca Raton, 515-36.

Wilson, D.E., Emi, M., and Iverius, P.-H. et al., 1990, Phenotypic expression of heterozygous lipoprotein lipase deficiency in the extended pedigree of a proband homozygous for a missense mutation, J. Clin. Invest. 86:735-50.

Zannis, V.I., Kardassis, D., and Zanni, E.E., 1993, Genetic mutations affecting human lipoproteins, their receptors, and their enzymes, in: "Advances in Human Genetics," Vol. 21. 145-319, H Harris, K. Hirschhorn eds, Plenum Press, New York,.

GENE THERAPY IN HEART DISEASE

Louis C. Smith,[1,2] Randy C. Eisensmith,[1] and Savio L. C. Woo[1,2,3]

[1]Department of Cell Biology
[2]Department of Medicine
[3]Howard Hughes Medical Institute
 Baylor College of Medicine
 Houston, TX 77030-3498

INTRODUCTION

Interest in gene therapy arises from the realization that, for many human diseases, current treatment of human disease, genetic or otherwise, is generally aimed at symptoms or secondary defects and almost never at the precise biochemical or genetic disease itself. The ideal treatment could involve actual change or replacement of defective genes (Friedman, 1983). In spite of remarkable surgical and pharmaceutical advances, therapy of the more than 5000 genetic disorders is not a clinical reality. The most effective drugs act, not only in the target organ, but also in other tissues, where they produce side effects. Moreover, drugs are expensive and must be continued for lifetime. Gene therapy is an alternative to a lifetime of medication, particularity when effective drugs do not exist.

Familial hypercholesterolemia (FH), caused by inherited defects in the receptor for low density lipoprotein (LDL) receptor which render it non functional (Goldstein and Brown, 1989), is associated with severe hypercholesterolemia and premature coronary artery disease. Several features of the disorder make it an ideal candidate for gene therapy. The homozygous form of FH is lethal at an early age and refractory to conventional therapy. An animal model, the Watanabe Heritable Hyperlipidemic (WHHL) rabbit, is well characterized (Buja et al., 1981; Yamamoto et al., 1984) and available. Serum lipid and lipoprotein values are informative and clinically relevant endpoints for measuring the response to therapy. Orthotopic liver transplantation in a human subject has shown that a functional LDL receptor in the liver is adequate therapy for the hypercholesterolemia (Bilheimer et al., 1984). Most of circulating LDL is degraded in the liver. It is also the only organ responsible for conversion of cholesterol to bile acids, which are the excreted end product of cholesterol metabolism.

Nutrition and Biotechnology in Heart Disease and Cancer
Edited by J.B. Longenecker *et al.*, Plenum Press, New York, 1995

79

APPROACHES TO GENE THERAPY

There are two general approaches to gene therapy. The first involves the preparation of cells from patients and the introduction of genes into them using viral vectors or other transfecting agents (Wolff et al., 1987; Wilson et al., 1988,1990; Anderson et al., 1989; Ponder et al., 1991). The transduced cells are then returned to the patients by autologous transplantation to reconstitute the missing functions in the patients. This *ex vivo* approach, while conceptually straightforward, suffers from the fact that tissue removal may require a major surgical procedure. The second approach involves the direct introduction of genetic material into specific tissue sites using a targeting mechanism. The *in vivo* approach, while conceptually simple, requires the ability to direct the injected genetic material into the proper tissues or organs to be effective. For correction of the metabolic defect in FH, the *ex vivo* utilization of recombinant retroviral vectors in WHHL rabbits (Wilson et al., 1990; Chowdhury et al., 1991) and in a human subject (Grossman et al., 1994), as well as the *in vivo* application of recombinant adenoviral vectors in LDL receptor knockout mice (Ishibashi et al., 1993) and WHHL rabbits (Li et al., 1994), demonstrate the technical feasibility and the present limitations of these approaches.

EX VIVO GENE THERAPY OF LDL RECEPTOR DEFICIENCY

The initial reports of correction of LDL receptor deficiency in WHHL rabbits utilized a recombinant virus that contained a functional human LDL receptor gene, as well as Moloney enhancer and promoter sequences (Wilson et al., 1990). When this protocol was repeated with a recombinant retrovirus containing a rabbit LDL receptor gene expresses by sequences from the chicken β-actin gene, there was long term improvement of hypercholesterolemia, a 30 to 50 percent reduction, after *ex vivo* gene therapy in WHHL rabbits (Chowdhury et al., 1991). These results provided the necessary background for *ex vivo* gene therapy directed to the liver of a patient with familial hypercholesterolemia (Grossman et al., 1994). A 29 year old woman with homozygous FH was treated by injection of her hepatocytes which had been cultured with a recombinant retrovirus containing a normal copy of the human LDL receptor gene. The surgical procedures were tolerated well. There was no apparent pathology in a liver biopsy sample at 4 months and *in situ* hybridization showed single cells expressing the gene. The baseline level of serum LDL after gene therapy was 17% lower than the pretreatment baseline. The patient also become more responsive to a cholesterol lowering agent after gene therapy. The most significant change is the change in the LDL:HDL ratio, which was 10-13 and 5-8, before and after treatment, respectively. Whether or not there will be detectable protection against vascular disease remains to be determined. Weatherall (1994) notes it will be important to evolve less traumatic approaches to introducing genes into liver and other cells.

IN VIVO GENE THERAPY OF LDL RECEPTOR DEFICIENCY

Thus far, the major problem with retrovirus-mediated approaches has been the low efficiency of therapeutic gene delivery to the target organs, and the requirement for dividing cells. By contrast, adenovirus-mediated gene delivery is highly efficient, can be used to transduce nondividing cells, and recombinant adenoviral vectors can be conveniently produced and administered to experimental animals. *In vivo* administration of a recombinant adenovirus containing the LDL receptor gene has recently been shown to increase the clearance of labeled VLDL in LDL receptor knock-out mice (Ishbashi et al., 1993). In our laboratory, a recombinant adenovirus was constructed which contained the rabbit low density lipoprotein receptor (rbLDLR) cDNA under the transcriptional control of the Rous sarcoma virus (RSV) long terminal repeat (LTR) promoter (Adv/RSV-rbLDLR) (Li et al., 1994). The functionality of the recombinant vector was assayed by measuring the binding, uptake and degradation of $[^{125}I]$-labeled LDL in LDL receptor-deficient Watanabe rabbit hepatocytes after viral transduction *in vitro*. While non-transduced Watanabe hepatocytes exhibited minimal uptake and degradation of ^{125}I-LDL, the Adv/RSV-rbLDLR vector successfully restored LDL receptor function in the LDL receptor-deficient hepatocytes in a dose-dependent manner.

This recombinant adenovirus containing the rabbit LDL receptor cDNA has been administered to Watanabe rabbits *in vivo* by splenic vein infusion. Plasma levels of total cholesterol were reduced by 75% at 6 days after treatment, while HDL-cholesterol and apoprotein A1 levels were increased subsequently by 3-4 fold. As a result, the LDL/HDL ratio exhibited a dramatic decrease. Three days after infusion, hepatocytes isolated from rabbits receiving the rbLDL receptor gene bound and internalized six times more $[^{125}I]$-LDL than hepatocytes isolated from animals treated with control virus Adv/RSV-hAAT. After reaching a minimum at 6 days, plasma total cholesterol values then gradually increased to control levels after three weeks. Thus, this recombinant vector efficiently reduced plasma cholesterol in the Watanabe rabbits, although the effect was transient as seen previously in other systems. While long-term therapeutic effects may be achievable with modified persistent adenoviral vectors, the current study illustrates the principle that hyperlipidemia can be ameliorated by a single inoculation of a vector containing a therapeutic gene.

SYNTHETIC DNA DELIVERY SYSTEMS

A conceptually different approach is to construct synthetic DNA complexes containing components that perform the roles of proteins in the viral vectors. The effectiveness of a synthetic DNA delivery system *in vivo* depends on the route of administration, uptake by a specific cell type, escape from lysosomal degradation, transit through the cytoplasm, transport through the nuclear membrane, recognition and utilization by nuclear enzymes and transcription factors for expression, and persistence of the DNA in an episomal form or by integration into genomic DNA for long term expression (Hoeben et al., 1993). There is little quantitative information about how efficiently the vector is processed through each stage of these complex processes. The steps in the overall process which account for the low efficiency observed for existing methods of gene delivery *in vivo* (Wu and Wu, 1988; Wu et al., 1989, 1991; Wilson et al., 1992) are not known. Viruses have developed highly efficient ways of delivering genes *in vivo*, and many delivery systems currently being examined are derivatives of various viruses. Thus, the incorporation of the surface features of different types of viruses may increase the efficiency and specificity of gene delivery mediated by synthetic DNA delivery systems. Roles assigned to viral and cellular proteins as important control elements for achieving high level expression of exogenous DNA are speculative.

Five components of synthetic DNA delivery systems are known at present, a DNA molecule of known sequence that contains a marker or therapeutic gene, a cationic DNA binding template to condense the DNA to a size small enough to undergo endocytosis, a ligand that recognizes and binds with high affinity to a cell surface receptor, a lytic peptide that ruptures the endosomes to release the complex directly into the cytoplasm, and a nuclear localization sequence (NLS) to enhance delivery of the DNA to the nucleus. At present, it is not clear which of the structural, thermodynamic, and kinetic properties of DNA complexes are important for high efficiency of DNA delivery. One strategy to obtain the information necessary is that of developing a chemically defined system that provides highly efficient delivery of genes into the nucleus of the targeted cells, first *in vitro* and then *in vivo*.

TARGETING MECHANISMS FOR DIRECT DNA DELIVERY

Direct delivery of genes to a given cell type requires a targeting mechanism. The observation that most, if not all, cells have specialized and sometimes unique cell surface receptors (Shephard, 1989; McGraw and Maxfield, 1992) provides the best strategy to achieve specific and exclusive delivery of genes. High level expression of genes delivered through this approach offers the potential for therapeutic intervention without the use of infectious agents and their cytopathic effects. A number of receptors expressed on the surface of the hepatocyte may be suitable for receptor-mediated transgene delivery to the liver. The asialoglycoprotein receptor is a part of the surveillance system in the circulation that removes proteins as they age by the spontaneous loss of sialic acid groups (Schwartz, 1989). This well-characterized cell surface receptor is found exclusively on the hepatocyte. The asialoglycoprotein receptor has been employed *in vitro* (Wu and Wu, 1987, 1988a; Neda et al., 1991; Cristiano et al., 1993a, 1993b) and *in vivo* (Wu and Wu, 1988b; Wu et al., 1989, 1991; Wilson et al., 1992). Other receptors used for gene delivery include the transferrin receptor *in vitro* (Cotten et al., 1990, 1992; Wagner et al., 1990, 1991a,b; 1992a,b; Zenke et al., 1990; Curiel et al., 1991,1992a,b; Plank et al., 1992; Michael et al., 1993) and the folate receptor *in vitro* (Leamon et al., 1991,1992; Turek et al., 1993; Gottschalk et al., 1994).

Ligands for these receptors have been covalently linked to poly-L-lysine which is then noncovalently complexed with negatively charged DNA through the positively charged template. Poly-L-lysine thus serves as the ligand template as well as the agent to condense the plasmid DNA. The resulting torus or doughnut-shaped particles are small enough to enter the cell through clathrin-coated pits containing the high affinity receptors (Goldstein et al., 1985). The clathin-coated pit places an upper limit on the size of DNA complex that can undergo endocytosis. Freeze fracture electron micrographs show diameters of clathin-coated pits as 100-200 nm (Goldstein et al., 1985; Robenek et al., 1991).

One of the rate limiting steps in achieving high levels of gene expression has been endosome lysis. Physiologically relevant levels can be obtained from DNA delivered into primary hepatocytes by combining receptor-mediated DNA delivery systems with the endosomal lytic abilities of adenovirus (Cristiano et al., 1993a,b). When asialoorosomucoid conjugated with poly-L-lysine was used to deliver the *E. coli* β-galactosidase gene into primary hepatocytes through binding with the hepatic asialoglycoprotein receptor, only low levels of β-galactosidase were detectable, with less than 0.1% of the hepatocytes being transfected. This level of activity was greatly enhanced by the cointernalization of the DNA/protein complex with the replication-defective adenovirus dl312; 100% of hepatocytes stained positive for ß-galactosidase activity (Cristiano et al., 1993a). Quantitative analysis of β-galactosidase expression also showed a 1000-fold enhancement of activity by coadministration adenovirus. To test the applicability of this DNA delivery system for the correction of phenylketonuria, a metabolic disorder that causes severe mental retardation in children, the human phenylalanine hydroxylase (PAH) cDNA was delivered to hepatocytes derived from the PAH-deficient mouse strain Pahenul. The level of PAH expression was raised to that of normal hepatocytes. The level of hPAH activity achievable by receptor-mediated gene delivery compared favorably to that produced by infection of PAH-deficient hepatocytes with a recombinant retrovirus that expresses the hPAH cDNA under the same CMV enhancer/promoter (Cristiano et al., 1993a).

The enhancement of gene delivery to hepatocytes depended on the addition of 3×10^6 adenoviral particles per cell. The high viral titer is needed since elevated gene expression is dependent on the chance that both an adenovirus and a DNA complex are internalized into the same endosome. In these experiments, virus titer was determined by spectrophotometric analysis, which is known to be 10- to 100-fold higher than viral titers determined by plaque assay. To include an adenoviral particle within the DNA complex, adenoviral particles were chemically conjugated to poly-L-lysine and bound ionically to DNA molecules (Cristiano et al., 1993b). Quantitative delivery of primary hepatocytes was achieved with significantly reduced viral titer when the asialoorosomucoid-poly-L-lysine conjugate was included in the complex. The conjugated adenovirus was used to deliver a DNA vector containing canine factor IX to mouse hepatocytes, resulting in the expression of significant concentrations of canine factor IX in the culture medium. The results suggest that receptor-mediated endocytosis and efficient endosomal lysis agent will give efficient targeted gene delivery.

The folate receptor, which is overexpressed in many tumors, was also used for targeted DNA delivery into KB epithelial cells (Gottschalk et al., 1994). This receptor internalizes folate through caveolae, a process named potocytosis (Anderson et al., 1992). Folate-labeled proteins are not degraded within 6 hours after intracellular uptake via this receptor, by contrast with endocytosis through clathrin-coated pits. Thirty minutes after injection of [^{125}I]-folyl-BSA into either the tail vein or the portal vein of the mouse, about 70% of the organ-associated radioactivity is found in the liver. Most of the injected material is in the extravascular space. When folate conjugated to poly-L-lysine was used to deliver the $E.$ $coli$ β-galactosidase gene into KB cells, little β-galactosidase activity was detectable, with less than 1% of cells transfected. The low level of gene expression appeared to be due to trapping of the DNA in multivesicular bodies after cellular uptake. The level of gene expression was greatly enhanced when a replication-defective adenovirus was coupled to the DNA/folate complex. This procedure increased the proportion of KB cells positive for ß-galactosidase from less than 1% to 20% to 30%, and produced a 300-fold increase in β-galactosidase activity (Gottschalk et al., 1994). Thus, for highly efficient DNA delivery and expression via caveolae, a membrane lysis agent is also needed as it is for clathrin-coated pit endocytosis. An alternative to coadministration of replication-defective adenovirus is to include a lytic peptide directly into the synthetic DNA delivery system, thereby ensuring that all endocytosed DNA will be released from the endosome, and at the same time, eliminating the adenovirus.

ENDOSOME LYSIS BY INFLUENZA VIRUS

An alternative approach to achieve endosome rupture is the lysis produced by specific peptides sequences in coat proteins of viruses (Ojcius and Young, 1991). The best studied example is that of the influenza virus. After binding to sialic acid-containing receptors on the plasma membrane, influenza virus is taken up into its host cell by receptor-mediated endocytosis (White, 1990,1992). The low pH of the endosomal compartment activates influenza virus haemagglutinin (HA), which produces fusion of viral and endosomal membrane and release of the viral nucleocapsid into the cytoplasm. The HA spike protein is a trimer of identical HA monomers, each of which is two subunits linked to each other by a disulfide bridge. Subunit HA$_2$ possesses an extremely hydrophobic N-terminus. The fusion activity of HA occurs only at pH 5-6. At 37°C, the half-times are generally <30 sec, and fusion is usually complete within 2 to 5 min. The fusion reactions of influenza viruses consist of two sequential reactions: a second order aggregation step followed by a first order fusion step. In the overall process, the aggregation step is rate-limiting. The irreversible conformational change exposes the hydrophobic N-terminus of HA$_2$ and subunit contacts within the HA trimer are partially lost. Short synthetic peptides from HA$_2$ subunit of influenza HA have been studied extensively with artificial lipid membranes (Wharton et al., 1988). Although the rates are slower, peptides with amino acid substitutions give both fusion and leakage of liposomal contents, similar to whole haemagglutinin molecules with the corresponding sequence changes.

Complexes containing plasmid DNA, transferrin-poly-L-lysine conjugates, and poly-L-lysine-conjugated peptides derived from the N-terminal sequence of the influenza hemagglutinin subunit HA_2 have been used for the transfer of luciferase or β-galactosidase marker genes (Wagner et al., 1992a). In HeLa cells, expression of β-galactosidase occurred in about 5-10% of all cells. This value is low compared to controls performed in the presence of free or DNA-bound adenoviruses, with frequencies >90%. Wagner et al. (1992a) note that transport complexes can be likened to artificial viruses that mimic the viral transfer of genes into cells but are devoid of all cytopathic and replicative functions of natural viruses.

Short synthetic peptides from HA_2 subunit of influenza HA have been studied extensively with artificial lipid membranes (Wharton et al., 1988). Although the rates are slower, peptides with amino acid substitutions give both membrane fusion and leakage of liposomal contents, similar to whole haemagglutinin molecules with the corresponding sequence changes. The low efficiency of gene expression with synthetic peptides as lytic agents may result from at least 4 factors. First, the spacing of the hydrophilic and hydrophobic amino acid residues along the α-helix prevents oligomer association of the peptides after their insertion into the membrane. Second, covalent attachment of the peptide to poly-L-lysine precludes oligomer formation and the necessary aggregation (Doms et al., 1991; Ramalho-Santos et al., 1993). All of the well-characterized viral fusion proteins are products of a single mRNA, and oligomerize in the rough endoplasmic reticulum (White, 1990), where *in vivo* conditions are radically different from those of the endosome. Third, sufficient aggregation of several oligomeric structures necessary to achieve lysis (Doms et al., 1991; Ramalho-Santos et al., 1993) may not have occurred because association of monomeric peptides is too slow. Finally, the synthetic peptides do not have the hydrophobic carboxyl terminal created by endosomal processing of intact virus (White, 1992). More than 100 different fusiogenic/lytic peptides have been described (Ojcius and Young, 1991). Some of these peptides may be useful for gene delivery. Once higher efficiencies of gene transfer have been obtained, such artificial viral complexes may become the method of choice.

CYTOPLASMIC DIFFUSION AND NUCLEAR TRANSPORT

In the design of synthetic reagents for delivery of genes, cytoplasmic diffusion of the exogenous DNA complex and its transport through nuclear pores remain poorly understood. The doughnut-shaped DNA complexes that are small enough to undergo efficient endocytosis are apparently too large to move in the cytoplasm and through the 24 nm pores of the nuclear membrane. The intriguing fact is that these apparent size constraints are overcome by as yet undefined cellular processes. When DNA/ligand complexes and adenovirus are co-internalized *in vitro*, it is possible to obtain 100% transduction and high level gene expression (Cristiano et al., 1993a,b). The DNA complex must undergo major structural and compositional changes in the cytoplasm after it is released by adenovirus mediated lysis of the endosome.

Living cytoplasm resembles a concentrated protein solution in a fibrous gel like polyacrylamide (Cohen and Paine, 1992). The gel-like components consist of complex networks of actin filaments, microtubules, intermediate filaments and other structural elements. The intracytoplasmic diffusion of proteins and small solutes is not influenced by size-sieving functions of the cytomatrix. As the size of tracers increases above 10 nm diameter, a strong size dependency becomes evident. Luby-Phelps et al. (1987) used a size-graded series of Ficoll, highly branched and cross-linked carbohydrates that approximate hard spheres in structure, to measure the ratio of their diffusion constants in the cytoplasm versus water. Particles with diameters greater than 50 nm were nondiffusable in the cytoplasm. Since high level gene expression is well documented, it is clear that interactions with cytoplasmic proteins occurred to transport the DNA, either free or in a complex, into the nucleus. Dissociation of the DNA/polypeptide complex must have occurred to allow the DNA to assume other structures small enough to penetrate the nuclear pores.

Silver (1991) summarizes how proteins enter the nucleus. Proteins destined for the nucleus contain a nuclear localization sequence (NLS), which interacts with other proteins located in the cytoplasm, on the nuclear envelope, and/or at the nuclear pore complex. Proteins are then translocated by processes requiring ATP. The model (Feldherr and Akin, 1990) of the nuclear pore complex derived from cryoelectron microscopy, image processing, and classification analysis, is comprised of two supramolecular iris-like assemblies which open asynchronously to provide an expanded pore for translocation while maintaining transport fidelity. The nuclear pore complexes visualized in isolated rat liver nuclei by fracture-flip/Triton-X (Fujimoto and da Silva, 1988) had an outer diameter of 120-130 nm, while the raised, "doughnut-like" nuclear pore complexes had an inner diameter of 40 nm.

A recent review (Dingwall and Laskey, 1991) has 38 entries as NLSs. The NLS have been identified by deletion and site-specific mutants of a number of proteins (Lanford et al., 1986,1987; Goldfarb et al., 1986; Garcia-Busto et al., 1991). They are generally less than 12 amino acids long and contain a high proportion of positively charged amino acids. The complexity of the transport processes underscore the need to identify the rate-limiting processes for delivery of exogenous genes to the nucleus.

The principal limitation is the inability to prepare reproducible proteinaceous DNA complexes in a consistent manner for gene delivery. The actual site of ligation is unknown. The covalent coupling of poly-L-lysine or ligands with proteins is nonspecific and gives a random mixture of conjugates. For example, commercially available poly-L-lysine used for direct DNA delivery is heterogeneous (Dolnik and Novotny, 1993). The derivatized poly-L-lysine sample was separated by capillary electrophoresis; more than 30 oligopeptides are present. Moreover, because the conformation of positively charged surfaces is formed by chance, binding of DNA to the charged template is also variable. High molecular weight and variable stoichiometry of the components of the complex has made it difficult to prepare the complexes either consistently well or in sufficient quantity for *in vivo* delivery. The complexity of the mixtures precludes a molecular definition of the biologically active reagent and may account for the meager *in vivo* results. A systematic study is necessary to provide a rational basis, rather than an empirical one, on which significant improvements can be made in the delivery systems.

SUMMARY

As the technology for gene therapy develops *in vitro* and *in vivo* in animal models, it is becoming clear that the three principal approaches--recombinant retroviruses, recombinant adenovirus, and direct DNA delivery--will ultimately have applications in specific therapeutic situations that take full advantage of the unique features of the specific delivery system: low level persistent expression after *ex vivo* recombinant retroviral therapy, high level transient expression after *in vivo* recombinant adenoviral therapy, or moderate level transient expression after *in vivo* administration of a synthetic DNA complex, which in principle could be repeated as desired.

ACKNOWLEDGMENTS

This work has been supported in part by HL-50422. S. L. C. W. is an Investigator of the Howard Hughes Medical Institute.

REFERENCES

Anderson, K.D., Thompson, J.A., Dipietro, J.M., Montgomery, K.T., Reid, L.M., Anderson, W.F., 1989, Gene expression in implanted rat hepatocytes following retroviral-mediated gene transfe, Somat. Cell Mol. Genet. **15**:215-27.

Anderson, R.G.W., Kamen, B.A., Rothberg, K.G., Lacey, S.W., 1992, Potocytosis: sequestration and transport of small molecules by caveolae, Science **25**:410-1.

Bilheimer, D.M., Goldstein, J.L., Grundy, S.C., Starzl, T.E., Brown, M.S., 1984, Liver transplantation provides low density lipoprotein receptors and lowers plasma cholesterol in a child with homozygous familial hypercholesterolemia, New Engl. J. Med. **311**:1658-1664.

Buja, L.M., Kita, T., Goldstein, J.L., Watanabe, Y., Brown, M.S., 1983, Cellular pathology of progressive atherosclerosis in the WHHL rabbit, an animal model of familial hypercholesterolemia, Arterioscler. **3**:87-99.

Chowdhury, J.R., Grossman, M., Gupta, S., Chowdhury, N.R., Baker, J.R. Jr., Wilson, J.M., 1991, Long term improvement of hypercholesterolemia after *ex vivo* gene therapy in LDLR-deficient rabbits, Science **254**:1802-1805.

Cohen, R.J., Paine, P.L., 1992, Biophysics of nucleocytoplasmic transport. in "Nuclear Trafficking," C.M. Feldherr, ed., Academic Press, New York.

Cotten, M., Langle-Rouault, F., Kirlappos, H., Wagner, E., Mechtler, K., Zenke, M., Beug, H., Birnstiel M.L., 1990, Transferrin-polycation-mediated introduction of DNA into human leukemic cells: stimulation by agents that affect the survival of transfected DNA or modulate transferrin receptor levels, Proc. Natl. Acad. Sci. USA **87**:4033-4037.

Cotten, M., Wagner, E., Zatloukal, K., Phillips, S., Curiel, D.T., Birnsteil, M.L., 1992, High-efficiency receptor-mediated delivery of small and large (48 kilobase) gene constructs using the endosome-disruption activity of defective or chemically inactivated adenovirus particles, Proc. Natl. Acad. Sci. USA **89**:6094-6098.

Cristiano, R., Smith, L.C., Woo, S.L.C., 1993a, Hepatic gene therapy: receptor-mediated gene delivery and elevated expression in primary hepatoctyes, Proc. Natl. Acad. Sci. USA **90**:2122-2127.

Cristiano, R.J., Smith, L.C., Kay, M.A., Brinkley, B., Woo, S.L.C., 1993b, Hepatic gene therapy: efficient gene delivery and expression in primary hepatocytes utilizing a conjugated adenovirus/DNA complex, Proc. Nat.l Acad. Sci. USA **90**:11548-11552.

Curiel, D.T., Agarwal, S., Wagner, E., Cotten, M., 1991, Adenovirus enhancement of transferrin polylysine mediated gene delivery, Proc. Natl. Acad. Sc. USA **88**:8850-8854.

Curiel, D.T., Wagner, E., Cotten, M., Birnstiel, M.L., Loechel, S., Hu, P.C., 1992a, High-efficiency gene transfer mediated by adenovirus coupled to DNA-polylysine complexes, Hum. Gene Ther. **3**:147-154.

Curiel, D.T., Agarwal, S., Romer, M.U., Wagner, E., Cotten, M., Birnstiel, M.L., Boucher, R.C., (1992b) Gene transfer to respiratory epithelial cells via the receptor-mediated endocytosis pathway, Am. .J Respir. Cell Mol. Bio.l **6**:247-252.

Dingwall, C., Laskey, R.A., (1991) Nuclear targeting sequences - a consensus?, Trends in Biol. Sci. **16**:478-481.

Dolník, V., Novotny, M.V., (1993) Separation of amino acid homopolymers by capillary gel electrophoresis, Anal. Chem. **65**:563-567.

Doms, R.W., White, J., Boulay, F., Helenius, A., 1991, Influenza virus hemagglutinin and membrane fusion, in "Membrane Fusion," J. Wilschut and D. Hockstra, eds., Marcel Dekker, Inc. New York.

Feldherr, C.M., Akin, D., 1990, EM visualization of nucleocytoplasmic transport processes, Electron Microsc Rev **3**:73-86.

T. Friedmann, 1983, "Gene Therapy-Fact and Fiction," Cold Spring Harbor Laboratory, New York.

Fujimoto, K., da Silva, P., 1988, Surface views of nuclear pores in isolated rat liver nuclei as revealed by fracture-flip/Triton-X. Eur, J. Cell. Biol **50**:390-397.

Garcia-Bustos, J., Heitman, J., Hall, M.N. 1991, Nuclear protein localization, Bioch. Biophys. Acta. **1071**:83-101.

Goldfarb, D.S., Gariepy, J., Schoolnik, G., Kornberg, R.D., 1986, Synthetic peptides as nuclear localization signals, Nature **322**:641-644.

Goldstein, J.L., Brown, M.S., Anderson, R.G.W., Russell, D.W., Schneider, W., 1985, Receptor-mediated endocytosis: concepts emerging from the LDL receptor system, Ann. Rev. Cell Biol. 1:1-39.

Goldstein, J.L., Brown, M.S., 1989, Familial hypercholesterolemia, in: "Metabolic Basis of Inherited Disease," C.R. Scriver, A.L. Beaudet, W.S. Sly, D. Valle, eds., McGraw-Hill, NewYork.

Grossman, M., Raper, S.E., Kozarsky, K., Stein, E.A., Engelhardt, J.F., Muller, D., Lupien, P.J., Wilson, J.M., 1994, Successful *ex vivo* gene therapy directed to liver in a patient with familial hypercholesterolemia, Nature Genetics 6:335-341.

Gottschalk, S., Cristiano, R.J., Smith, L.C., Woo, S.L.C., 1994, Folate-mediated gene delivery to tumor cells, Gene Ther. 2:1-7.

Hoeben, R.C., Fallaux, F.J., van Tilburg, N.H., Cramer, S.J., van Ormondt, H., Briet, E., van der E.b., A.J., 1993, Toward gene therapy for hemophila A: long-term persistence of factor VIII-secreting fibroblasts after transplantation into immunodeficient mice, Hum. Gene Ther. 4:179-186.

Ishibashi, S., Brown, M.S., Goldstein, J.L., Gerard, R.D., Herz, H.J., 1993, Hypercholesterolemia in low density lipoprotein receptor knockout mice and its reversal by adenovirus-mediated gene delivery, J Clin. Invest. 92:883-891.

Lanford, R.E., Kanda, P., Kennedy, R.C., 1986, Induction of nuclear transport with a synthetic peptide homologous to the SV40 T antigen transport signal, Cell 46:575-582.

Lanford, R.E., White, R.G., Dunham, R.G., Kanda, P., 1988, Effect of basic and nonbasic amino acid substitutions on transport induced by simian virus 40 T-antigen synthetic peptide nuclear transport signals, Mol. Cell Biol. 8:2722-2729.

Leamon, C.P., Low, P.S., 1991, Delivery of macromolecules into living cells: a method that exploits folate receptor endocytosis, Proc. Natl. Acad. Sci. USA 88:5572-5576.

Leamon, C.P., Low, P.S., 1993, Cytoxicity of momordin-folate conjugates in cultured human cells, J. Biol. Chem. 267:24966-24971.

Li, J., Fang, B., Eisensmith, R.C., Li, X.H.C., Nasonkin, I., Lin-Lee, Y.C., Mims, M., Hughes, A., Montgomery, C., Roberts, J., Parker, T., Levine, D., Woo, S.L.C., 1994, In vivo gene therapy for hyperlipidemia: phenotypic correction in Watanabe rabbits by hepatic delivery of the rabbit LDL receptor gene, J. Clin. Invest., accepted.

Luby-Phelps, K., Castle, P.E., Taylor, D.L., Lanni, F., 1987, Hindered diffusion of inert tracer particles in the cytoplasm of mouse 3t3 cells, Proc. Natl. Acad. Sci. USA 84:4910-13.

McGraw, T.E. and Maxfield, F.R., 1992, Internalization and sorting of macromolecules: endocytosis: in "Targeted Drug Delivery," RL Juliano, ed., Springer-Verlag.

Michael, S.I., Huang, C., Romer, M.U., Wagner, E., Curiel, D.T., 1993, Binding-incompetent adenovirus facilitates molecular conjugate-mediated gene transfer by the receptor-mediated endocytosis pathway, J. Biol. Chem. 268:6866-6869.

Neda, H., Wu, C.H., Wu, G.Y., 1991, Chemical modification of an ecotropic murine leukemia virus results in redirection of its target cell specificity, J. Biol. Chem. 266:14143-14146.

Ojcius, D.M., Young, J.D.E., 1991, Cytolytic pore-forming proteins and peptides: is there a common structural motif?, Trends in Biol. Sci. 16:225-229.

Plank, C., Zatloukal, K., Cotten, M., Mechtler, K., Wagner, E., 1992, Gene transfer into hepatocytes using asialoglycoprotein receptor mediated endocytosis of DNA complexes with an artificial tetra-antennary galactose ligand. Bioconjug, Chem. 3:533-9.

Ponder, K.P., Gupta, S., Leland, F., Darlington, G., Finegold, M., Demayo, J., Ledley, F., Chowdhury, J., & Woo, S.L.C., 1991, Mouse Hepatocytes Migrate to Liver Parenchyma & Function Indefinitely After Intrasplenic Transplantation, Proc. Natl. Acad. Sci. USA 88:1217-1221.

Ramalho-Santos, J., Nir, S., Düzgünes, N., Carvalho, A.P. de, Lima, M.C.P. de, 1993, A common mechanism for influenza virus fusion activity and inactivation, Biochemistry 32:2771-9.

Robenek, H., Harrach, B., Severs, N.J., 1991, Display of low density lipoprotein receptors is clustered, not dispersed, in fibroblast and hepatocyte plasma membranes, Arterioscl Thromb 11:261-271.

Schwartz, A.L., 1989, The Hepatic Asialoglycoprotein Receptor. CRC Crit. Rev. Biochem. 16:207-233.Silver PA., 1991, How proteins enter the nucleus, Cell 64:489-497.

Shepherd, V.L., 1989, Intracellular mechanisms of sorting in receptor-mediated endocytosis, Trends in Physiol. Sci. 10:458-462.

Silver, P.A., 1991, How proteins enter the nucleus, Cell 64:489-497.

Turek, J.J., Leamon, C.P., Low, P.S., 1993, Endocytosis of folate-protein conjugates: ultrastructural localization in KB cells, J. Cell Sci. 106:4223-4230.

Wagner, E., Zenke, M., Cotten, M., Beug, H., Birnstiel, M.L., 1990, Transferrin-polycation conjugates as carriers for dna uptake into cells, Proc. Natl. Acad. Sci. USA 87:3410-4.

Wagner, E., Cotten, M., Mechtler, K., Kirlappos, H., Birnstiel, M.L., 1991a, DNA-binding transferrin conjugates as functional gene-delivery agents: synthesis by linkage of polylysine or ethidium homodimer to the transferrin carbohydrate moiety, Bioconjugate Chem. 2:226-231.

Wagner, E., Cotten, M., Foisner, R., Birnstiel, M.L., 1991b, Transferrin-polycation-DNA complexes: effect of polycations on the structure of the complex and dna delivery to cells, Proc. Natl. Acad. Sci. USA **88**:4255-4259.

Wagner, E., Plank, C., Zatloukal, K., Cotten, M., Birnstiel, M.L., 1992, Influenza virus hemagglutinin HA$_2$-N-terminal fusogenic peptides augment gene transfer by transferrin-polylysine-DNA complexes: toward a synthetic virus-like gene-transfer vehicle, Proc. Natl. Acad. Sci. **89**:7934-7938.

Wagner, E., Zatloukal, K., Cotten, M., Kirlappos, H., Mechtler, K., Curiel, D.T., Birnstiel, M.L., 1992, Coupling of adenovirus to transferrin-polylysine/DNA complexes greatly enhances receptor-mediated gene delivery and expression of transfected genes, Proc. Natl. Acad. Sci. USA **89**:6099-6103.

Weatherall, D., 1994, Heroic gene surgery, Nature Genetics **6**:325-326.

Wharton, S.A., Martin, S.R., Ruigrok, R.W.H, Skehel, J.J., Wiley, D.C., 1988, Membrane fusion by peptide analogues of influenza virus haemagglutinin, J. Gen. Virol. **69**:1847-1857.

White, J.M., 1990, Viral and cellular membrane fusion proteins, Ann. Rev. Physiol. **52**:675-697.

White, J.M., 1992, Membrane fusion, Science **258**:917-924.

Wilson, J.M., Jefferson, D.M., Chowdhury, J.R., Novikoff, P.M., Johnston, D.E., and Mulligan, R.C., 1988, Retrovirus-mediated Transduction of Adult Hepatocytes, Proc. Natl. Acad. Sci. USA **85**:3014-3018.

Wilson, J.M., Chowdhury, N.R., Grossman, M., Wajsman, R., Epstein, A., Mulligan, R.C., and Chowdhury, J.R., 1990, Temporary Amelioration of Hyperlipidemia in Low Density Lipoprotein Receptor-deficient Rabbits Transplanted with Genetically Modified Hepatocytes, Proc. Natl. Acad. Sci. USA **87**:8437-8441.

Wilson, .JM., Grossman, M., Wu, C.H., Chowdhury, N.R., Wu, G.Y., Chowdhury, J.R., 1992, Hepatocyte-Directed gene transfer *in vivo* leads to transient improvement of hypercholesterolemia in low density lipoprotein receptor-deficient rabbits, J. Biol. Chem. **267**:963-967.

Wolff, J.A., Yee, J.-K., Skelly, H.F., Moores, J.C., Respess, J.G., Friedmann, T., and Leffert, H., 1987, Expression of Retrovirally Transduced Genes in Primary Cultures of Adult Rat Hepatocytes, Proc. Natl. Acad. Sci. USA **84**:3344-3348.

Wu, G.Y., Wu, C.H., 1987, Receptor-mediated *in vitro* gene transformation by a soluble DNA carrier system, J. Biol. Chem. **262**:4429-4432.

Wu, G.Y., Wu, C.H., (1988a) Evidence for targeted gene delivery to HepG2 hepatoma cells in vitro, Biochemistry **27**:887-892.

Wu, G.Y., Wu, C.H., 1988b, Receptor-mediated gene delivery and expression *in vivo*, J. Biol. Chem. **263**:14621-14624.

Wu, C.H., Wilson, J.M., Wu, G.Y., 1989, Targeting genes: delivery and persistent expression of a foreign gene driven by mammalian regulatory elements *in vivo*, J. Biol. Chem. **264**:16985-16987.

Wu, G.Y., Wilson, J.M., Shalaby, F., Grossman, M., Shafritz, D.A., Wu, C.H., 1991, Receptor-mediated gene delivery *in vivo*, J. Biol. Chem. **266**:14338-14342.

Yamamoto, T., Bishop, R.W., Brown, M.S., Goldstein, J.L., Russel, D.W., 1986, Deletion in cysteine-rich region of LDL receptor impedes transport to cell surface in WHHL rabbit, Science **232**:1230-1237.

Zenke, M., Steinlein, P., Wagner, E., Cotten, M., Beug, H., Birnstiel, M.L., 1990, Receptor-mediated endocytosis of transferrin polycation conjugates:an efficient way to introduce DNA into hematopoietic cells, Proc. Natl. Acad. Sci. USA **87**:3655-3659.

POSSIBLE ROLE OF VIRUSES IN ATHEROSCLEROSIS

Jason C. H. Shih and Donald W. Kelemen

Biotechnology Laboratory
Department of Poultry Science
North Carolina State University
Raleigh, NC 27695-7608

INTRODUCTION

Much progress has been made in last two decades in research, prevention, and treatment of atherosclerosis and subsequent coronary heart diseases. Hypotheses have been proposed to define a unifying concept for the basic mechanisms of atherogenesis. The response-to-injury hypothesis, as recently updated (Ross, 1993a,b), suggests that various forms of insults to the endothelium and arterial cells initiate a chronic, inflammatory response featuring the subendothelial infiltration and activation of macrophages and T-cells. Upregulation of expression and secretion of cytokines and growth factors associated with these activated leukocytes result in smooth muscle cell (SMC) migration and proliferation into the intima of the arterial wall. Simultaneous with this process is thought to be the modification of lipoproteins in the arterial wall. In one such process low-density lipoproteins (LDL) can be oxidatively modified and taken up by macrophages to become foam cells (Steinberg et al., 1989). Because it is cytotoxic, oxidized LDL may kill macrophages and endothelial cells. In response to endothelial injury, platelets adhere, aggregate and secrete platelet-derived growth factor (PDGF), which further stimulates the proliferation of SMCs and thrombosis as proposed in the original response-to-injury hypothesis (Ross and Glomset, 1976).

Another interesting hypothesis is the monoclonal origin of atherosclerosis (Benditt and Benditt, 1973; Benditt, 1977), which proposes that the formation of atherosclerotic plaques is the result of SMC proliferation derived from transformation or a stable mutation of SMCs. This hypothesis originated from the observation that the SMCs of many atherosclerotic plaques were monotypic based on isozyme analysis (Benditt and Benditt, 1973). Independently, the discovery of the transforming potential of DNA isolated from plaque cells (Penn et al., 1986) seemed to support this hypothesis. If this hypothesis is proved to be true, the basic mechanisms of the initiation of atherogenesis and tumorigenesis may be similar.

Intriguing as these theories may be, the injuring agents or the genetic transforming elements have yet to be defined. One of the possible candidates is viruses. Viral infections

Nutrition and Biotechnology in Heart Disease and Cancer
Edited by J.B. Longenecker *et al.*, Plenum Press, New York, 1995

89

are known to cause immune and inflammatory responses and cell death. Some viruses carrying transforming genes are tumorigenic. The implications and evidence for the association of viruses with atherosclerosis in human patients and experimental animals, will be reviewed and discussed in this chapter.

HUMAN STUDIES

Using *in situ* DNA hybridization techniques, Benditt and coworkers (1983) first detected the nucleic acids of a herpesvirus in human arterial samples with thickened intima. Thin sections of the tissues were hybridized with DNA probes for herpes simplex virus-2 (HSV-2), cytomegalovirus (CMV), and Epstein-Barr virus (EBV). The cells in the lesion areas were found to be positive for the presence of HSV-2, but not for the other two viruses tested. However, CMV, also a herpesvirus, has been found in the vascular tissues and SMC cultures by other investigators.

A series of studies were carried out by Melnick and his colleagues in search of viruses in aortic tissues of the patients with advanced atherosclerosis. Using immunocytochemistry, they were unable to detect the viral antigens of HSV-1, HSV-2 or CMV directly in the tissues, but CMV-specific antigens, not HSV's, were found in the explant cell cultures (Melnick *et al.*, 1983, 1990; Gyorkey *et al.*, 1984). The presence of nucleic acids of CMV were subsequently recognized in SMC cultures by DNA hybridization. But CMV antigens and CMV DNA sequence were found in cell cultures from both involved and non-involved tissues (Petrie *et al.*, 1987). In plasma antibody analysis, higher levels of antibodies to CMV were associated with patients undergoing vascular surgery for atherosclerosis than those of the control group. Interestingly, the patient group was found to have lower levels of plasma cholesterol than the control group (Adam *et al.*, 1987). In these studies, although viral infection was diagnosed as positive in arterial tissues by different methods, direct isolation of CMV from tissues and cell cultures was not successful.

The results of two other clinical studies (Grattan *et al.*, 1989; MacDonald *et al.*, 1989) also strongly support the link between atherosclerosis and viral infection. They found that the incidence of recurrence of atherosclerosis among heart transplant patients correlated positively with the prevalence of CMV infection. In another study, the presence of antigens and nucleic acids of CMV or HSV were demonstrated in the coronary arteries and thoracic aortas of young trauma victims without any manifestation of atherosclerosis (Yamashiroya *et al.*, 1988). Histology of the specimens indicated that the viral DNA and antigen were associated with focal clusters of "foamy" cells. Again the investigators attempted but failed to isolate active viruses. Infection of CMV in the blood vessels was therefore believed to be in a latent state.

In Holland, Hendrix *et al.* (1989) studied the presence of CMV in the arterial samples of patients with and without atherosclerosis. Nucleic acids of CMV were detected in both the atherosclerotic and control samples by *in situ* DNA hybridization and dot blot hybridization. However, virus isolation and immunocytochemical tests were negative in all samples. In the following year, they used the sensitive polymerase chain reaction (PCR) to analyze for trace amounts of CMV DNA in the tissue samples (Hendrix *et al.*, 1990). They found 90% of the specimens with atherosclerosis were positive; while only 53% of normal specimens were positive. A clear correlation of the presence of CMV or CMV genes with atherosclerosis was demonstrated, suggesting a possible role for CMV in the pathogenesis of atherosclerosis (Bruggemen and van Dam-Mieras, 1991).

A unique and comprehensive epidemiological study was conducted in China (Chen *et al.*, 1990). Data from a 1976 retrospective mortality survey for causes of death between 1973-75 and an ecologic survey were combined to study the relationship between various mortality rates and several dietary, lifestyle, and environmental characteristics in 65 mostly rural counties in China. A total of approximately 6500 adults aged 35-64 were included in the survey. In examining the data, it was discovered that the cumulative mortality from stroke, hypertensive heart disease, and myocardial infarction/coronary heart disease correlated positively with the presence of antibody against HSV in the plasma (Campbell, personal communication). The correlation was weak (r=0.38), but statistically significant (p<0.01). Antibodies against CMV were not measured in this study. Furthermore, the mortality of the diseases was not associated with plasma levels of cholesterol, where the range of cholesterol concentration among rural Chinese was generally 100-200 mg/dl.

In summary, infection of herpesviruses have been detected in human aortic tissues, and the infection seems to correlate with atherosclerosis in many studies. However, it has not been determined in these studies whether the viruses are the cause of the plaques or enter the arterial wall after the plaques are formed. Is the virus indeed the etiopathogenic agent? This question does not have an easy answer for two reasons. First, herpesviruses are ubiquitous and mostly latent among the human population. Second, isolation of the virus from arterial cells has not been successful to date. These facts have made it difficult to apply Koch's Postulates to this problem.

THE CHICKEN MODEL

Paterson and Cottral reported in 1950 the co-development of neurolymphomatosis and atherosclerosis in chickens that had been inoculated with tracheal washings from diseased chickens. In 1967, Churchill and Biggs identified the etiologic agent of neurolymphomatosis or Marek's disease as an avian herpesvirus called Marek's disease virus (MDV).

In 1973, Burch and coworkers suggested that viral infection may be associated with early atherosclerotic changes in humans. Benditt and Benditt, in the same year, published their findings of the monoclonal origin of plaque cells in atherosclerotic human patients and suggested that proliferative mutation of SMCs may be due to the effects of chemicals or viruses. However, experimental evidence for viral atherogenesis did not come until Fabricant and her colleagues conducted a series of studies on atherosclerosis in chickens caused by MDV.

The first observation made by Fabricant *et al.* (1973) was in the course of studies on feline cell cultures infected by a feline herpesvirus. The infected cell cultures were found to accumulate appreciable amounts of intracellular and extracellular lipids including crystalline cholesterol. This serendipitous observation led them to hypothesize a role for herpesvirus in the pathogenesis of atherosclerosis. At this time, MDV was isolated and identified as a herpesvirus which causes proliferative disease in the chicken. Specific pathogen-free (SPF) chickens susceptible to MDV were developed. They became a perfect host-and-pathogen system for Fabricant to test this hypothesis (Fabricant *et al.*, 1973; Fabricant, 1975).

In their experiments, SPF-chickens were divided into four groups (Fabricant *et al.*, 1978; Minick *et al.*, 1979). All of them were fed a normal low-cholesterol diet, but two of the groups were infected with a low-virulence strain of MDV at 2 days of age. After 15 weeks, one infected and one uninfected group were fed a diet supplemented with 2% cholesterol; the other infected and uninfected groups remained on the normal diet. After an additional 15 weeks, aortas were examined for atherosclerotic lesions. It was found that only MDV-infected chickens developed atherosclerosis whether they were normocholesterolemic or hypercholesterolemic. High levels of cholesterol appeared to have a synergistic effect. Viral infection did not influence serum cholesterol concentration. Uninfected chickens, even hypercholesterolemic, did not develop the disease. Grossly visible atherosclerotic lesions were observed in large coronary arteries, aortas, and major arterial branches. Frequently they were occlusive. Microscopically, the arterial lesions were fatty and proliferative with fibrous caps overlying areas of atheromatous change. They closely resembled human atherosclerosis in character and distribution. Furthermore, specific viral internal antigens of MDV were found in medial layers of arteries of infected birds, as detected by an immunofluorescence technique.

Immunization with herpes virus of turkey (HVT) has been a common commercial practice to prevent Marek's disease in chickens. It was found that the development of atherosclerosis in the MDV-infected chickens was markedly reduced when they were first immunized with HVT (Fabricant, 1985). This is further evidence that MDV infection may be the direct cause of atherosclerosis.

Aortic tissue from the normocholesterolemic, MDV-infected chickens had a significantly higher content of free and esterified cholesterol, triglycerides and phospholipids than did uninfected controls (Fabricant *et al.*, 1981). Infection of chicken SMCs with MDV *in vitro* greatly increased the accumulation of cholesterol and cholesterol esters (CE). Detailed analysis of enzyme activation demonstrated that the CE cycle is altered, resulting in cytoplasmic accumulation of CE, not only in MDV-infected chicken aortic SMCs (Hajjar *et al.*, 1985; Hajjar, 1986), but also in HSV-infected human arterial SMCs (Hajjar *et al.*, 1987, 1989).

In chickens, experimental results have strongly indicated an etiologic role of MDV in atherosclerosis. The results include: 1) only MDV-infected chickens develop arterial lesions; 2) MDV-specific internal antigens are found in the medial SMCs of atherosclerotic arteries; 3) immunization with HVT protects chickens against atherosclerosis as well as Marek's disease; 4) MDV alters lipid metabolism of both *in vivo* and *in vitro* infected arterial SMCs and induces cytosolic accumulation of cholesterol and cholesterol esters. To date, the chicken is the only animal species in which the causal effect of a herpesvirus for atherosclerosis has been demonstrated. This host-and-pathogen system merits further investigation to unfold the basic mechanism of the initiation of atherogenesis by viruses.

THE QUAIL MODEL

The Japanese quail (*Coturnix coturnix japonica*) is an ideal laboratory animal, especially for atherosclerosis research. It is small (110-130 g body weight), low in feed consumption (14 g of feed per day), sexually mature in 40 days, omnivorous, hardy and resistant to most infectious diseases (National Academy of Science, 1969). Most importantly, it is susceptible to spontaneous and cholesterol-induced atherosclerosis. It has been used for studies of atherosclerosis in many laboratories (reviewed by Shih, 1983). Genetically bred Japanese quail in the authors' and other laboratories are extremely sensitive to cholesterol feeding and develop aortic lesions in 10-12 weeks. This model has made it possible to study a chronic disease such as atherosclerosis for its development, prevention and treatment within a relatively short period of time.

Nutritional studies have shown that several kinds of nutrients and dietary ingredients prevent atherosclerosis in Japanese quail. These include vitamin A (Baker *et al.*, 1972, 1976), polyunsaturated fat (Smith and Hilker, 1973), fish oil (Pullman and Shih, 1987) and high intake of soy protein (McClelland and Shih, 1988). Several chemicals and drugs such as lipoic acid (Shih, 1983), aspirin, nicotinic acid, gemfibrozil, clofibrate (Shih and Pyrzak, 1987), and cholestyramine (unpublished data) were shown to be effective in prevention or regression of the disease in quail. Because of its small size, low feed intake, short life cycle, and rapid atherosclerosis development, the genetically selected Japanese quail is one of the most sensitive and cost-effective animal species for nutritional and pharmaceutical studies of atherosclerosis.

A genetic strain of SEA (susceptible to experimental atherosclerosis) quail was first genetically developed in the laboratory of Day and his colleagues (Chapman *et al.*, 1976;

Day *et al.*, 1977) and further characterized by McCormick *et al.* (1982). In our department, the original stock of Japanese quail were generally susceptible to atherosclerosis but responded to dietary cholesterol with great individual variations (Morrisey and Donaldson, 1977). A breeding program was undertaken with 40 pairs of male and female quail as the starting flock (Shih *et al.*, 1983). After four generations of selection, the segregation into susceptible (SUS) and resistant (RES) lines was evident. The SUS birds developed severe lesions in 9-10 weeks on a diet containing 0.5% cholesterol; the RES birds developed none to moderate atherosclerosis on the same atherogenic diet. Histological studies of the aortic lesions indicated that they were similar to human atherosclerosis. The aortic plaque was characterized by intimal thickening, the presence of foam cells, proliferation of SMC and fibroblast cells, and formation of scar tissue with collagen deposition.

More recently, immunocytochemical and scanning electron microscopic studies on the progression of atherosclerosis were conducted in this laboratory (Casale *et al.*, 1992). Both proliferation of SMCs and infiltration of macrophages were characteristics of the development of the disease. However, we found that SMC proliferation began at a very early stage, contrary to other studies in which macrophage infiltration was thought to be the initiating event (Ross, 1993a,b). The development of genetically distinct SUS and RES quail has provided us with two advantages: the availability of sensitive SUS birds for a variety of studies on atherosclerosis and the availability of RES birds for comparative studies. The second advantage became apparent when we investigated the link between viral infection and atherosclerosis. (Now both genetic lines of the quail are maintained and available at the Center of Genetic Stocks of Japanese Quail, Department of Animal Science, University of British Columbia, Vancouver, Canada.)

In light of Fabricant's discovery MDV-induced atherosclerosis in chickens, a series of studies were undertaken to determine the possibility of MDV infection in Japanese quail (Pyrzak and Shih, 1987; Shih *et al.*, 1989). Virus isolation was attempted by direct cultivation of kidney cells from quail with and without atherosclerosis and by inoculation of tissue suspensions onto the chorioallantoic membrane (CAM) of embryonated chicken eggs. Cell cultures and embryos were observed for cytopathic effects, embryo mortality, lesions on CAM, virus particles, and immunofluorescent reaction with MDV-specific antigens. All of these tests were negative for MDV. To confirm the absence of MDV, MDV-susceptible sentinel chicks were put in contact with SUS quail in the same cages for 12 weeks. Throughout the exposure period, none of the chickens were infected or appeared sick. Serological tests and virus isolation were also negative in these chickens.

DNA hybridization techniques were then used to investigate the presence of the MDV genome in quail. A gene library consisting of 14 EcoRI fragments of MDV DNA cloned into plasmid pBR328 was used to prepare the probe representing 30% of the total MDV genome. Results of Southern blot analysis demonstrated the presence of MDV homologous sequences in the DNA of quail atherosclerotic tissues, but the restriction patterns were entirely different from that of MDV. Since some specific sequences were complementary with the MDV probe, it is believed that the quail may be infected by a putative quail herpes virus (QHV) related to MDV.

Is the viral DNA incorporated into the germline and involved in the genetic selection of SUS and RES quail? To answer this question, a dot blot method was used to screen embryo DNA from SUS and RES quail (Pyrzak and Shih, 1987). DNA samples were isolated and tested for 68 SUS embryos and 126 RES embryos. All DNA samples from SUS embryos demonstrated a strong positive hybridization with the MDV probe, equivalent to 10 or more virus copies per cell. Among the samples of RES embryos, only 16% showed the same strong hybridization, 43% intermediate (1-10 virus copies per cell) and 41% negative (less than 1 virus copy per cell). It is quite possible that viral genes had been incorporated into the host DNA and thus were co-selected by genetic breeding. A small group of RES quail positive for viral genes was consistent with the fact that some quail in the RES group always developed atherosclerosis when fed cholesterol. Even though the presence of viral genes in SUS quail had been demonstrated, dietary cholesterol was still required to induce the disease. An unknown interaction between cholesterol and viral genes must exist.

Since SMC proliferation is a key feature in atherosclerosis and it occurs in the early stage of the disease in the quail, a study of quail aortic SMCs in culture has been conducted (Shih *et al.*, manuscript in preparation). Aortic SMCs from SUS quail were cultured by the same procedures developed for pigeon arterial SMCs (Smith *et al.*, 1979). SMCs which grew normally in monolayer **(Figure 1A)** were characterized and passaged weekly. After 4 or 5 passages of normal growth, the cultured SMCs began to develop cytopathic effects (CPE) **(Figure 1B)**, including the formation of giant cells, multinucleated syncytia, highly vacuolated cells and eventually cell death. This scenario was similar to active CMV infection of human arterial SMC cultures (Tulmilowicz, 1990). Cells grown on formar-coated gold mesh grids were subsequently processed for whole mount electron microscopy **(Figure 1C)** and revealed electron-dense virus-like particles approximate the same size as CMV inside the cytoplasm and nuclei. Isolation of the virus-like particles and inoculation to quail embryo fibroblasts (QEF) for propagation were not successful. The cell cultures were positive for DNA hybridization with the MDV probe. It was speculated that latent QHV may be activated upon passages of SMCs *in vitro*.

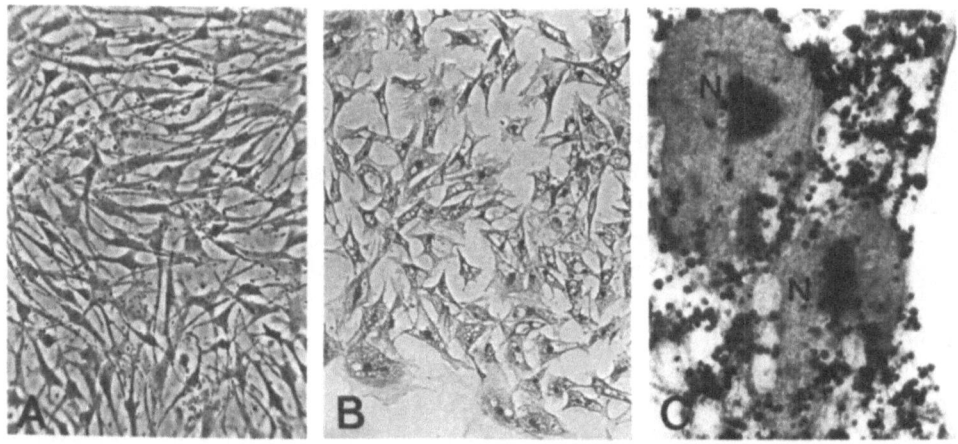

Figure 1. Aortic SMC cultures from SUS quail. A: passage 2; B: passage 4 of the same culture; C: whole mount electron microscopy of passage 4 cells showing virus-like particles; N: nucleus.

If SUS quail are indeed latently infected by a putative QHV, as suggested by our experimental results, how can a latent virus be involved in atherogenesis? One possibility is the latent provirus may become reactivated under certain conditions and cause an active viral infection. Active herpesvirus can be an injuring agent on endothelium (Kefalides, 1992) that initiates the atherogenic process as proposed by Ross (1976; 1993). The other possibility is that the virus may introduce an exogenous transforming gene to the host or, alternatively, the insertion of viral genes may activate an endogenous proto-oncogene or deactivate a suppressor gene in the host genome (Bishop and Varmus, 1982; Bishop, 1991). Whatever the genetic mechanism may be, cell transformation or mutation may initiate atherogenesis as suggested by Benditt's monoclonal hypothesis (Benditt and Benditt, 1973). Since MDV is an oncogenic virus and carries genes with transforming potential (Bradley *et al.*, 1989; Jones *et al.*, 1992), transformation of arterial cells may initiate the disease in the chicken. This hypothesis has been supported by the discovery of DNA with transforming activity in human arterial plaque cells (Penn *et al.*, 1986). To test this hypothesis in quail, similar experiments as described by Penn *et al.* (1986) were conducted in search of a gene with transforming activity (Kelemen *et al.*, manuscript in preparation).

Because of the small size of quail aortas, they were pooled into two groups for the isolation of high molecular weight DNA. NIH3T3 cells were transfected with the quail DNA by the standard calcium phosphate protocol (Chen and Okayama, 1987). The transformed cells were trypsinized, washed and resuspended for injection (5×10^5 cells/inj.) into athymic nude mice. Nude mice were injected subcutaneously on their backs and monitored weekly for tumor development. Animals positive for tumors were sacrificed and the tumor mass collected for DNA isolation. Tumor DNA samples were used for Southern analysis and for a secondary tumorigenicity test of the NIH3T3 cells-nude mice system. The results are summarized in **Table 1**.

Table 1. NIH3T3 Cells-Nude Mice Assay for tumorigenicity of quail DNA

DNA	Tumor positive/ Mice injected	Latency (wks)	Quail sequence
Primary:			
Quail Gp.1	4/8	7-13	Yes[1]
Quail Gp.2	2/4	9-14	Yes
NIH3T3 cells[2]	0/4	>14	No
T24	4/4	2-3	No
Secondary:			
Tumor Gp.1	2/7	7-12	Yes
Tumor Gp.2	7/12	9-13	No
NIH3T3 cells	1/4	10	No
T24	4/4	2-3	No

[1] Hybridization positive with quail genomic DNA probe.
[2] NIH3T3 cells: negative control; T24: positive control.

It was evident that quail aortic DNA transformed NIH3T3 cells that subsequently produced tumors in athymic nude mice. Based on Southern analysis, these primary (6/6) and secondary (2/9) tumors contained sequences which hybridized with a quail genomic probe (data not shown). Positive results of a secondary transfection confirmed that the transformed phenotype was induced initially by quail aortic DNA. The quail sequences retained in the secondary tumor DNA are candidates for the transforming gene(s) required for tumorigenesis. Cloning and sequence analysis of these sequences are in progress.

The availability of genetic lines of Japanese quail susceptible and resistant to atherosclerosis will provide a powerful animal model for the studies of the pathogenesis, prevention and treatment of atherosclerosis. The discovery of the association of atherosclerosis with a latent viral infection in the quail should provide an opportunity to further explore the potential role of viral mediated transforming genes in the development of atherosclerosis. If this can be demonstrated, it would further add to the concept of the multifactorial nature of atherosclerosis.

SUMMARY

Relative to the role of viruses in the pathogenesis of atherosclerosis a number of questions remain to be answered. Being ubiquitous, are the herpesvirus infections the norm or a diseased state in humans and animals? In all cases studied, including MDV induced atherosclerosis, direct isolation of viruses from arterial tissues has not been successful. Are herpesviruses always latent or dormant in vascular tissues? What is the relationship between provirus and disease development? Are they setting the stage for the pathogenic process triggered by certain environmental factors? Is hypercholesterolemia to trigger the development of atherosclerosis in the presence of proviruses? Is the reactivated and infectious virus the injuring agent that initiates atherogenesis? Or, are the proviral genes activated to transform arterial cells? In the latter case, are these proviral genes equivalent to "proto-atherogenes"? Hopefully, further study on Japanese quail will help clarify many of these questions.

ACKNOWLEDGMENT

The authors wish to thank NIH, USDA, N.C. Biotechnology Center, Burroughs Wellcome Pharmaceutical Co., and the Pew Charitable Trusts (Pew Fellowship to JCHS) for grant support, and the Chemical Industry Institute of Toxicology, Dr. Cheryl Walker, Dr. Roman Pyrzak, Dr. Jackie McClelland, and Ms. Liz Pullman for their collaboration and technical assistance. A special appreciation is given to Dr. Richard St. Clair, with whom the author (JCHS) spent a year of sabbatical study on this subject, for his support of my work and review of this manuscript.

REFERENCES

Adam, E., Melnick, J.L., Probstfield, J.L., Petrie, B.L., Burek, J., Bailey, K.R., McCollum, C.H., and DeBakey, M.E., 1987, High levels of cytomegalovirus antibody in patients requiring vascular surgery for atherosclerosis, Lancet, 2:291.

Baker, R.C., Ringer, R.K., and Cogger, E.A., 1972, The influence of dietary vitamin A on atherosclerosis in Japanese quail, Poultry Sci. 51:925.

Baker, R.C., Smith, M.A., and Ringer, R.K., 1976, The influence of vitamin A deficiency on aortic endothelium of Japanese quail. Artery, 2:423.

Benditt, E.P., 1977, Implications of the monoclonal character of human atherosclerotic plaques, Am. J. Pathol. 86:693.

Benditt, E.P., Barrett, T., and McDougall, J.K., 1983, Viruses in the etiology of atherosclerosis, Proc. Nat. Acad. Sci. U.S.A. 80:6386.

Benditt, E.P., and Benditt, J.M., 1973, Evidence for a monoclonal origin of human atherosclerotic plaques, Proc. Nat. Acad. Sci. U.S.A. 70:1753.

Bishop, J.M., 1991, Molecular themes in oncogenesis, Cell, 64:235.

Bishop, J.M., and Varmus, H.E., 1982, Functions and origins of retroviral transforming genes, in "Molecular Biology of Tumor Viruses. Part III, RNA Tumor Viruses", R. Weiss, N. Teich, H. Varmus, and J. Coffin, eds., Cold Spring Harbor Laboratory Press, New York.

Bradley, G., Hayashi, M., Lancz, G., Tanaka, A., and Nonoyama, M., 1989, Structure of the Marek's disease virus BamHI-H gene family: genes of putative importance for tumor induction, J. Virol. 63:2534.

Bruggeman, C.A., and van Dam-Mieras, M.C.E., 1991, The possible role of cytomegalovirus in atherogenesis, Prog. Med. Virol. 38:1.

Burch, G.E., Harb, J.M., Hiramoto, Y., and Shewey, L., 1973, Viral infection of the aorta of man associated with early atherosclerotic changes, Am. Heart J. 86:523.

Casale, E.S., Qureshi, M., and Shih, J.C.H., 1992, Immunocytochemical and scanning electron microscopic studies of atherosclerosis in Japanese quail, Poultry Sci. 71:141.

Chapman, K.P., Stafford, W.W., and Day, C.E., 1976, Animal model for experimental atherosclerosis produced by selective breeding of Japanese quail, in "Atherosclerosis Drug Discovery", C.E. Day, ed., Plenum, New York.

Chen, J., Campbell, T.C., Li, J., and Peto, R., 1990, "Diet, Lifestyle and Mortality in China. A Study of the Characteristics of 65 Chinese Counties", Joint publication of Oxford University Press, Cornell University Press and China People's Publishing House, Oxford, U.K.

Chen, C., and Okayama, H., 1987, High-efficiency transformation of mammalian cells by plasmid DNA, Mol. Cell. Biol. 7:2745.

Churchill, A.E., and Biggs, P.M., 1967, Agent of Marek's disease in tissue culture. Nature, 215:528.

Day, C.E., Stafford, W.W., and Schurr, P.E., 1977, Utility of a selected line (SEA) of the Japanese quail (Coturnix coturnix japonica) for the discovery of new anti-atherosclerosis drugs, Lab. Anim. Sci. 27:817.

Fabricant, C.G., 1975, Herpesvirus induced cholesterol - an added dimension in the pathogenesis, prophylaxis, or therapy of atherosclerosis, Artery, 1:361.

Fabricant, C.G., 1985, Atherosclerosis: the consequence of infection with a herpesvirus, Adv. Vet. Sci. Comp. Med. 30:39.

Fabricant, C.G., Fabricant, J., Litrenta, M.M., and Minick, C.R., 1978, Virus-induced atherosclerosis, J. Exp. Med. 148:335.

Fabricant, C.G., Hajjar, D.P., Minick, C.R., and Fabricant, J., 1981, Herpesvirus infection enhances cholesterol and cholesteryl ester accumulation in cultured arterial smooth muscle cells, Am. J. Pathol. 105:176.

Fabricant, C.G., Krook, L., and Gillespie, J.H., 1973, Virus-induced cholesterol crystals, Science, 181:566.

Grattan, M.T., Moreno-Cabral, C.E., Starnes, V.A., Oyer, P.E., Stinson, E.B., and Shumway, N.E., 1989, Cytomegalovirus infection is associated with cardiac allograft rejection and atherosclerosis, JAMA, 261:3561.

Gyorkey, F., Melnick, J.L., Guinn, G.A., Gyorkey, P., and DeBakey, M.E., 1984, Herpes viridae in the endothelial and smooth muscle cells of the proximal aorta of atherosclerosis patients, Exp. Mol. Pathol. 40:326.

Hajjar, D.P., 1986, Herpesvirus infection prevents activation of cytoplasmic cholesteryl esterase in arterial smooth muscle cells, J. Biol. Chem. 261:7611.

Hajjar, D.P., Falcone, D.J., Fabricant, C.G., and Fabricant, J., 1985, Altered cholesterol ester cycle is associated with lipid accumulation in herpesvirus infected avian arterial smooth muscle cells, J. Biol. Chem. 260:6124.

Hajjar, D.P., Nicholson, A.C., Hajjar, K.A., Sando, G.N., and Summers, B.D., 1989, Decreased messenger RNA translation in herpesvirus-infected arterial cells: effect on cholesteryl ester hydrolase. Proc. Nat. Acad. Sci. U.S.A. 86:3366.

Hajjar, D.P., Pomerantz, K.B., Falcone, D.J., Weksler, B.B., and Grant, A.J., 1987, Herpes simplex virus infection in human arterial cells: implications in atherosclerosis, J. Clin. Invet. 80:1317.

Hendrix, M.G.R., Dormans, P.H.J., Kitslaar, P., Bosman, F., and Bruggeman, C.A., 1989, The presence of cytomegalovirus nucleic acids in arterial walls of atherosclerotic and nonatherosclerotic patients, Am. J. Pathol. 134:1151.

Hendrix, M.G.R., Salimans, M.M.M., van Boven, C.P.A., and Bruggeman, C.A., 1990, High prevalence of latently present cytomegalovirus in arterial walls of patients suffering from grade III atherosclerosis, Am. J. Pathol. 136:23.

Jones, D., Lee, L., Liu, J.L., Kung, H.J., and Tillotson, J.K., 1992, Marek disease virus encodes a basic-leucine zipper gene resembling the fos/jun oncogenes that is highly expressed in lymphoblastoid tumors, Proc. Nat. Acad. Sci. U.S.A. 89:4042.

Kefalides, N.A., 1992, Response of human vascular cells to viral infection, in "Endothelial Cell Dysfunctions", N. Simionescu and M. Simionescu, eds., Plenum, New York.

MacDonald, K., Rector, T.S., Braulan, E.A., Coubo, S.H., and Olivari, M.T., 1989, Association of coronary artery disease in cardiac transplant recipients with cytomegalovirus infection, Am. J. Pathol. 64:359.

McClelland, J.W., and Shih, J.C.H., 1988, Prevention of hypercholesterolemia and atherosclerosis in Japanese quail by high intake of soy protein, Atheroscler., 74:127.

McCormick, D.L., Radcliff, J.D., Metha, R.G., Thompson, C.A., and Moon, R.C., 1982, Temporal association between arterial cholesterol deposition, thymidine incorporation into DNA and atherosclerosis in Japanese quail fed an atherogenic diet, Atheroscler., 42:1.

Melnick, J.L., Adam, E., and DeBakey, M.E., 1990, Possible role of cytomegalovirus in atherogenesis, JAMA, 263:2204.

Melnick, J.L., Petrie, B.L., Dreesman, G.R., Burek, J., McCollum, C.H., and DeBakey, M.E., 1983, Cytomegalovirus antigen within human arterial smooth muscle cells, Lancet, 2:644.

Minick, C.R., Fabricant, C.G., Fabricant, J., and Litrenta, M.M., 1979, Atheroarteriosclerosis induced by infection with a herpesvirus, Am. J. Pathol. 96:673.

Morrissey, R.B., and Donaldson, W.E., 1977, Rapid accumulation of cholesterol in serum, liver and aorta of Japanese quail, Poultry Sci. 56:2003.

National Academy of Sciences, 1969, "Coturnix", Washington, D.C.

Paterson, J.C., and Cottral, G.E., 1950, Experimental coronary sclerosis III. Lymphomatosis as a cause of coronary sclerosis in chickens, Arch. Pathol. 49:699.

Penn, A., 1990, Mutational events in the etiology of arteriosclerotic plaques, Mutation Res. 239:149.

Penn, A., Garte, S.G., Warren, L., Nesta, D., and Mindich, B., 1986, Transforming gene in human atherosclerotic plaque DNA, Proc. Nat. Acad. Sci. U.S.A. 83:7951.

Petrie, B.L., Melnick, J.L., Adam, E., Burek, J., McCollum, C.H., and DeBakey, M.E., 1987, Nucleic acid sequences of cytomegalovirus in cells cultured from human arterial tissue, J. Infect. Dis. 155:158.

Pullman, E.P., and Shih, J.C.H., 1987, Cholesterol-lowering and atherosclerosis-preventing effect of fish oil in quail, Federation Proc. 46:1171.

Pyrzak, R., and Shih, J.C.H., 1987, Detection of specific DNA segments of Marek's disease herpesvirus in Japanese quail susceptible to atherosclerosis, Atheroscler., 68:77.

Ross, R., 1993a, Atherosclerosis: a defense mechanism gone awry, Am. J. Pathol. 143:987.

Ross, R., 1993b, The pathogenesis of atherosclerosis: a perspective for the 1990s, Nature, 362:801.

Ross, R., and Glomset, J.A., 1976, The pathogenesis of atherosclerosis, New Eng. J. Med. 295:369.

Shih, J.C.H., 1983, Atherosclerosis in Japanese quail and the effect of lipoic acid, Federation Proc. 42:2494.

Shih, J.C.H., and Pyrzak, R., 1987, Atherosclerosis and viral gene in Japanese quail, in "Cardiovascular Disease", L.L. Gallo, ed., Plenum, New York.

Shih, J.C.H., Pyrzak, R., and Guy, J.S., 1989, Discovery of noninfectious viral genes complementary to Marek's disease herpesvirus in quail susceptible to cholesterol-induced atherosclerosis, J. Nutr. 119:294.

Shih, J.C.H., Pullman, E.P., and Kao, K.J., 1983, Genetic selection, general characterization, and histology of atherosclerosis-susceptible and -resistant Japanese quail, Atheroscler., 49:41.

Smith, B.F., St. Clair, R.W., and Lewis, J.C., 1979, Cholesterol esterification and cholesteryl ester accumulation in cultured pigeon and monkey arterial smooth muscle cells, Exp. Mol. Pathol. 30:190.

Smith, R.L., and Hilker, D.M., 1973, Experimental dietary production of aortic atherosclerosis in Japanese quail, Atheroscler., 17:63.

Steinberg, D., Parthasarathy, S., Carew, T.E., Khoo, J.C., and Witztum, J.L., 1989, Modifications of low-density lipoprotein that increase its atherogenicity, New Eng. J. Med. 320:915.

Tumilowicz, J.J., 1990, Characteristics of human arterial smooth muscle cell cultures infected with cytomegalovirus, In Vitro Cell. Dev. Biol. 26:1144.

Yamashiroya, H.M., Ghosh, L., Yang, R., and Robertson, A.L., 1988, Herpesviridae in the coronary arteries and aorta of young trauma victims, Am. J. Pathol. 130:71.

IMPACT OF BIOTECHNOLOGY IN THE DIAGNOSTIC AND THERAPEUTIC
MANAGEMENT OF CARDIOVASCULAR DISORDERS

Jawed Fareed, Debra Hoppensteadt, Lalitha Iyer, Michael Koza,
Jeanine M. Walenga and Edward Bermes, Jr.

Loyola University Medical Center
Departments of Pathology and Pharmacology
2160 S. First Avenue
Maywood, IL 60153

INTRODUCTION

Biotechnology has already exerted a major impact in the management of cardiovascular disorders by providing such drugs as tissue plasminogen activator (t-PA), urokinase, platelet targeting antibodies, recombinant anticoagulant agents such as hirudin and various modulatory proteins. On the diagnostic front, biotechnology has provided newer, more sensitive methods to detect minute changes in blood for the proper diagnosis and management of cardiovascular disorders. Several major breakthroughs in both the diagnosis and treatment of cardiovascular disorders have recently occurred. The measurement of troponin T and I in circulating blood provides a reliable measure of myocardial damage. Other analytes have also been identified for the rapid diagnosis of cardiovascular disorders. Such agents as tissue factor pathway inhibitor (TFPI), protein C and antithrombin III have been developed and may prove to be useful in the management of antithrombotic disorders. Although the use of transgenic animals is still at experimental stages, the initial findings are promising and will provide important leads into the development of therapeutic proteins. Through biotechnology, recombinant inhibitors of fibrinolysis have also been developed and may be useful in the management of bleeding disorders. Nucleic acid derived drugs have been used for sometime. Several newer nucleic acid derivatives have been recently developed. An important consideration in the use of biotechnology derived drugs is the availability of a US Pharmacopoeial monograph on specific drugs. The subject of providing a monograph on biotechnology derived drugs is now being dealt with by the USP. Several other considerations in the development, clinical applications and the results of clinical trials on new biotechnology derived drugs are being addressed at this time. It is hoped that the impact of biotechnology on the diagnosis and treatment of cardiovascular disorders will result in a significant change in the management of these disorders.

Biotechnology has contributed significantly during the past decade to the diagnostic and therapeutics of cardiovascular disorders (Adams, 1993; Bauer, 1987; Broze, 1988). The

Nutrition and Biotechnology in Heart Disease and Cancer
Edited by J.B. Longenecker *et al.*, Plenum Press, New York, 1995

99

development of recombinant technology leading to the cloning of t-PA has revolutionized the treatment of myocardial infarction. On the other hand, utilization of monoclonal antibodies for the specific detection of various markers of myocardial damage such as troponin-I and troponin-T have added a new diagnostic spectrum to the diagnosis of cardiovascular disorders. Continued developments in both the therapeutics and diagnostics have added to the early diagnosis and effective treatment of cardiovascular disorders. Many of the cardiovascular disorders are primarily due to the activation of platelets or a dysfunction of the vascular surface. A list of some of the disorders due to this pathophysiologic syndrome are listed in **Table 1.** Coronary artery disease resulting in ischemic disorders such as unstable angina is primarily due to the occlusive process which results in the impediment of blood flow. The rupture of plaque in the stressed or occluded coronary artery results in acute myocardial infarction. Acute occlusion during such cardiovascular interventional procedures as coronary angioplasty and atherectomy are dependent on many predisposing factors. Some of these factors can be readily identified by blood analysis and specific measures can be taken to avoid these events. Almost 50% of the patients undergoing coronary angioplasty experience late restenosis within a three month period of time. This phenomenon is rather complex and this pathophysiology is not clearly understood (Faxon, 1984). Cerebrovascular disorders such as thrombotic stroke and other cerebral ischemic disorders also result from the activation of platelets or microvascular spasm. The activation of platelets in diabetics also contributes to the ischemic disorder of lower legs and Raynaud's disease. These disorders of platelet activation of vascular spasm can be diagnosed by using various newer biotechnologic procedures. The pathophysiology of these disorders is multifactorial and is determined by several plasmatic and cellular factors. Thus, measurement of cytokines, lipoproteins profiling and other specific tests aid in this diagnosis.

Table 1. Arterial thrombosis

1.	Cardiovascular disorders	
	a.	Coronary artery diseases
	b.	Myocardial infarction
	c.	Unstable angina
	d.	Acute occlusion during interventional cardiologic procedures
2.	Cerebrovascular disorders	
	a.	Thrombotic stroke
	b.	Cerebral ischemic disorders
3.	Peripheral vascular disorders	
	a.	Peripheral arterial occlusive disorders
	b.	Raynaud's disease

PATHOPHYSIOLOGY OF CARDIOVASCULAR DISORDERS

A simplified diagram representing the pathophysiology of cardiovascular disorders leading to ischemia and occlusion is shown in **Figure 1.** Vascular thickening, spasm and narrowing of the lumen contribute to the pathophysiology leading to cardiovascular disorders. The interactions of cellular and humoral components of blood and their activation products with the vascular surface also play a crucial role in the ischemic/occlusive phenomenon. While platelets play a significant role in both the clotting and localized narrowing, many other cellular activation processes also participate in this complex pathology. This concept is very important for the proper development of newer diagnostic methods and new drugs for the treatment of cardiovascular disorders.

Our understanding of the important role that is played by prostanoids in various aspects of vascular function and dysfunction continue to increase but are still not completely understood.

Thromboxane is generated during platelet activation and produces marked constriction of coronary vessels resulting in ischemia. Platelet derived growth factors released during activation also contribute to the pathophysiology of myocardial infarction. Platelet serotonin increases the coronary vascular tone. Knowing this process, a serotonin antagonist has been developed for the treatment of myocardial infarction. Leukocyte activation results in the formation of leukotrienes and granular proteases which may produce several effects resulting in vasoconstriction and localized necrosis. Leukocytes also interact with the pathologic vascular surface to extend myocardial ischemia by altering coronary artery tone and thrombus formation. Leukotrienes and platelet activating factor contribute to sustained arterial spasm. Endothelial cells are known to provide various mediators of vascular function. Newly developed methods based on the use of specific monoclonal antibodies and flow cytometry have added a new dimension to the diagnosis of vascular and cardiovascular disorders.

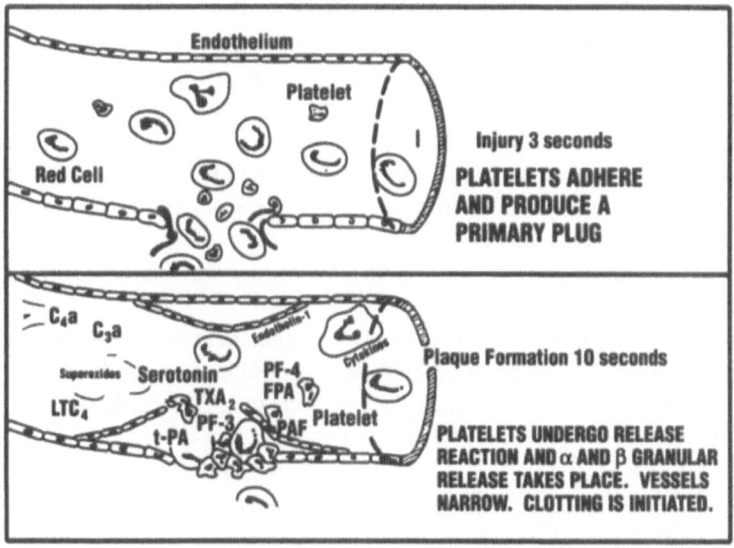

Figure 1. Pathophysiology of thrombus and occlusive lesion formation in the coronary arteries. Atherosclerotic masking or endothelial injury results in the activation of platelets. Activated platelets mediate several direct or signal transduction induced processes on other cellular responses. Cellular activation results in the release of various mediators which amplify vascular spasm and the coagulation process. Thus, anaphylatoxins (C_3a, C_4a and C_5a), superoxide, leukotrienes, thromboxane B_2, serotonin, platelet factor 3, platelet factor 4, platelet activating factor, endothelin-1 and cytokines play a major role in the overall pathophysiology of myocardial infarction.

Release of such proteins as angiotensin converting enzyme and von Willebrand's factor is known to produce a pathologic environment. More recently, a new mediator of vasoconstriction, namely endothelin, has been described. Toxic oxygen products such as superoxides, proteases and thromboplastic material produced by cellular activation also contribute to endothelial detachment and clot formation. In addition, free radical species are known to effect the release and function of endothelium derived relaxing factor (EDRF) and also predispose the endothelium towards a procoagulant tendency. Having a knowledge of this phenomenon, such therapeutic enzymes as superoxide dismutase have been developed. Thrombin mediated endothelial damage, protease generation, platelet activation and endothelial spasm may precede or follow further narrowing of atherosclerotic coronary arteries and facilitate the progression of infarction. This complex pathophysiologic event is now considered as one of the main trigger mechanisms in the progressive development of many of the cardiovascular disorders. However, cytokines, cellular mediators of vascular responses and cell signalling play a key role in the overall pathophysiology of cardiovascular disorders. The knowledge of the molecular events leading to a subclinical or clinical event is, therefore, useful in the development of new methods and new drugs. Biotechnology provides a useful tool and will be very helpful in the management of cardiovascular disorders.

DIAGNOSTIC PROFILING OF CARDIOVASCULAR DISORDERS

Laboratory methods greatly facilitate the diagnosis of cardiovascular disorders. Many new methods are being continually developed (Fareed, 1981; Sidoli, 1990). A list of various laboratory markers used in the diagnostic profiling of high risk patients is shown in **Table 2.** Fibrinogen is the sole clottable protein in plasma. High levels of this protein in plasma are indicative of risk for cardiovascular disorders. Factor VII and factor VIII related antigen are also increased in this group of patients. Cholesterol, low density lipoproteins, and apolipoprotein a have been found to provide a reliable index of high risk patients. More recently, two specific markers of myocardial injury, namely troponin-I and troponin-T have been identified. These two markers are useful in the diagnosis of myocardial damage (Hake, 1993). Troponins represent specific proteins which are found in the myocardial tissue and the presence of these in circulating blood is a clear indication of myocardial damage. Antiphospholipid antibodies resulting from the derangement of vascular membrane also reflect high risk for cardiovascular disorders. Various peptides such as endothelin and atrial natriuretic factor are also increased in patients with cardiovascular disorders.

Biotechnology has provided specific methods to quantitate most of these risk factors in the blood of high risk patients. A profile based on these analytes aids immensely in the diagnosis of cardiovascular disorders. With the advances in newer technologies, a cardiovascular risk profile can be readily determined.

Table 2. Diagostic profiling of high risk patients

1.	Fibrinogen
2.	Factor VII
3.	Factor VIII relted antigen
4.	Cholesterol
5.	LDL/HDL ratio
6.	Apolipoprotein a
7.	Troponin I and T
8.	Antiphospholipid antibodies
9.	Atrial natriuretic factor
10.	Endothelin

Table 3 provides an overview of the biotechnology markers for the diagnosis of cardiovascular disorders. The development of specific monoclonal antibodies to target the fibrin clot (Fibriscint[R]), platelet rich thrombi (Thrombiscint[R]) and glycoproteins in a platelet rich thrombi have enabled the identification of specific sites by using imaging methods. Myoscint represents another antibody which can be used to identify myosin-like filaments in damaged areas of the heart muscle. Thus, these site targeting antibodies are of major value in the identification of cardiovascular disorders. Flow cytometry has also contributed to the diagnosis of vascular disorders. Activated platelets can be readily identified by using specific antibodies which can identify the activation of platelets. Leukocytes also contribute to the pathophysiology

of cardiovascular disorders. Receptor density measurement on leukocytes can also provide useful information on the pathophysiology of vascular disorders. Troponin-I and troponin-T have already been described. Newer more sensitive methods have recently become available to measure these two markers in the blood of patients with cardiovascular risk profiles. Vascular stress results in the generation of various markers. The presence of these markers provides a useful diagnostic index on the vascular function. Several endothelial markers have been recently identified in plasma. These newer markers include thrombomodulin, endothelin and plasminogen activator inhibitor (PAI). Newer endothelial distress markers will be useful in the profiling of patients with cardiovascular disorders.

Table 3. Biotechnology markers for the diagnosis of cardiovascular disorders

1.	Vascular imaging
	Fibriscint
	Thromboscint
	Glycoprotein targeting antibodies
2.	Flow cytometric analysis of activated cells
3.	Cardiac markers
	Troponin I
	Troponin T
4.	Hemostatic markers of vascular function
	Endothelial markers
	Platelet markers
	Plasmatic markers

The impact of biotechnology in the diagnosis of cardiovascular disorders is briefly summarized in **Table 4.** Through biotechnology, highly specific monoclonal antibodies have become available for the determination of plasma markers and vascular imaging. Through the use of these antibodies, immunohistochemical mapping can be performed. This technology provides a specific approach in the understanding of the pathophysiology of vascular disorders. Biotechnology has also provided various specific reagents such as tissue factor, TFPI, purified coagulation factors and purified antibodies to the coagulation factors to investigate the molecular mechanisms involved in the pathogenesis of cardiovascular disorders. Through biotechnology, various cytokines, growth factors and other mediators of the pathogenesis of restenosis and related occlusive disorders have been identified. These developments have led to a multiparametric approach to the diagnosis of cardiovascular disorders.

Table 4. Impact of biotechnology in the diagnosis of cardiovascular disorders

1.	Availability of highly specific monoclonal antibodies for lesion specific targeting of the coronary arteries.
2.	Immunohistochemical mapping of the coronary lesion to determine the role of various mediators.
3.	Availability of recombinant tissue factor, tissue factor pathway inhibitor, coagulation factor activators and inhibitors to study the molecular pathogenesis of vascular occlusion.
4.	Development of highly specific and ultrasensitive methods to determine the markers of cardiac dysfunction.
5.	Identification of factors responsible for post-PTCA restenosis.
6.	Development of multiparametric test battery to identify cardiac risks.

One of the major breakthroughs in this area is the subclinical diagnosis of cardiovascular disorders. With the advent of these molecular markers of vascular dysfunction, subclinical events can be readily identified. Thus, prior to a clinical event, risk groups can be identified and various preventive measures can be taken to avoid a pathologic event.

Table 5 shows a list a diagnostic and therapeutic targets in the management of cardiovascular disorders. A fibrin clot is the main target in the management of acute myocardial infarction. Many of the thrombolytic agents are capable of digesting the fibrin clot. Biotechnology has continued to provide newer thrombolytic agents for the management of fibrin clots for both the coronary and peripheral sites. Platelets and subendothelial lesions are also targeted in the management of various cardiovascular disorders. More recently, surface glycosaminoglycans and all specific cytoskeletal structure have also been targeted. The development of these agents has added a new dimension to the diagnosis and treatment of cardiovascular disorders. Drugs can be readily delivered at specific pathologic sites by utilizing these methods.

Table 5. Diagnostic and therapeutic targeting of cardiovascular disorders

1. Fibrin target
2. Platelet cytoprotein targets
3. Subendothelial lesion target
4. Glycosaminoglycan targeting
5. Cell specific cytoskeletal targets

BIOTECHNOLOGY DERIVED ANTICOAGULANT DRUGS

Anticoagulant drugs play an important role in the management of cardiovascular disorders (Fareed, 1988, 1989, 1991, 1993). Since most of the ischemic and occlusive disorders are due to the activation of clotting or platelet thrombus formation, anticoagulant and antithrombotic drugs can be very useful in the prevention and treatment of cardiovascular disorders. A list of some of the drugs which are obtained by biotechnology means is given in **Table 6.**

Table 6. Biotechnology derived anticoagulant drugs used in the management of cardiovascular disorders

Disorder	Agents
Unstable angina	Hirulog, hirudin
Myocardial infarction	T-PA, hirudin
Post-PTCA atherectomy restenosis reduction	Hirudin, TFPI, growth factor inhibitor
Thrombotic stroke	Hirudin
Transplantation restenosis prevention	Hirudin, hirulog

Hirudin and hirulog represent site-directed thrombin inhibitors which can be used in the treatment of unstable angina. Both of these agents are potent inhibitors of thrombin and retard clot formation. Several clinical trials are currently being carried out to test the usefulness of these thrombin inhibitors in this indication (Fareed, 1994).

Recombinant tissue plasminogen activator has been used for the treatment of myocardial infarction. More recently, hirudin is used as an adjunct drug for the management of myocardial infarction. Hirudin, TFPI, and growth factor inhibitors have been used for the reduction of post-PTCA restenosis. However, to date the results have not been very promising. Additional studies may be required for the validation of the effect of these agents in cardiovascular indications. Anticoagulants such as hirudin and hirulog have also been used in the management of thrombotic stroke and transplantation restenosis reduction. It has been suggested that thrombin may play a role in these pathologic events. Most of the drugs listed in **Table 6** have been obtained by using biotechnology methods. It is very likely that additional drugs will become available for the management of cardiovascular disorders.

Figure 2 shows a comparison of the structures of recombinant hirudin and tissue factor pathway inhibitor. Both of these agents are strong inhibitors of the hemostatic activation processes which may contribute to the pathophysiology of cardiovascular disorders. Hirudin represents a 65 amino acid protein which was first isolated from the medicinal leech, *hirudo medicinalis*. This inhibitor of thrombin is a 3-5 times stronger anticoagulant in comparison to heparin. Several variants of recombinant hirudin have recently become available. Many pharmaceutical companies are currently developing this agent for several cardiovascular indications. Hirudin can also be used in heparin compromised patients as it does not activate platelets. TFPI represents a polydomain inhibitor of tissue factor, tissue factor VIIa complex and factor Xa. This inhibitor can also inhibit other proteases such as elastase released from the macrophage. Thus, it has multiple targets. This inhibitor was originally isolated from the blood. Currently, it is produced by recombinant technology. Several variants of this inhibitor are currently being developed for various therapeutic purposes. Because of the pathophysiologic activation processes in cardiovascular disorders, TFPI is likely to provide therapeutic effects in various pathophysiologic states. At the present time, there is only limited data on clinical trials with this agent. However, there is experimental data indicating that TFPI is a potent inhibitor of tissue factor in various settings. **Table 7** shows that TFPI targets platelets, macrophages and endothelial cells. Thus, it may be of value in the inhibition of the pathology of events leading to vascular stress and eventually to cardiovascular disorders.

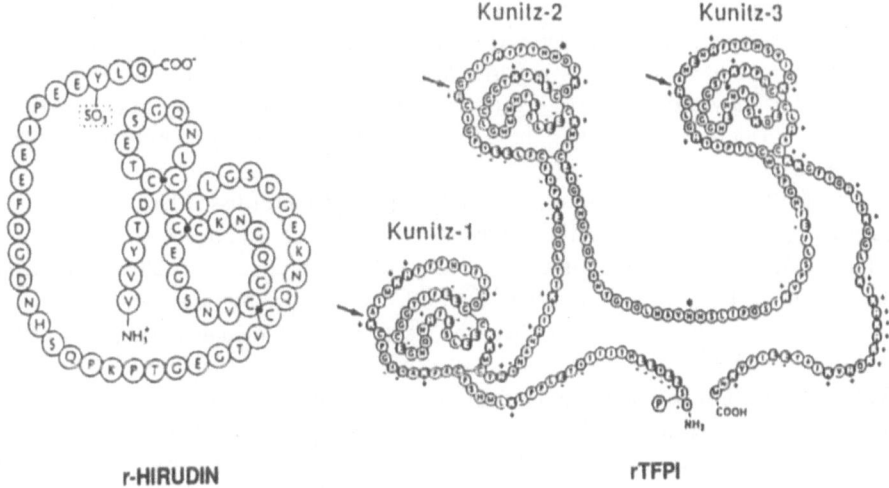

Figure 2. A comparison of the structures of recombinant hirudin and recombinant TFPI. Both of these proteins can now be produced with biotechnologic procedures.

Table 7. Possible cellular targets for TFPI

1. Platelets
2. Macrophages
3. Endothelial cells

Table 8 depicts some of the methods by which tissue factor can be modulated by various methods. Monoclonal and polyclonal antibodies can be used to neutralize the effects of tissue factor. These antibodies can be produced by biotechnology based processes. Various forms of recombinant TFPI have already been developed. These recombinant proteins can also neutralize the actions of TFPI. Various truncated analogues of TFPI have also been developed. These analogues can target tissue factor. Based on the neutralizing sites, specific peptides have also been developed to neutralize the actions of tissue factor. The development of tissue factor neutralizing agents, thus represents a new approach to the control of cardiovascular disorders.

Table 8. Modulation of tissue factor

1. Antibodies to tissue factor
2. TFPI
3. Truncated analogues of tissue factor
4. Peptide targets

Figure 3 shows that TFPI can effectively inhibit the activation of platelets by tissue factor as measured by flow cytometric methods. This data clearly suggests that TFPI may be useful in the modulation of activated platelets, thrombotic and occlusive events since receptors to platelets can be identified. The top panel shows the effect of saline on platelets. The bottom left hand panel shows the effects of tissue factor on platelet activation, whereas the bottom right hand panel shows the effect of TFPI on tissue factor induced activation of platelets. These data also suggest that cells other than platelets can also be mediated by TFPI.

Heparin and related polyelectrolytes are also capable of releasing TFPI from the vascular sites. A comparison of TFPI levels in various anticoagulant states with heparin therapy is shown in **Figure 4.** It can be seen that the degree of heparinization is proportional to the level of TFPI released. Therapeutic anticoagulation and low molecular weight heparin are capable of releasing TFPI, however, to a lesser extent. A remarkable rise in the TFPI levels is observed during angioplasty and cardiovascular bypass surgery. In both methods, large amounts of heparin are used for anticoagulation purposes. Thus, these levels of TFPI may actually contribute to the level of anticoagulants observed in these states.

Beside TFPI, several other means to modulate the function of tissue factor can be used. Antibodies to tissue factor, truncation of the tissue factor molecule and development of peptide targets for the modulation of tissue factor are some of the approaches to alter tissue factor function. Tissue factor is unquestionably one of the most important mediators in the pathophysiology of cardiovascular disorders and its therapeutic modulation may provide a reliable means to treat these disorders.

Tables 9-11 show a comparison of TFPI with some other biotechnology produced antithrombotic drugs such as protein Ca, antithrombin III and hirudin. It is likely that TFPI individually or in combination with some of these agents may prove to be extremely useful in the management of cardiovascular disorders.

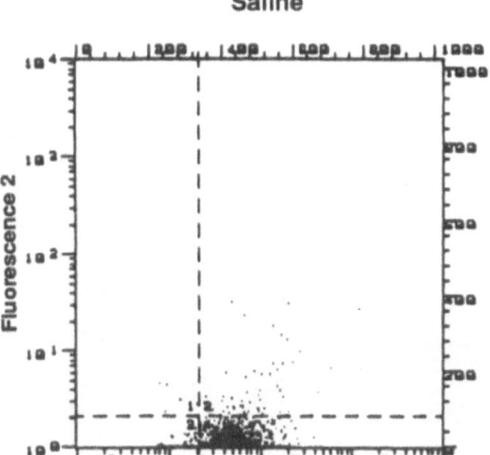

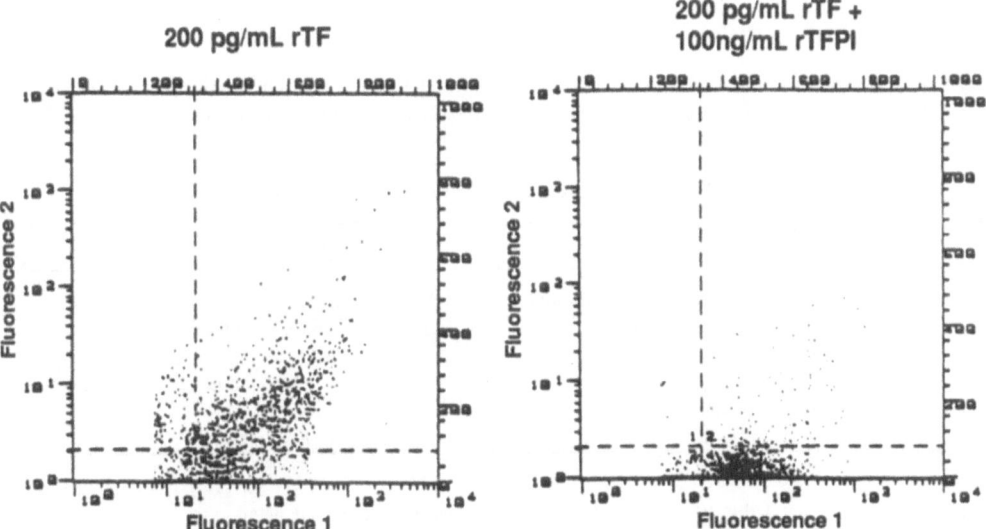

Figure 3. Tissue factor mediated activation of platelets and its inhibition by TFPI. The top panel shows the saline control with minimal platelet activation. Tissue factor at 200 ng/ml (bottom left) activates platelets and TFPI is capable of inhibiting this activation (bottom right).

DISCUSSION

Protein Ca is able to digest factor VIII coagulant and factor V activity. This protein also releases t-PA in circulation. Thus, it produces a cumulative anticoagulant state. The infusion of this agent may, therefore, be useful in the management of various thrombotic disorders. Antithrombin III has already been used in the management of various hypercoagulable states and disseminated intravascular coagulation. In both of these states, TFPI may also be very useful. TFPI inhibits the extrinsic pathway of coagulation activation. Furthermore, it also inhibits the action of some of the proteases from white cells. Recombinant hirudin is also a very potent inhibitor of the coagulation process. Currently, this agent is undergoing clinical trials. In comparison to hirudin, protein C and antithrombin III, TFPI is a polydomain antithrombotic

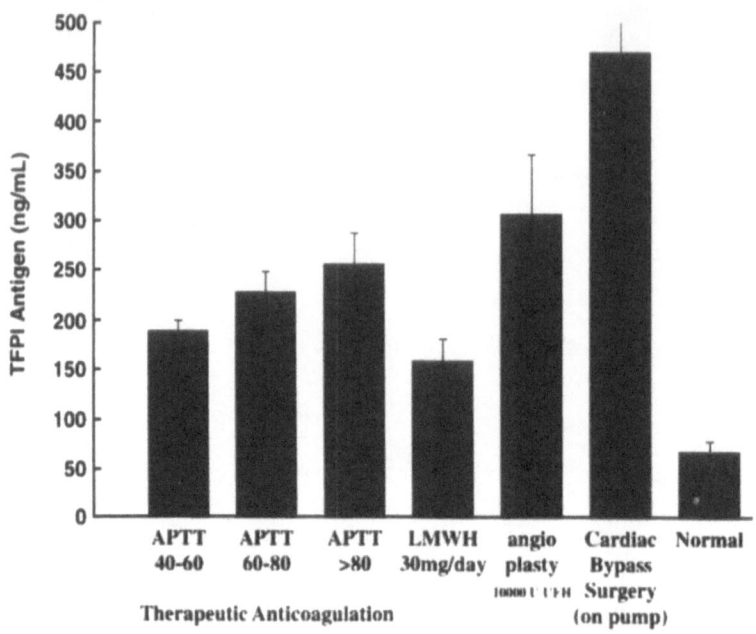

Figure 4. A comparison of the TFPI levels in various heparinization states. Twenty to thirty patients in each group were tested for the TFPI antigen level using an ELISA method. All results represent a mean ± 1 SD.

Table 9. TFPI vs. Protein Ca as an antithrombotic agent

TFPI	Protein Ca
Polydomain protease and cellular inhibition	Serine protease with inhibitory action on factors V and VIII
Independent of any endogenous cofactors	Requires protein S and newly discovered protein Ca activators for its function

Table 10. TFPI vs. Antithrombin III as an anticoagulant agent

TFPI	Antithrombin III
Polydomain protease and cellular inhibitor	Heparin cofactor, inhibitor of factors IIa and Xa
MW 34-42 KDA	MW 60 KDA
Partially activated by heparin	Strong activation by heparin

Table 11. Hirudin vs. TFPI as an antithrombotic agent

Hirudin	TFPI
Recombinant protein with a MW of 6.3 KDA	Recombinant protein with a MW of 35-45 KDA
Monospecific potent inhibitor of thrombin	Polydomain inhibitor of tissue factor and serine proteases
Relatively poor inhibitor of protease generation	Potent inhibitor of protease generation

agent which may be extremely useful in producing antithrombotic states. As the present time, some clinical trials are being carried out on this agent.

Vascular disorders offer a major challenge to laboratory medicine as their diagnosis is rather difficult. Only radiologic and invasive methods have been conventionally used to assess the vascular status of a given patient. With the advances in biotechnology, several newer approaches have been developed. Contrary to the original belief that the vascular lining is inert, it is now known that vascular endothelium plays a pivotal role in the regulation of hemostasis, fibrinolysis, complement pathways and vascular tone. Blood vessels under various conditions are capable of releasing specific markers whose measurement provides a useful monitor of the pathophysiologic status of vascular and cardiac function. With the availability of ultrasensitive immunoassays, minute levels of substances released under endothelial stress such as endothelin, prostaglandins, thrombomodulin, tissue factor, von Willebrand's factor, t-PA, u-PA, PAI-1, PAI-PA complexes and other markers can be readily quantitated. Similarly, target specific peptides to detect endothelial stress/damage have been developed to map the sites where endothelial function is impaired. These tests can be performed in resting and activated states. Clearly, biotechnology will play a key role in the development of newer approaches to assess vascular function which may be associated with atherosclerotic, thrombotic (stroke, myocardial infarction) or bleeding disorders.

Recombinant technology has added a new dimension to the development of antithrombotic and thrombolytic agents. Such drugs as plasminogen activators and their variants, protein C/C_a and derivatives, thrombomodulin and its variants, antibodies and peptide targeting platelet membrane glycoproteins, plasma inhibitors, eg. antithrombin III and heparin cofactor II and anticoagulant proteins such as lipoprotein associated coagulation inhibitor (LACI) are currently under development. Similarly, targeting of platelets and subendothelial lesions has been accomplished using cell or cytoskeletal specific antibodies. Cloning of genes for protein C/C_a protein S and thrombomodulin has resulted in the development of therapeutic grade products for the management of thrombotic disorders. Genes for antithrombin III and heparin cofactor II have been cloned and the expressed proteins exhibit therapeutic properties. The introduction of recombinant hirudin has added a new dimension to the management of such indications as cardiovascular bypass surgery, unstable angina, prophylaxis of DVT and coating of biomaterials. While the developments in this area are dramatic, only a few products have entered into clinical use at this time. Currently, protein Ca, hirudins and glycoprotein targeting antibodies are undergoing clinical trials and will potentially have a major impact in the future management of thrombotic disorders. Thus, the recent developments in this area offer a challenge for both basic science as well as clinical investigators which must be met in an objective and stepwise fashion.

SYNOPSIS

Although biotechnology has been useful in the development of new diagnostic methods and drugs for the management of cardiovascular disorders, there are several issues which raise certain questions on the global use of biotechnology based drugs and diagnostic methods (Piascik, 1991; Fareed, 1993a; Fareed, 1994a). The cost is rather prohibitive in the development of this type of technology. Most diagnostic methods and drugs developed utilizing biotechnology based methods are relatively expensive. The second important consideration is the equivalence of the newer biotechnology derived drugs to the natural products. Many of the biotechnology derived drugs are obtained in prokaryotic systems (E. coli). Post-transcriptional modifications such as glycosylation are often important in determining the function of various proteins. On the other hand, biotechnology based diagnostic methods exhibit somewhat different specificity in comparison to conventional methods. Thus, it is rather important to assess the developments in this area in a careful manner. Furthermore, validation of the clinical, diagnostic and therapeutic efficacy of biotechnology derived diagnostic devices and drugs is a prerequisite for their use in cardiovascular medicine.

ACKNOWLEDGEMENT

The authors appreciate the skillfulness of Mrs. E. Grzeda in preparing this manuscript.

REFERENCES

Adams, J.E., Bodor, G.S., Davila-Roman, V.G., Delmex, J.A., Apple, F.S., Hadenson, J.H. and Jaffe, A.S., 1993, Cardiac Troponin I - A marker with high specificity for cardiac injury, Circulation 88:101-106.

Bauer, K.A. and Rosenberg, R.D., 1987. The pathophysiology of the prothrombotic state in humans: Insights gained from studies using markers of hemostatic system activation, Blood 70:343-350.

Broze, G., Warren, J. and Novotny, L.A., 1988. The lipoprotein associated coagulation inhibitor that inhibits the factor VIIa-tissue factor complex, also inhibits factors Xa: Insight into its possible mechanism of action, Blood 71:335-343.

Fareed, J., 1981, New methods in hemostatic testing. 1981, Perspectives in hemostasis, in "Perspectives in Hemostasis", J. Fareed. ed., Pergamon Press, New York, NY.

Fareed, J., Walenga, J.M. and Breddin, K., 1989. Newer avenues in antithrombotic therapy, Semin. Thromb. Hemost. 15.

Fareed, J., Walenga, J.M. and Hoppensteadt, D.A., 1988. Pharmacologic profiling of defibrotide in experimental models. Semin., Thromb. Hemost. 14:27-37.

Fareed, J., Walenga, J.M. and Iyer, L., 1991. An objective perspective on recombinant hirudin: A new anticoagulant and antithrombotic agent, Blood Coag. Fibrinoly 2:135-148.

Fareed, J., Hoppensteadt, D., Walenga J.M. and Pifarre, R., 1993. New antithrombotic drugs for myocardial infarction in "Anticoagulation, Hemostasis, and Blood Preservation in Cardiovascular Surgery", R. Pifarre, ed., Hanley & Belfus, Philadelphia.

Fareed, J., 1993a, Recombinant antithrombotic drugs. Impact on the management of thrombotic disorders in the 1990s. Proceedings in the International Biotechnology Exposition and Scientific Conference (IBEX 1993), San Francisco, CA, pp. 50, 51, Cartlidge Associates, Belmont, CA.

Fareed, J., Hoppensteadt, D., Walenga, J.M. and Bick, R.L., 1994, Current trends in the development of anticoagulant and antithrombotic drugs, Med. Clin. No. America 78:713-731.

Fareed, J., 1994a, Proceedings of the International Biotechnology Exposition and Scientific Conference (IBEX 1994), San Francisco, CA, PP. 44, 46, 49, 50, 55, 67, Cartlidge Associates, Belmont, CA.

Faxon, D.P., Sanborn, T.A. and Haudenschild, C.C., 1984. Effect of antiplatelet therapy on restenosis after experimental angioplasty. Am. J. Cardiol. 53:72C-76C.

Hake, U., Schmid, F.X., Iversen, S., Dahm, M., Mayer, E., Hafner, G. and Oelert, H., 1993, Troponin T - a reliable marker of perioperative myocardial infarction? Eur. J. Cardiothorac. Surg. 7:628-633.

Piascik, M.M., 1991. Research and development of drugs and biologic entities. Am. J. Hosp. Pharm. 48(10):S4-13.

Sidoli, A., Galliani, S. and Baralle, F.E., 1990. DNA based diagnostic tests: Recombinant DNA and cardiovascular disease risk factors. Br. Med. Bulletin 46(4):941-959.

NUTRITION AND CARCINOGENESIS: HISTORICAL
HIGHLIGHTS AND FUTURE PROSPECTS

R. K. Boutwell

Department of Oncology
University of Wisconsin Medical School
1400 University Avenue
Madison, WI 53706

INTRODUCTION

Gene therapy for cancer is on the horizon and gene prophylaxis is likely under specific circumstances. However, prevention, or at least a lowered risk of developing cancer is possible now. The knowledge to accomplish prevention is based primarily on laboratory research with experimental animals. The basic principles of carcinogenesis, nutrition, and intermediary metabolism do not differ in most cases between nonhuman animals and people. There is every reason to be certain that most of the facts established in laboratories are effective for the prevention of cancer in people.

Although food stuffs contain carcinogens, both naturally occurring and man made (IARC Monographs, 1993; Miller and Miller, 1979; Miller et al., 1979; Pariza and Felton, 1990), these will not be considered; emphasis will be solely on the role of both diet and nutrition in the prevention of cancer. In fact, if one thinks about it, the effective balance between naturally occurring inhibitors of cancer development and the carcinogenic potency of food stuffs is far-and-away in favor of a dominant role for the inhibitors, for it is only by adding natural foods to control diets that the existence of inhibitors is demonstrable in laboratory experiments. Of course, there are specific food stuffs that contain naturally occurring carcinogens in excess of any protective factors and pose a serious risk of cancer.

Volumes could be written on the history of the role of nutrition in cancer prevention. This review, of necessity, will be selective. The items that have been selected are important but many that are equally important have been omitted. Entry to the literature on nutrition and cancer may be found in the following references (Hartman and Shankel, 1990; Hayashi et al., 1986; Kuroda et al., 1990; Milner, J. A., 1989; Rowland, 1991).

At least two types of events are involved in the genesis of most cancers: mutational and gene activating. Specific mutations (e.g., in oncogenes) initiate the processes that constitute carcinogenesis. Multiple gene activations that are highly pleiotropic reveal the mutational events, ultimately resulting in neoplasia; these latter processes are generally known as promotion. Progression of a neoplasm to more malignant states is accomplished

Nutrition and Biotechnology in Heart Disease and Cancer
Edited by J.B. Longenecker *et al.*, Plenum Press, New York, 1995

111

by specific mutations. Each of these processes can be modified by nutrition, either enhancing or inhibiting the process.

CALORIC CONTROL

It has been known since the early years of this century that the caloric balance of an animal has a minimal influence on the fatal outcome of animals bearing established malignancies (e.g. Moreschi, 1909; Rous, 1914; Bischoff and Long, 1938). Cancer is truly parasitic, living off the body for its source of energy and the molecules to synthesize its protein, nucleic acid, etc. (LePage *et al.*, 1952).

In contrast, the incidence of cancer, whether induced or of spontaneous origin, may be greatly reduced by caloric restriction. The earliest experiments involved reduction in the amount of the total diet made available to the animals (Sivertsen and Hastings, 1938; Tannenbaum, 1940). In their experiments, the incidence of spontaneous mammary cancer in C3H mice was recorded (Table 1).

Table 1. The Inhibition of Tumor Incidence by Caloric Restriction

Investigator	Tumor Type	Incidence, %	
		Fully Fed (Obese)	Calorie Restricted
Sivertsen and Hastings, 1938	Spontaneous: mammary	88	16
Tannenbaum, 1940	Spontaneous: mammary	40	2
Visscher *et al.*, 1942	Spontaneous: mammary	67	0
Saxton *et al.*, 1944	Spontaneous: leukemia	65	10
Larsen, 1945	Spontaneous: lung	50	30
Rusch *et al.*, 1945	Ultraviolet light: skin	87	7
Boutwell *et al.*, 1948	Hydrocarbon: skin	82	18
Weindruch and Walford, 1982	Spontaneous: lymphoma	47	4

However, it is essential to study the effect of each dietary component separately. In 1942 Visscher *et al.* reported that simple caloric restriction (consumption of minerals, vitamins, and amino acids/protein was held constant) of C3H mice resulted in a lower incidence of spontaneous mammary cancer. In the following 6 years, observations of the protective effect of caloric restriction was extended to the occurrence of spontaneous malignancy of several organ systems as well as to induced neoplasms **(Table 1).**

It was recognized that the protective effect of caloric restriction tended to be over-whelmed by large doses of carcinogens. That fact is generally true of studies of all factors capable of modifying carcinogenesis. Carcinogen doses that allow a long tumor induction time and/or less than 100% incidence show a greater spread between the incidence of neoplasms in the control group and the group subjected to the variable under investigation.

As laboratory husbandry of experimental animals improved and experimental animals were free of pathogens, the incidence of degenerative diseases including cancer increased in the *ad libitum*-fed animals. They died at a younger age than fully-fed animals in earlier experiments. The *ad libitum* animals became pathologically obese; they were no longer normal. In contrast, controlling caloric intake at a level that maintained the young adult body weight throughout life, resulted in healthy animals with extended life spans. McCay's group at Cornell University was the first to establish clearly the beneficial effects of restricted calorie diets adequate in essential nutrients on longevity (McCay *et al.*, 1935) and on tumor incidence (McCay *et al.*, 1939). Rats were observed for periods up to 4 years.

Controlled studies that attempt to examine directly the relationship between caloric intake (energy balance) and human cancer incidence are fraught with complications (Weindruch *et al.*, 1991). However, the public should be urged vigorously to maintain a healthy body weight in order to lower the risk of developing a variety of degenerative diseases including cancer.

A review of the history of research on the mechanism by which caloric restriction reduces cancer incidence reveals important advances in cancer prevention. Several lines of evidence support the conclusion that moderate caloric restriction maintains the activity of the adrenal cortex. In contrast, the adrenal cortex becomes inactive as the *ad libitum*-fed animal ages. Thus, the ratio of adrenal weight to body weight is low, the liver glycogen level falls rapidly when food is withheld, and the ability to store a test dose of glucose as glycogen after a fast is lost. In contrast, in the case of the calorically controlled mice, a much larger proportion of the test dose of glucose is stored in the liver as glycogen, indicative of an active adrenal cortex and consequent increase in the level of circulating cortisone (**Table 2**, Boutwell *et al.*, 1948). Based on this observation, the effect on skin tumor induction of cortisone was investigated; about 0.5 mg of cortisone per mouse per day added to the diet of *ad libitum*-fed mice inhibited carcinogenesis in mice as effectively as caloric restriction (**Table 3**, Boutwell, 1964).

Table 2. Loss of Adrenal Function in Obese Mice

	Mg Glycogen/100 g liver	
		Calorie Restricted
	Obese	Normal
After 32 hr fast	41	931
After 32 hr fast plus 30 mg glucose	209	2376

Table 3. Inhibitory Effect of Cortisone and Caloric Control on Tumor Promotion

Group	Papillomas per mouse	Carcinoma Incidence %
Ad libitum	5.9	50
Ad libitum cortisone diet	1.8	6
Calorie controlled	1.8	14

At the same time that the corticoid (substances that inhibit cell division) levels are increased in calorie restricted animals, growth stimulating polypeptide hormone levels are controlled at lower levels. This is manifest by a decrease in gonadotropes (Boutwell *et al.*, 1948), in lower plasma insulin levels (42 mU/ml in mice restricted by 30% compared to 122 mU/ml in *ad libitum*-fed mice) (Klurfeld *et al.*, 1989), and in plasma IL-6 levels (Volk *et al.*, 1994). Volk *et al.* provide evidence that the decrease in IL-6 levels with aging is related to the decreased incidence of spontaneous lymphoma in mice (Weindruch and Walford, 1982) and in humans.

In summary, the protective effect of caloric restriction is the result of increased levels of growth inhibitory antiinflammatory adrenal steroids and a decrease in the level of the growth promoting polypeptide hormones.

The fact that the antiinflammatory steroids inhibited carcinogenesis led to studies on the nonsteroidal antiinflammatory agents. It was found that antiinflammatory agents, such as aspirin, indomethacin, and flufenamic acid not only inhibited tumor promotion (Verma *et al.*, 1980), they also inhibited the synthesis of prostaglandins (Ashendel and Boutwell, 1979) and the induction of ornithine decarboxylase activity (Verma *et al.*, 1977), changes that are essential for tumor formation (Takigawa *et al.*, 1983).

It is clear that progress in elucidating biochemical mechanisms underlying carcinogenesis evolved directly from research on the role of the level of caloric intake on carcinogenesis.

DIETARY FAT LEVEL

The earliest observation of the effect of the level of fat on carcinogenesis was made by Watson and Mellanby (1930). They reported that the incidence of tar cancer of the skin was greater in mice fed a diet of powdered bread and ground oats to which fat was added. The systematic investigation of the effect of the level and kind of dietary fat began with the studies of Baumann, Rusch, and their colleagues in 1937. They reported that diets high in fat accelerated the induction of mouse skin tumors by ultraviolet light (Baumann and Rusch, 1939) or by benzpyrene (Baumann *et al.*, 1939). They found that the enhancing effect of fat occurred in several strains of mice and was manifest with three different hydrocarbon carcinogens. Furthermore, the effect of fat was found to be unique; variation in carbohydrate or protein failed to alter tumor incidence (Jacobi and Baumann, 1940). Although the model system for tumor promotion was not developed until 1944 (Mottram, 1944), the concept was utilized in an experiment reported in 1941 in which the carcinogen was applied to the skin twice weekly for 2 months to mice fed a 2% fat diet. At 2 months, half the mice were switched to a 15% fat diet for the next 4 months. The incidence of skin tumors in the mice switched to the 15% fat diet was 43% at 5 1/2 months whereas those left on the 2% fat diet was 11% (Lavik and Baumann, 1941).

The conclusion of that study was that the major influence of the high fat diet was not on the action of the carcinogen but rather on the later processes in tumor induction. The observation that high fat diets enhance the appearance of neoplasms was confirmed (Tannenbaum, 1942) and reconfirmed countless times since and so the extant dogma is that high fat diets increase the risk of cancer (**Table 4**).

Table 4. The Effect of the Level of Fat in the Diet on the Incidence of Neoplasms

Investigators	Tumor Type	Neoplasms, %	
		Stock diet	Stock diet 70% Fat 30%
Baumann and Rusch, 1939	Ultraviolet light: skin	16	48
Baumann *et al.*, 1939	Benzpyrene: skin	6	75
Tannenbaum, 1942	Spontaneous: mammary	13	32

The results of an insightful and pivotal yet largely overlooked experiment was published over 50 years ago (Lavik and Baumann, 1943). They established that mice fed diets high in fat at a lowered level of caloric intake developed fewer hydrocarbon-induced skin tumors than those mice that were fed a low fat diet *ad libitum*. At the *ad libitum* feeding level for the 5% fat diet, the mice consumed 0.06 g of fat per mouse per day on average over the duration of the experiment and 54% developed skin carcinomas in response to methylcholanthrene applications. In contrast, mice fed a 15% fat diet restricted to 66% of the caloric intake of the *ad libitum*-fed mice consumed 0.24 g of fat per mouse per day (four-fold more fat) yet the carcinoma incidence was reduced by one-half to 28% (**Table 5**). Clearly, the incidence of skin carcinomas was not dependent on the percentage of fat in the diet nor on the amount of fat consumed. Rather, the incidence of skin carcinomas was determined by the level of calories consumed.

Table 5. Restriction of Caloric Intake Modulates the Enhancing Effect of Fat on Mouse Skin Carcinogenesis (Lavik and Baumann, 1943)

Caloric Intake (% of *ad lib.*)	Fat Level (% of diet)	Fat Intake (g/mouse/day)	Skin Carcinomas (%)
100	5	0.06	54
66	15	0.24	28

Data confirming the results of Lavik and Baumann were published in 1949 (Boutwell *et al.*, 1949). The degree of restriction of the mice fed the high fat diet was moderate (83% of *ad libitum*) and the average gain at the end of the experiment was 8.4 g per mouse beginning at the young adult weight of 24.5 g. The *ad libitum*-fed mice on the low fat diet gained 10.2 g per mouse over the 24 week experiment. Although the mice fed the high fat diet consumed almost 8-fold more fat (calculated), the incidence of benzo[a]pyrene-induced skin cancer was 72% vs. 82% of the mice fed the lower fat diet.

The fact that the level of calories and not the level of fat is the determinant of cancer incidence was extended to the rat mammary cancer model. The relationship between the level of dietary fat and the caloric intake of rats on the incidence of mammary cancer caused by a single dose of 7,12-dimethylbenz[a]anthracene at 7 weeks of age was reported (Boissonneault *et al.*, 1986). The high fat diet (30%) was fed to rats at a level of caloric intake reduced to 81% of the amount that was consumed by the animals fed the lower fat (5%) diet *ad libitum*. The *ad libitum*-fed rats consumed an average of 0.6 g of fat per day and the incidence of mammary cancer was 43%. In contrast, the rats fed the diet high in fat but calorically restricted to 81%, consumed 2.2 g of fat per day and only 7% developed mammary neoplasms at 24 weeks after administration of the carcinogen (**Table 6**). Essentially identical results were obtained by Kritchevsky and Klurfeld (1987) and by Welsch *et al.* (1990).

The aforedescribed experiments clearly establish that the major determinant of the risk of cancer in specific model systems is attributable to diets high in calories and that the quantity of fat that is consumed (i.e., the % fat in the diet) is not important if caloric intake (i.e., caloric balance) is controlled at an appropriate level. In these experiments the animals were fed diets adequate in protein, minerals, and vitamins. Human diet-related issues are complicated (Willett, 1994) whereas in animal studies single variables can be evaluated. There is little doubt that the principles established in animal experiments are applicable to people; the major human concern should be caloric balance but because fat contains over twice as many calories per unit weight as carbohydrate and protein, eliminating fat-rich foods (candies, desserts, and snack foods) is the appropriate approach to a controlling caloric balance together with exercise.

Table 6. Moderate Restriction of Caloric Intake Modulates the Enhancing Effect of Fat on Mammary Carcinogenesis in the Rat (Boissonneault *et al.*, 1986)

Caloric Intake (% of *ad lib.*)	Fat Level (% of diet)	Fat Intake (g/rat/day)	Mammary Neoplasms (%)
100	5	0.6	43
81	30	2.2	7

Although the risks associated with saturated fats in the human diet are emphasized to the public, recent experimental animal data point to polyunsaturated fatty acids as risk factors for cancer (Reddy and Maeura, 1984). The incidence of colon cancer was high in rats treated with azoxymethane and fed a diet containing 23.5% fat if the fat was corn or safflower oil compared to a lower incidence if the fat was olive (monounsaturated) or coconut oil (predominantly saturated) (**Table 7**).

Table 7. Incidence of Total Colon Neoplasms at 48 Weeks Induced by a Single Subcutaneous Injection of Azoxymethane in Rats (Reddy and Maeura, 1984)

Group	Dietary Fat %	Type	Total Neoplasms %
1	5.0	Corn	17
2	23.5	Corn	46[a]
3	5.0	Safflower	13
4	23.5	Safflower	36[a]
5	5.0	Olive	10
6	23.5	Olive	13
7	23.5	Coconut	13

[a] Significantly higher incidence than the other 5 groups.

The data obtained in the colon cancer model by Reddy and Maeura were corroborated in a rat mammary cancer model (Braden and Carroll, 1986) and in pancreatic cancer (Roebuck *et al.*, 1981). The onus for the increased cancer incidence of each of those high fat diets is clearly on an overabundance of the polyunsaturated fatty acid content of the fats. There is an optimum level for the essential fatty acid, linoleic acid (Ip, 1987). It is likely that the adverse effects of excessive levels of polyunsaturated fatty acids is related to overproduction of the eicosanoids via cyclooxygenase that enhances the risk of cancer (Ashendel and Boutwell, 1979; Verma *et al.*, 1977, 1980). However, in all of the reports cited above (e.g., Tables 6 and 7) showing that the enhancing effect of fat is a phenomenon dependent on *ad libitum* feeding, the fat was corn oil. The enhanced risk of cancer associated with the polyunsaturated fat-containing diets (Table 8) is precluded by controlled caloric balance.

MICRONUTRIENTS

The role of inorganic micronutrients in the development of cancer continues under investigation. In many cases, no consensus exists between epidemiological data and animal studies. Only one example of an effective inhibitor will be cited. In 1943 selenium (as selenide) was reported to cause liver tumors in rats (Nelson et al., 1943). This observation was discredited on the basis of defective histological identification of the lesions (Scott, 1973). It is now recognized that inorganic forms of selenium are capable of protecting against carcinogenesis of the liver (Griffin and Jacobs, 1977), colon (Jacobs et al., 1977), mammary gland (Ip, 1981), and skin (Wilt et al., 1979). Levels of sodium selenite at levels of 5 ppm as selenium in the diet were effective. Organoselenium compounds are even more powerful inhibitors than inorganic forms in animal tumor models and several organic selenium-containing compounds are the major source of selenium in foods (Ip and Ganther, 1990, 1992). In fact, little or no dietary selenium is in the inorganic form. The mechanism by which selenium inhibits carcinogenesis is not known; it appears that it may be multi-modal. Selenite has been shown to inhibit DNA-DMBA adduct formation in mammary epithelial cells (Liu et al., 1991).

VITAMINS

An anecdote that is unique in the annals of the Nobel prize is relevant. In 1907, a Danish investigator by the name of Johannes Fibiger began his studies of neoplastic stomach lesions that he found in wild rats roaming free in a sugar refinery. He interpreted the lesions as carcinoma of the stomach and he concluded that a microscopic nematode was causative with the cockroach as the intermediate host. Because this conclusion was deemed important to eliminating a very common human cancer, Fibiger was awarded the Nobel prize in 1926. Subsequently, the causative role for a nematode could not be substantiated in controlled laboratory experiments. Today, one might guess that wild rats in a sugar refinery consumed nutritionally deficient diets, particularly in vitamin A. If the stomach wall was damaged by bits of wood that wild rats characteristically chew on, the lesions might remain undifferentiated in vitamin A deficient rats and appear neoplastic under the microscope. If this interpretation of Fibiger's observation is correct, it is the first example of a role for a vitamin deficiency in the development of a neoplastic lesion.

Validation for this interpretation of Fibiger's observation is based on the pivotal discovery of a relationship between vitamin A and neoplasia (Wolbach and Howe, 1925). They concluded that "... the deficiency results in loss of specific chemical functions of the epitheliums concerned, while the power of growth becomes augmented... [This] suggests the acquisition of neoplastic properties ..."

Vitamin A and carotene deficiencies have been associated by epidemiologists with established human malignancy (the alpha-tocopherol, beta-carotene cancer prevention study group, 1994; Willett, 1994). Experimental animal models for evaluating the effect of vitamin A on carcinogenesis were equivocal (Saffiotti et al., 1967). The state of affairs was best summarized by Cohen et al. (1976) based on their studies of bladder carcinogenesis: "Although a deficiency of vitamin A accelerated the carcinogenic process in the bladder, hypervitaminosis did not inhibit neoplasia compared to rats receiving normal levels of vitamin A together with the bladder carcinogen."

Because of the conviction that the effects of vitamin A on carcinogenesis merited additional study, advantage was taken of the advances in establishing the metabolic pathways of vitamin A. Hector DeLuca's group established that retinoic acid was a metabolite of vitamin A; retinoic acid is an active form of the vitamin in epithelial tissues (Emerick et al., 1967). Bollag was also involved with vitamin A metabolism and studying the effect of vitamin A and its synthetic derivatives, retinoic acid, on experimental cancer therapy (Bollag, 1971). He was the first to report the unequivocal ability of small amounts of retinoic acid to inhibit the induction of benign and malignant neoplasms (Bollag, 1972). The potency of a variety of synthetic retinoids for the inhibitions of skin carcinogenesis was established by Verma and Boutwell (1977). The induction of the enzyme, ornithine decarboxylase, has been established as an essential (but not sufficient) component of tumor promotion (Takigawa *et al.*, 1983) and the assay for the inhibition of the induction of the polyamine biosynthetic enzyme, ornithine decarboxylase was utilized as a rapid assay to screen retinoid derivatives for prophylaxis of cancer (Verma and Boutwell, 1977; Verma *et al.*, 1979). There is little doubt but that control of the production of polyamines by controlling the activity of the enzyme, ornithine decarboxylase, is the underlying metabolic function of vitamin A in determining the role of the vitamin in maintaining the differentiated state of epithelial tissues. The polyamines, putrescine, spermidine, and spermine function to control growth and differentiation.

It is clear that adequate levels of vitamin A intake protect against the appearance of cancer and that control is mediated through the tissue-active form of the vitamin, retinoic acid. However, levels of vitamin above nutritionally adequate levels do not assure protection against cancer and vitamin A toxicity is an ever-present risk. In contrast, synthetic derivatives of retinoic acid are potent chemoprophylactic agents against cancer (Moon *et al.*, 1983) at dose levels that are free of the toxic effects of high dietary levels of vitamin A. Thus, although not naturally occurring, recognition of the highly effective class of inhibitors known as retinoids evolved logically from research on vitamin A.

NON-NUTRITIVE COMPONENTS OF NATURAL FOOD STUFFS

A most important, heterogeneous group of compounds are highly effective agents against cancer and occur naturally as minor components of a number of foods, mostly fruits, vegetables, and cereal grains. Recognition of their existence has come slowly, in part masked by overemphasis on the importance of carotenoids and fiber in cancer prevention. However, as long ago as 1949 it was observed that mice fed semipurified diets developed more tumors sooner than those whose diet was reinforced with natural substances (Boutwell *et al.*, 1949). This fact was confirmed and extended (Silverstone *et al.*, 1952). They investigated the relative influences of natural (Purina Laboratory Chow) and semipurified diets on the development of spontaneous hepatomas in male DBA and C3H mice as well as on induced skin neoplasms. The incidence of hepatomas in mice fed the stock diets was 0, 18, and 26% in three experiments while the incidence in mice fed the purified diet was, respectively, 26, 54, and 45%. As was found in the experiment of Boutwell *et al.*, Silverstone and Tannenbaum reported a smaller effect of diet on induced skin tumors.

Subsequently, evidence that stock diets contain protective factors has been demonstrated in a number of model systems (Engle and Copeland, 1952; Reddy *et al.*, 1974; Roebuck *et al.*, 1981; Hecht *et al.*, 1989). As a specific example, Sprague-Dawley rats fed DMBA at a level of 0.03% in a purified diet developed mammary cancer in 90% of the animals by 27 weeks and ear duct and liver tumors in 50% and 60% of the rats. In contrast, all of those rats fed DMBA in a stock diet were still tumor free at 27 weeks (Engle and Copeland, 1952). They concluded "... that purified or semipurified diets are preferable to diets of natural food stuffs for studies in which a rapid and uniform production [of tumors] is desired." The difference could not be accounted for based on body weight (caloric intake or known essential nutrients) and therefore "... none of the dietary ingredients seem to be responsible for any particular anticarcinogenic properties." There was no follow-up to ascertain what constituents of the natural diet provided such powerful protection against cancer.

Wattenberg (1971) pioneered in searching for substances in natural diets that conferred protection against induced cancer. His rationale was based on the emerging knowledge of the molecular mechanisms of carcinogenesis: activation, detoxification, interaction with cellular DNA, and repair systems (Miller and Miller, 1979).

It has long been known that all plants contain countless numbers of xenobiotic compounds and Wattenberg reasoned that some of these natural compounds would account for the protective effect of diets containing natural constituents. In the initial studies, dried pieces of cabbage and other vegetables were ground, dried, and incorporated at a level of 25% in a purified diet. The diet was readily eaten by the rats and hydroxylase activity was induced (Wattenberg, 1971). Later it was reported that cabbage and broccoli inhibited mammary tumor formation caused by DMBA and subsequently it was ascertained that benzylisothiocyanate was one of the effective agents (Wattenberg, 1977). The inhibitory effect of benzylisothiocyanate was established later for carcinogenesis of the forestomach and lungs of mice given diethylnitrosamine or benzo[a]pyrene (Wattenberg, 1987). It is now reported that the aromatic isothiocyanates inhibit carcinogen activation (Wattenberg, 1992a).

Because butylated hydroxy anisol (BHA, a synthetic phenolic antioxidant) had been demonstrated to be a versatile inhibitor of carcinogen-induced neoplasia, Wattenberg investigated the inhibitory effect of three phenols that are found in fruits and vegetables (Wattenberg, Coccia, and Lam, 1980). They reported that o-hydroxycinnamic acid, 3,4-dihydroxycinnamic acid (caffeic acid), and 4-hydroxy-3-methoxycinnamic acid (ferulic acid) suppressed benzo[a]pyrene-induced neoplasia of mice. Dietary antioxidants such as BHA had been reported to increase hepatic glutathione S-transferase activity of mice as much as 5- to 10-fold (Benson et al., 1978). They suggested that the protective effects of BHA might be accounted for, at least in part, by the elevated glutathione S-transferases. The concept was adopted by Wattenberg and he suggested that enhancement of glutathione S-transferase activity might be used as a method of identifying compounds likely to inhibit carcinogenesis by detoxifying the ultimate electrophilic form (Sparnins and Wattenberg, 1981). The value of a rapid assay for inducers of phase 2 carcinogen detoxication enzymes (e.g., glutathione S-transferase, NADP-H:quinone reductase) was illustrated recently by Talalay's group to isolate another highly effective inhibitor, sulforaphane, from broccoli (Zhang et al., 1994). Thus, the assay for the induction of a phase 2 enzyme activity may serve as a rational, rapid assay for dietary components that inhibit neoplasia and for the isolation of the active compounds as well as for assessing molecular structures that may prove more active than the naturally occurring inducers belonging to the Brassicaceae family.

Another class of naturally occurring compounds that inhibit carcinogenesis by inhibiting tumor promotion was discovered using the following rationale (Belman, 1983). Vanderhoek et al. (1980) showed that onion and garlic oils inhibit fatty acid oxygenases and Verma et al. (1977) established a role for oxidation of fatty acids by cyclooxygenase to prostaglandins in tumor promotion. These observations led Belman to test the effect of onion and garlic oil on tumor promotion; he reported protection (Belman, 1983; Belman et al., 1989). In addition, organo-sulfur compounds found in onion and garlic inhibit glutathione-S-transferase activity, an enzyme that detoxifies the activated, electrophilic form of carcinogens (Sparnins et al., 1988). Thus both the mutagenic and gene activating (growth stimulating) activity of carcinogens are inhibited by these dietary components.

A number of nonnutrient minor components of food stuffs have been identified and each has an interesting history. Some of these compounds together with a dietary source of each is listed in **Table 8**.

Table 8. Some Dietary Sources of Compounds that Inhibit Carcinogenesis[a,b]

Source	Active Compounds	Reference
Allium species (garlic, onion, etc.)	Organosulfur compounds such as allylmethylsulfide	Wargovich et al., 1988 Belman, 1983
Cruciferous vegetables (cabbage, broccoli, etc.)	Indoles, isothiocyanates, dithiolethiones	Wattenberg, 1977
Certain dairy products	Conjugated linoleic acid	Ip et al., 1991
Citrus fruits	d-Limonene	Elson et al., 1988
Beans and grains	Protease inhibitors, phytic acid, isoflavones	Troll et al., 1987 Shamsuddin et al., 1988 Messina et al., 1994
Vegetables	Sulforaphane	Zhang et al., 1994
Green tea, fruits, vegetables	Many phenolic compounds	Huang et al., 1988
Tumeric, curry[c]	Curcumin	Huang et al., 1988
Fruit	Tannic acid	Gali et al., 1992

[a] An incomplete list; see Wattenberg (1992a, 1992b).

[b] Many relevant references may be found in the bibliographies of the cited papers.

[c] Essential oils obtained from a number of spices contain a variety of inhibitors.

SUMMARY AND CONCLUSION

Historical reviews were presented of several selected nutritional factors that are determinants of cancer incidence in laboratory experiments utilizing animals. An all-inclusive review of nutrition as it impacts cancer incidence was not done. Rather, the selection of subjects was based on a combination of several factors. (1) The efficacy of the factor as an inhibitor. (2) Current interest in the factor. (3) The extent to which the mechanisms of the inhibition is known and that knowledge may facilitate future studies. (4) The relevance to the human problem.

The future of research on cancer prevention is bright. There are now mechanism-based rapid assays to detect food stuffs that prevent cancer and to assay for the active compounds therein. The list of inhibitors shown in **Table 8** will continue to grow. The challenge is to achieve universal application to the human population of appropriate dietary practices that include foods that provide the protective factors shown in **Table 8**.

REFERENCES

Ashendel, C.L., and Boutwell, R.K., 1979, Prostaglandin E and F levels in mouse epidermis are increased by tumor-promoting phorbol esters, Biochem. Biophys. Res. Commun. 90:623.

Baumann, C.A., Jacobi, H., and Rusch, H.P., 1939, The effects of diet on experimental tumor production, Am. J. Hyg. 30:1.

Baumann, C.A., and Rusch, H.P., 1939, Effect of diet on tumors induced by ultraviolet light, Am. J. Cancer 35:213.

Belman, S., 1983, Onion and garlic oils inhibit tumor promotion, Carcinogenesis 4:1063.

Belman, S., Solomon, J., Segal, A., Block, E., and Barany, G., 1989, Inhibitors of soybean lipoxygenase and mouse skin tumor promotion by onion and garlic components, J. Biochem. Toxicol. 4:151.

Benson, A.M., Batzinger, R.P., Ou, S.-Y.L., Bueding, E., Cha, Y.-N., and Talalay, P., 1978, Elevation of hepatic glutathione S-transferase activities and protection against mutagenic metabolites of benzo[a] pyrene by dietary antioxidants, Cancer Res. 38:4486.

Bischoff, F., and Long, M.L., 1938, The influence of calories *per se* upon the growth of sarcoma 180, Am. J. Cancer 32:418.

Boissonneault, G.A., Elson, C.A., and Pariza, M.W., 1986, Net energy effects of dietary fat on chemically induced mammary carcinogenesis in F-344 rats, J. Natl. Cancer Inst. 76:335.

Bollag, W., 1971, Therapy of chemically induced skin tumors of mice with vitamin A palmitate and vitamin A acid [retinoic acid], Experimentia 27:90.

Bollag, W., 1972, Prophylaxis of chemically induced benign and malignant epithelial tumors by vitamin A acid (retinoic acid), Eur. J. Cancer 8:689.

Boutwell, R.K., 1964, Some biological aspects of skin carcinogenesis, Prog. Exptl. Tumor Res. 4:207.

Boutwell, R.K., Brush, M.K., and Rusch, H.P., 1948, Some physiological effects associated with chronic caloric restrictions, Am. J. Physiol. 154:517.

Boutwell, R.K., Brush, M.K., and Rusch, H.P., 1949, The influence of vitamins of the B-complex on the induction of epithelial tumors in mice, Cancer Res. 9:747.

Braden, L.M., and Carroll, K.K., 1986, Dietary polyunsaturated fat in relation to mammary carcinogenesis in rats, Lipids 21:285.

Cohen, S.M., Wittenberg, J.F., and Bryan, G.T., 1976, Effect of avitaminosis A and hypervitaminosis A on urinary bladder carcinogenicity of N-[4-(nitro-2-furyl)-2-thiazolyl] formamide, Cancer Res. 36:2334.

Elson, C.E., Maltzman, T.H., Boston, J.L., Tanner, M.A., and Gould, M.N., 1988, Anticarcinogenic activity of d-limonene during the initiation and promotion/progression stages of DMBA-induced rat mammary carcinogenesis, Carcinogenesis 9:331.

Emerick, R.J., Zile, M., and DeLuca, H.F., 1967, Formation of retinoic acid from retinol in the rat, Biochem J. 102:606.

Engle, R.W., and Copeland, D.H., 1952, Protective action of stock diets against cancer-inducing action of 2-acetylaminofluorene in rats, Cancer Res. 12:211.

Gali, H.V., Perchellet, E.M., Klish, D.S., Johnson, S.M., and Perchellet, J.-P., 1992, Antitumor-promoting activities of hydrolyzable tannins in mouse skin, Carcinogenesis 13:715.

Griffin, A.C., and Jacobs, M.M., 1977, Effects of selenium on azo dye hepatocarcinogenesis, Cancer Lett. 3:177.

Hartman, P.E., and Shankel, D.M., 1990, Antimutagens and anticarcinogenesis: a survey of putative interceptor molecules, Environmental Mol. Mutagenesis 15:145.

Hayashi, Y., Nagao, M., Sugimura, T., Takayama, S., Tomatis, L., Wattenberg, L.W., and Wogan, G.N., 1986, Diet, Nutrition and Cancer, Princess Takamatsu Symposium, vol. 16, Japan Societies Press, Tokyo.

Hecht, S.S., Morse, M.A., Shantu, A., Stoner, G.D., Jordan, K.G., Choi, C.-I., and Chung, F.-L., 1989, Rapid single-dose model for lung tumor induction in A/J mice by 4-(methylnitrosoamino)-1-(3-pyridyl)-1-butanone and the effect of diet, Carcinogenesis 10:1901.

Huang, M.-T., Smart, R.C., Wong, C.-Q., and Conney, A. H., 1988, Inhibitory effect of curcumin, chlorogenic acid, caffeic acid, and ferrulic acid on tumor promotion in mouse skin by 12-O-tetradecanoylphorbol-13-acetate, Cancer Res. 48:5941.

IARC Monographs on the Evaluation of Carcinogenic Risk to Humans, 1993, Some Naturally Occurring Substances: Food Items and Constituents, Heterocyclic Aromatic Amines, and Mycotoxins, vol. 56.

Ip, C., 1981, Prophylaxis of mammary neoplasia by selenium supplementation in the initiation and promotion phase of chemical carcinogenesis, Cancer Res. 41:4386.

Ip, C., 1987, Fat and essential fatty acid in mammary carcinogenesis, Am. J. Clin. Nutr. 45:218.

Ip, C., Chin, S.F., Scimeca, J.A., and Pariza, M.W., 1991, Mammary cancer prevention by conjugated dienoic derivative of linoleic acid, Cancer Res. 51:6118.

Ip, C., and Ganther, H.E., 1990, Activity of methylated forms of selenium in cancer prevention, Cancer Res. 50:1206.

Ip, C., and Ganther, H.E., 1992, Comparison of selenium and sulfur analogs in cancer prevention, Carcinogenesis 13:1167.

Jacobi, H.P., and Baumann, C.A., 1940, The effect of fat on tumor formation, Am. J. Cancer 39:338.

Jacobs, M.M., Jansson, B., and Griffin, A.C., 1977, Inhibitory effects of selenium on 1,2-dimethylhydrazine and methylazoxymethanol acetate induction of colon tumors, Cancer Lett. 2:133.

Klurfeld, D.M., Welch, C.B., Davis, M.J., and Kritchevsky, D., 1989, Determination of degree of energy restriction necessary to reduce mammary tumorigenesis in rats during the promotion phase, J. Nutr. 119:286.

Kritchevsky, D., and Klurfeld, D.M., 1987, Caloric effects in experimental mammary tumorigenesis, Am. J. Clin. Nutr. 45:236.

Kuroda, Y., Shankel, D.M., and Waters, M.D., 1990, "Antimutagenesis and Anticarcinogenesis, Mechanisms II" Basic Life Sciences, vol. 52, Plenum Press, New York and London.

Larsen, C.D., 1945, Effects of cystine and calorie restriction on the incidence of spontaneous pulmonary tumors in strain A mice, J. Natl. Cancer Inst. 6:31.

Lavik, P.S., and Baumann, C.A., 1941, Dietary fat and tumor formation, Cancer Res. 1:181.

Lavik, P.S., and Baumann, C.A., 1943, Further studies on the tumor promoting action of fat, Cancer Res. 3:749.

LePage, G.A., Potter, V.R., Busch, H., Heidelberger, C., and Hurlbert, R.B., 1952, Growth of carcinoma implants in fed and fasted rats, Cancer Res. 12:153.

Liu, J., Gilbert, K., Parker, H.M., Hashek, W.M., and Milner, J.A., 1991, Inhibition of 7,12-dimethylbenz[a]anthracene-induced mammary tumors and DNA adducts by dietary selenite, Cancer Res. 51:4613.

McCay, C.M., Crowell, M.F., and Maynard, L.A., 1935, The effect of retarded growth upon the length of the life span and upon the ultimate body size, J. Nutr. 10:63.

McCay, C.M., Ellis, G.H., Barnes, L.L., Smith, C.A.H., and Spewling, G., 1939, Chemical and pathological changes in aging and after retarded growth, J. Nutr. 18:15.

Messina, M.J., Persky, V., Setchell, K.D.R., and Barnes, S., 1994, Soy intake and cancer risk: a review of *in vitro* and *in vivo* data, Nutrition and Cancer 21:113.

Miller, E.C., and Miller, J.A., 1979, Miletones in Chemical Carcinogenesis, Seminars in Oncology 6:445.

Miller, E.C., and Miller, J.A., 1979, Naturally occurring chemical carcinogens that may be present in food, in "Biochemistry of Nutrition 1A," A. Neuberger and T.H. Jakes, eds., University Park Press, Baltimore.

Miller, E.C., Miller, J.A., Hirono, I., Sugimura, T., and Takayama, S., 1979, Naturally Occurring Carcinogens–Mutagens and Modulators of Carcinogenesis, Japan Scientific Societies Press, Tokyo.

Milner, J.A., 1989, Mechanism for nutritional inhibition of cancer, in "Nutrition and Cancer Prevention, Investigating the Role of Micronutrients," T.E. Moon and M.S. Micozzi, eds., Marcel Decker, New York.

Moon, R.C., McCormick, D.L., and Mehta, R.G., 1983, Inhibition of carcinogenesis by retinoids, Cancer Res. 43:2469.

Moreschi, C., 1909, Beziehungen zwischen ernährung und tumorwachstum, Z. Immunitsforsh. 2:651.

Mottram, J.C., 1944, A developing factor for experimental blastogenesis, J. Path. Bact. 56:181.

Nelson, A.A., Fitzhugh, O.G., and Calvery, H.O., 1943, Liver tumors following cirrhosis caused by selenium in rats, Cancer Res. 3:230.

Pariza, M.W., and Felton, J.S., 1990, Mutagens and Carcinogens in the Diet, Wiley-Liss, New York.

Reddy, B.S., and Maeura, Y., 1984, Tumor promotion by dietary fats in azoxymethane-induced colon carcinogenesis in female F344 rats: influence of amount and source of dietary fat, J. Natl. Cancer Inst. 72:745.

Roebuck, B.D., Yager, J.D., Longnecker, D.S., and Wilpone, S.A., 1981, Promotion by unsaturated fat of azaserine-induced pancreatic carcinogenesis in the rat, Cancer Res. 41:3961.

Rous, P., 1914, The influence of diet on transplanted and spontaneous mouse tumors, J. Exptl. Med. 20:433.

Rowland, I.R., 1991, Nutrition, Toxicity, and Cancer, CRC Press, Boca Raton, FL.

Rusch, H.P., Kline, B.E., and Baumann, C.A., 1945, The influence of caloric restriction and of dietary fat on tumor formation with ultraviolet radiation, Cancer Res. 5:431.

Saffiotti, U., Montesano, R., SellaKumar, A.R., and Borg, S.A., 1967, Experimental cancer of the lung. Inhibition by vitamin A of the induction of trachealbronchial metaplasia and squamous cell tumors, Cancer 20:857.

Saxton, J.A., Boon, M.C., and Furth, J., 1944, Observations on the inhibition of the development of spontaneous leukemia in mice by underfeeding, Cancer Res. 4:401.

Scott, M.L., 1973, The selenium dilemma, J. Nutr. 103:803.

Shamsuddin, A.M., Elsayed, A.M., and Ullah, A., 1988, Suppression of large intestinal cancer in F344 rats by inositol hexaphosphate, Carcinogenesis 9:577.

Silverstone, H., Solomon, R.D., and Tannenbaum, A., 1952, Relative influences of natural and semipurified diets on tumor formation in mice, Cancer Res. 12:750.

Sivertsen, I., and Hastings, W.H., 1938, A preliminary report on the influence of food and function on the incidence of mammary gland tumor in A stock albino mice, Minn. Med. 21:873.

Sparnins, V.L., Barany, G., and Wattenberg, L., 1988, Effects of organo-sulfur compounds from garlic and onions on benzo[a]pyrene-induced neoplasia and glutathione-S-transferase activity, Carcinogenesis 9:131.

Sparnins, V.L., and Wattenberg, L.W., 1981, Enhancement of glutathione S-transferase activity of the mouse forestomach by inhibitors of benzo[a]pyrene-induced neoplasia of this anatomic site, J. Natl. Cancer Inst. 66:769.

Takigawa, M., Verma, A.K., Simsiman, R.C., and Boutwell, R.K., 1983, Inhibition of mouse skin tumor promotion and of promoter-stimulated epidermal polyamine synthesis by α-difluoromethylornithine, Cancer Res. 43:3732.

Tannenbaum, A., 1940, The initiation and growth of tumors. I. Effects of underfeeding, Am. J. Cancer 38:335.

Tannenbaum, A., 1942, The genesis and growth of tumors. III. Effects of a high fat diet, Cancer Res. 2:468.

The Alpha-tocopherol, Beta-carotene Cancer Prevention Study Group, 1994, The effect of vitamin E and beta-carotene on the incidence of lung cancer and other cancers in male smokers, New Eng. J. Med. 330:1029.

Vanderhoek, J. Y., Makheja, A. N., and Bailey, J. M., 1980, Inhibition of fatty acid oxygenases by onion and garlic oils, Biochem. Pharmacol. 29:3169.

Verma, A.K., 1989, Inhibition of phorbol ester-induced ornithine decarboxylase gene transcription by retinoic acid: a possible mechanism of antitumor promoting action of retinoids, Prog. Clin. Biol. Res. 259:245.

Verma, A.K., Ashendel, C.L., and Boutwell, R.K., 1980, Inhibition by prostaglandin synthesis inhibitors of the induction of epidermal ornithine decarboxylase activity, the accumulation of prostaglandins, and tumor promotion caused by 12-0-tetradecanoylphorbol-13-acetate, Cancer Res. 40:308.

Verma, A.K., and Boutwell, R.K., 1977, Vitamin A acid (retinoic acid), a potent inhibitor of 12-0-tetradecanoylphorbol-13-acetate-induced ornithine decarboxylase activity in mouse epidermis, Cancer Res. 37:2197.

Verma, A.K., Rice, H.M., and Boutwell, R.K., 1977, Prostaglandins and skin tumor promotion: inhibition of tumor promoter-induced ornithine decarboxylase activity in epidermis by inhibitors of prostaglandin synthesis, Biochem. Biophys. Res. Commun. 79:1160.

Verma, A.K., Shapas, B.G., Rice, H.M., and Boutwell, R.K., 1979, Correlation of the inhibition by retinoids of tumor promoter-induced epidermal ornithine decarboxylase activity and of skin tumor promotion, Cancer Res. 39:419.

Visscher, M.B., Ball, Z.B., Barnes, R.H., and Sivertsen, I., 1942, The influence of caloric restriction upon the incidence of spontaneous mammary carcinoma in mice, Surgery 11:48.

Volk, M.J., Pugh, T.D., Moon, J.K., Frith, C.H., Daynes, R.A., Ershler, W.B., and Weindruch, R., 1994, Dietary restrictions from middle age attenuates age-associated lymphoma development and IL-6 deregulation in C57BL/6 mice, Cancer Res. 54:May 15 issue.

Wargovich, M.J., Woods. C., Eng, V.W.S., Stephens, L.C., and Gray K., 1988, Chemoprevention of N-nitrosomethylbenzylamine-induced esophageal cancer in rats by the naturally occurring thioether, diallylsulfide, Cancer Res. 48:6872.

Watson, A.F., and Mellanby, E., 1930, Tar cancer in mice II: the condition of the skin when modified by external treatment or diet, as a factor influencing the cancerous reaction, Brit. J. Exptl. Path. 11:311.

Wattenberg, L.W., 1971, Studies of polycyclic hydrocarbon hydroxylases of the intestine possibly related to cancer. Effect of diet on benzpyrene hydroxylase activity, Cancer 28:99.

Wattenberg, L.W., 1977, Inhibition of carcinogenic effects of polycyclic hydrocarbons by benzylisothiocyanate and related compounds, J. Natl. Cancer Inst. 58:395.

Wattenberg, L.W., 1987, Inhibitory effects of benzylisothcocyanate administered shortly before diethylnitrosamine or benzo[a]pyrene on pulmonary and forestomach neoplasia in A/J mice, Carcinogenesis 8:1971.

Wattenberg, L.W., 1992a, Inhibition of carcinogenesis by minor dietary constituents, Cancer Res. (Suppl.) 52:2085s.

Wattenberg, L.W., 1992b, Chemoprevention of cancer by naturally occurring and synthetic compounds, in "Cancer Chemoprevention," L. Wattenberg, M. Lipkin, C.W. Boone, and G.J. Kelloff, eds., CRC Press, Inc., Boca Raton, FL.

Wattenberg, L.W., Coccia, J.B., and Lam, L.K.T., 1980, Inhibitory effects of phenolic compounds on benzo[a]pyrene-induced neoplasia, Cancer Res. 40:2820.

Weindruch, R., Albanes, D., and Kritchevsky, D., 1991, The role of calories and caloric restriction in cancer, Hematol./Oncol. Clinics No. America 5:79.

Weindruch, R., and Walford, R.L., 1982, Dietary restriction in mice beginning at 1 year of age: effects on life span and spontaneous cancer incidence, Science 215:1415.

Welsch, C.W., House, J.L., Herr, B.L., Eliasberg, S.S., and Welsch, M.A., 1990, Enhancement of mammary carcinogenesis by high levels of dietary fat: a phenomenon dependent on *ad libitum* feeding, J. Natl. Cancer Inst. 82:1615.

Willett, W.C., 1994, Diet and health: what should we eat?, Science 264:532.

Wilt, S., Pereira, M., and Couri, D., 1979, Selenium effect on initiation and promotion of tumors by benzo[a]pyrene and 12-0-tetradecanoylphorbol, Proc. Am. Assoc. Cancer Res. 20:21.

Wolbach, S.B., and Howe, P.R., 1925, Tissue changes following deprivation of fat soluble A vitamin, J. Exptl. Med. XLII:753.

Zhang, Y., Kensler, T.W., Cho, C-G., Posner, G.H., and Talalay, P., 1994, Proc. Natl. Acad. Sci. 91:3147.

EPIDEMIOLOGY OF ANTICARCINOGENS IN FOOD

Lenore Kohlmeier

Departments of Nutrition and Epidemiology
School of Public Health and School of Medicine
University of North Carolina at Chapel Hill
Chapel Hill, NC 27599-7400

INTRODUCTION

Ever since Doll and Peto suggested that up to 70% of all cancers might be related to our diet, there has been a mad dash to try to discover exactly what components of our diet and which active ingredients are acting where and how to affect carcinogenesis. The cancers considered to be nutritionally related include not only the cancers of the gastrointestinal tract, but also those of the reproductive organs and the lung. Initially, interest was focused on how diet causes cancer. More recent research is focusing on exploration of the ways in which diet can prevent cancer. Epidemiologic research remains the most powerful tool in understanding of the role of nutrition in the etiology of cancer in human populations. The challenges of this field of research include the problems of measurement error, exposure-disease lag times, collinearity and weak associations.

This paper addresses the important epidemiologic issues within the topic area of nutrition and cancer, and, in particular, chemoprotective agents in foods.

Glucosinolates, antioxidants and phenols are discussed as potential protection against cancer. Studies which form the foundation of our knowledge are critiqued. General weaknesses of effect in epidemiologic studies on diet and cancer are addressed, and strategies to overcome them proposed.

ANTICARCINOGENS IN PLANTS

Epidemiologic studies examining relationships between the intake of fruits and vegetables provide consistent evidence of important protective effects. A review of the 156 studies examining intakes of fruits and vegetables (Block et al., 1992) showed lower rates of cervical, ovarian, lung, esophageal, oral cavity, larynx, pancreatic, stomach, colon, rectal, bladder and breast cancer in people consuming fruits or vegetables relatively more frequently than those who did not. The vegetable epidemiology has until now added little to our understanding of the active ingredients in these foods.

Nutrition and Biotechnology in Heart Disease and Cancer
Edited by J.B. Longenecker *et al.*, Plenum Press, New York, 1995

125

A number of substances are drawing attention because of these potential anticarcinogenic effects. These include phenolic compounds, glucosinolates and naturally occurring antioxidants. Care is needed, however, in analyzing the results of epidemiologic studies with the same rigorous standards demanded of basic scientific experiments with regard to design, measurement, and internal and external validity.

Phenolic Compounds

Among phenolic compounds, three categories of substances are considered to be particularly biologically active. They are the flavonoids, the proanto- and anto-cyanins, and the catechins. The flavon family includes flavonols, and flavonol glycosides. Among the catechins, tea catechins, epicatechin gallate and quercitin are shown to be active in animal experiments. Tea catechins are seen to reduce tumorogenesis and tumor growth in mice. Epicatechin gallate inhibits free radical chain reactions of cell membrane lipids and can influence mutagenicity and DNA damaging activity. Quercitin is the most common and biologically active flavonoid. In a Dutch population, intakes of flavonoids were calculated to be 25.9 mg/day, with the majority of it derived from tea (Hertog et al., 1993).

Ingestion of quercitin by rats provided evidence of a reduced bioavailability of Benzo(a)pyrene, but not the carcinogen IQ (2-amino-3-methyl-imidazol quinoline), when this was added to the diet. This was assessed by measurement of radioactivity from marked compounds in the bile of the rats, bile being the excretory route of these substances. (Stavrie et al., 1992).

The action of some flavonoids is mediated by binding to estrogen receptors. It has been argued that by this binding they act on the regulation of gene transcription and may protect against certain cancers (Baker, 1992).

The effect of phenolic compounds on Phase II enzymes has also been unequivocally proven. The Phase II enzymes are the enzymes responsible for catalyzing the conjugation of endogenous ligands to functionalized xenobiotics. The metabolically inactive products formed can then be processed for excretion. Induction of Phase II enzymes has been demonstrated *in vitro* from a number of glucosinolate products (Prochaska et al., 1992). Induction of quinone reductase activity in murine hepatoma cells was the measure of effect. The active ingredients were isolating using reverse-phase HPLC (Zhang et al., 1992). It has not yet been shown, however, that this mechanism directly effects carcinogenesis *in vivo*.

Blockage of carcinogenic electrophiles may occur in a number of ways. Ultimate carcinogens, such as episodes of polycyclic aromatic hydrocarbons may bind with phenolic groups and produce inactive adducts. Oxidative damage can also be suppressed by phenolic compounds, inhibition of active oxygen species and of the generation and scavenging of superoxide anions (Huang and Farraro, 1992).

The ability of phenolic compounds to reduce proliferation may be caused through the modulation of protein kinase C activity. Protein kinase C is a TPA (tissue plasminogen activator) receptor believed to regulate cellular proliferation, (Castagna, 1982; Niedel et al., 1983). Several phenolic compounds inhibit the activity of rat brain protein kinase C (Castgna, 1982).

The mechanisms hypothesized here as influencing carcinogenicity have not yet been examined directly in epidemiologic studies at the substance level. Despite the wealth of studies on food groups such as fruits and vegetables and cancer, there is a dearth of studies on active ingredients in foods. This is largely because of the lack of reliable food compositional data on these substances.

Glucosinolates

Glucosinolates are present exclusively in vegetables of the Cruciferae family, especially in the genus Brassicae. They have been implicated as contributing to the lower cancer rates for a number of sites in East Germany, as compared with West Germany (Kohlmeier and Dortschy, 1991). Differences in cabbage consumption were among the most striking differences in food intakes across Europe (Kohlmeier, 1993).

The modulation of carcinogenesis by the consumption of Brassica vegetables has been investigated in laboratory animals challenged with carcinogens. The comparison of tumor development in animals with and without cabbage supplements provides evidence of an effect on the production of some tumors. Mammary tumors, particularly in mice and rats, seem to be reduced upon the addition of 5-20% of cabbage by weight to the diet (Bresnick et al., 1990; Stoewsand et al., 1988).

Glucosinolates are cleaved by the enzyme myrosinase, which the plant synthesizes for this purpose. The myrosinase is activated by mechanic disruption of the plant cell (Fenwick et al., 1983). As a result of more extensive enzymatic degradation, diets with uncooked Brassica vegetables contain more active metabolites of glucosinolates than cooked products (Betz and Obermeyer, 1993). Upon cleavage, isothiocyanates, thiocyanates, and nitriles are formed.

Modification of Phase I enzymes has been demonstrated by various phenols. Phase I enzymes are the cytochrome P-450 or mixed function oxidases (microsomal enzymes) which insert an oxygen atom between carbon or nitrogen-hydrogen bonds. Their modification can result in the activation of endogenous compounds and xenobiotics to mutagenic metabolites. How this may reduce carcinogenicity remains unclear, but it has been suggested that detoxification of enzymes which accelerate elimination of carcinogens may be the mechanism (Wood et al., 1986).

One of the phase I enzyme inducers, indole-3-carbinol, has been shown to affect estrogen metabolism in a way which may reduce carcinogenesis. It is known that the 2-hydroxy-estrone shows minimal estrogenic activity whereas the 16-hydroxy-estrone is both genotoxic and exerts full estrogenic potency (Doris et al., 1993). The *in vivo* effect of many compounds implied in the carcinogenic process on this pathway have been demonstrated (Michnoviez and Bradlow, 1991). So, for example, DMBA and Benzo(a)pyrene reduce the 2-OH metabolite levels in urine, whereas indole-3-carbinol, a compound found in Brassica, enhances excretion. Alcohol and the polyunsaturated fatty acids linoleic and arachidonic, on the other hand, increase the excretion of 16-OH metabolites. Induction of genotoxic damage by 16-hydroxy-estrone has been demonstrated in mammary epithelial cells of the mouse (Balang et al., 1992; Bradlow et al., 1991).

Other Naturally Occurring Antioxidants

Many naturally occurring phenolic antioxidants are found in foods. Chili peppers, ginger, green tea, pepper, oregano, rosemary, sesame seeds, soybean, and thyme contain particularly high concentrations of these substances per gram dry weight. however, other foods, consumed in greater quantities are equally important sources of naturally occurring antioxidants. Antioxidants reduce the auto-oxidation of lipids, and probably also reduce lipo-oxygenase activity. The metabolism of arachidonic acid to lipid peroxides is influenced by antioxidants, and inhibition of arachidonic acid metabolism inhibits tumor promotion (Fischer et al., 1982).

Specific substances from plant sources under study because of their known physiologic importance, high consumption levels and antioxidant potential include tocopherols (vitamin E); carotenoids; ascorbic acid. These are the so-called "ACE vitamins" although vitamin A is not among them. The provitamin A carotenoids are referred to here, with β-carotene receiving the most attention as a chemoprotective agent among the carotenoids. β-carotene is one in a large family of over 500 carotenoids (Pfander, 1987).

All carotenoids contain extensive conjugated double-bonds. Because of this, they absorb light in the visible spectrum, and serve as natural pigments, providing color to flamingoes, lobster, tomatoes, many flowers, and yellow vegetables (Olson, 1993). Individual caratenoids differ in regard to their antioxidant potential in man (Simic, 1992). Some have no measurable antioxidant potential *in vitro*.

Epidemiologic findings have led to the suggestion that β-carotene and other carotenoids are preventive in the carcinogenesis process (Peto et al., 1981). Of the approximately 10,000 oxidative hits to DNA per cell per day estimated in man, most of the damage is repaired by glycosylases, which excise the lesion and excrete the free base in the urine. However the accumulation of unrepaired damage over time may influence the risk of cancer. Antioxidative plant constituents may play a role in the cellular defense against oxidative stress, and reduce mutagenic and carcinogenic DNA alterations.

The results of studies of the effects of β-carotene on tumor formation in animals is inconsistent. An overview of experimental findings regarding the effects of β-carotene on chromosome breaks, sister chromatid exchange, and tumor growth is presented by Krinsky (1993). He attributes the mixed results to the problem that most animals are extremely poor absorbers of carotenoids, and therefore doses used are necessarily in the pharmacological arena. There is also the question of whether β-carotene or its retinoic derivatives are acting in these tests. The administration of carotenoids has however shown impressive results regarding a decrease in tumor incidence and even regression of tumors in hamster. The final outcome on carotenoid intake and cancer in man has not yet been determined.

Proposed Mechanisms of Effect

The hypothesized activities of the various anticarcinogens discussed are listed in **Table 1**. They include activity at the levels of absorption of toxins, activation of detoxifying enzymes, blockage of active carcinogens directly affecting gene transcription and reduction of cell proliferation (Huang and Ferraro, 1992).

Table 1. Potential mechanisms of action of anticarcinogens in food

• Reduce Bioavailability of Mutagens from Intestine
• Modify Phase I enzymes
• Induce Phase II enzymes
• Block Carcinogenic Electrophiles in Cell
• Scavenge active oxygen species Reduce Proliferation
• Influence gene transcription
• Reduce proliferation
• Alter Estrogen Metabolism

Food Concentrations of Anticarcinogens

Little data exists on availability of phenolic compounds in foods. The information which does exist is in various publications, but not drawn together in edited food compositional databases. Therefore, estimates on the amounts consumed daily vary widely. Based on usual intakes, tea provides the greatest amounts of flavonoid (61%); onions and apples are the next greatest sources, providing 13% and 10% respectively in a Dutch population (Hertog et al., 1993). Tea leaves may have catechin concentrations which represent up to 30% of the dry weight of the tea leaf (HO, 1992). Allium vegetables (leeks, shallots, scallions, garlic and onions) range in their flavonol content from none to more than 1g/kg of vegetable. Shallots have uniformly high concentrations, but onions range widely, with no measurable amounts in white onions and high levels in yellow onions (Leighton et al., 1992).

The glucosinolate content of foods varies from species to species, from crop to crop, and from lab to lab. The amount of thiocyanate, for example, in various Brassica ranges as much as seven-fold within a species. Their average concentrations in mg/kg fresh weight vary up to four-fold between species (Langer, 1983). Among the subspecies, late sowing and younger samples show a larger percentage of acetonitrile extracts from these vegetables, ranging from 4.6 to 15.6% of dry weight (Huang and Ferraro, 1992).

Different foods contribute different levels of carotenoids. Carrots contain primarily and β-carotene, broccoli contains lutein and β-carotene. Tomatoes are a primary source of lycopene. Raw guava and watermelon also provide large concentrations of lycopene. Lutein can be found in very high amounts in chicory, chives, kale, collards, cress and other greens. The major sources of carotenes in the American diet are carrots, tomatoes, sweet potatoes, yellow squash, spinach and cantaloupe (Byers and Perry, 1992). These six foods account for 70% of carotenoids in the diet. Romaine lettuce, broccoli, spinach and iceberg lettuce contribute approximately another 10%.

All in all, information on concentrations of anticarcinogens in food is not widely available. Furthermore, reported specie concentrations are not predictive of the concentrations of samples from different cultivars.

EPIDEMIOLOGIC STUDIES ON ANTICARCINOGENS

Animal studies are inconclusive about the effects of many dietary anticarcinogens. *In vitro* evidence of anticarcinogenic potential of these compounds are inadequate for prediction of effects on cancer development in man. Epidemiologic studies alone can indicate whether these exposures are showing a measurable influence in the population at large within the range of doses normally consumed.

Epidemiologic research has established a strong relationship between β-carotene consumption and lung cancer risk. At least 32 studies have examined relationships between consumption of foods rich in carotenoids or B carotene levels in serum and lung cancer risk, with 30 of these showing a protective relationship (Block et al., 1992). Seven of these studies examined serum levels of beta carotene in cases and controls, and six of them found strong protective effects of higher circulating carotenoid levels. This "scorecard" approach to epidemiologic findings , without rigorous examination of the study designs, measurement instruments, or analyses can be misleading, and is the source of mistrust of epidemiologic findings. Despite differences in design, populations, and measures of intake, the consistency of effect in this area is impressive. However, the nature of the activity of carotenoids on carcinogenesis is not known; carotenoids may actually serve as indicators of consumption of foods containing some factors which may actually be responsible for the protection seen.

More than 25 intervention studies, largely supported by the NIH, are currently underway which test whether the addition of one or a combination of naturally occurring antioxidants at various doses affect disease development in risk populations. One intervention study with a semi-factorial design, conducted in a population in China with high risk of esophageal and stomach cancer studied the effect of combinations of nutrients, as compared with placebos on cancer incidence. The study failed to show a significant reduction of esophageal cancer in 5 years of intervention. It did show reductions in total mortality, cancer mortality and, in particular, stomach cancer mortality when a combination of β-carotene, vitamin E and selenium were supplemented. Supplementation of retinol and ascorbic acid in this period of time was without effect (Blot et al., 1993). This trial is limited in its interpretability because of the combination of nutrients which were supplemented. The poor nutritional status of the population before supplementation restricts generalizability of these findings to well-nourished populations.

Supplementation has been limited in its dosage and restricted almost exclusively to the carotenoid, β-carotene. Research on phenolic compounds and potential protective agents in Cruciferous vegetables is still in its infancy. Epidemiologic studies can help identify whether effects are seen in man, and direct the focus on substances of interest. Two types of epidemiologic studies lend credence to the hypothesis of a true population influence: ecologic studies across countries, and case-control studies within countries.

Ecologic Studies

Ecologic studies, when conducted properly, with adjustment of confounding factors, have consistently drawn the attention of scientists to previously unnoticed relationships and driven the basic research on diet and disease. So, for example, the strong associations seen between fiber consumption and colon cancer, salt and hypertension, and dietary fat and breast cancer in ecologic research have been the primary reasons for etiologic research on these exposures. The limitations in this type of study are many. Analyses need to be carefully designed and interpretation of findings tempered by the path that the data reflects average yearly intake over populations and ignores the underlying variation in the populations under study. Furthermore, measurements at this level are questionable as far as food intake is concerned, since per capita statistics on the disappearance of commodities are generally used. These are collected for economic purposes, and beyond their being of questionable comparability between countries (Kelly et al., 1992), they do not account for factors such as home production and waste, which can be considerable when one considers foods of plant origin. Another source of measurement error with per capita statistics can result when the geographical units of comparison differ, as for example comparison of regional cancer incidence with national per capita food disappearance statistics.

Most of these measurement errors tend to reduce the ability to discover a relationship, even if one exists. Thus, notice is generally given to strong associations from ecologic analyses. If an ecologic relationship is seen, the danger however remains that observed associations are due to an underlying latent variable relationship to the dietary exposures, such as economic factors.

An ecologic exploration of relationships between chemoprotective foods, such as cabbage and citrus fruits and breast cancer mortality, or mortality from stomach cancer and colon and rectal cancer death rates do reveal preventive associations. In an analysis of cabbage and citrus fruit disappearance across 20 European countries over a twenty year time span, it is seen that citrus fruit intakes are associated in a protective fashion with colon cancer and stomach cancer. Cabbage intake appears to be protective against colon and breast cancer, when alcohol and tobacco intakes are controlled for. The strength of this association weakens as the time between exposure and disease increases (Kohlmeier and Dortschy, submitted).

Case Control Studies on Nutrition and Cancer

More than 14 epidemiologic studies provide information on the topic of consumption of cruciferous vegetables and the risk of colorectal cancer (Block et al., 1992). They differ in their selection of cases and controls, study size and power, and in their exposure measurements. Exploration of differences in design of these case control studies from around the globe, most of which were conducted to study the effect of dietary fiber on colon cancer, illustrates the potential effects of design and exposure measurement differences on study findings.

DESIGN ISSUES

Two case-control studies on colon cancer and diet conducted in Japan reported opposite results regarding a protective effect of cabbage. Haenzel et al. (1980), conducted a hospital based case control study in three areas of Japan, and pooled the results. In this study, hospital based controls were patients drawn with a variety of diagnoses from a number of clinics. The exposure to food in this study was assessed with a food frequency questionnaire using food consumption frequency, not amounts, as the basis of the analyses.

The median consumption of the foods of interest in these Japanese was four times per month. Risk of developing colon cancer among those consuming Hakusai (pickled cabbage) more frequently than four times per month were compared with those whose consumption was less frequent. The odds ratio was protective here too, with an odds ratio of 0.76 ; a 24% lower risk for those with more frequent Hakusai consumption. In the second Japanese study (Tajima and Taminaga, 1985) cabbage consumption was reported to carry a two-fold risk of colon cancer development (odds ratio =2.07) when consumed four times or more per week.

The similarities in design between these 2 studies include use of a hospital based case-control study design, and the assessment of dietary exposure by food frequency questionnaire. A closer look at intrinsic differences helps to explain some of the differing results between these two studies.

Study Power

As mentioned, no significant protective effect of Hakusai was seen in the Tajima and Taminaga (1985) study, and the point estimate of risk for consumption of other cabbage was greater than one, but not statistically significant. A reason for this is the low power of this study. Haenzel et al. (1988) analyzed 583 cases and 1173 controls. Tajima and Taminaga (1985) collected information on 42 colon and 51 rectal cancer patients. Few epidemiology studies of this size can contribute concretely to etiologic studies. A significant result would require excellent exposure measurements and relative risks greater than 4 to achieve statistical significance at the $p=0.05$ level.

When risks are weak, three epidemiologic tools can be employed to enhance them, and to make the study more powerful: one of these is to increase validity of measure; a second is to increase the range of exposures; and the third is to use a larger sample. Two major limitations in current epidemiology are that most studies are on relatively homogeneous populations, and weak measures of dietary exposure are being used.

Interviewer Bias

Since eating behavior is laden with judgment of peers, it is easily subject to bias. A potential flaw in the Tajima and Taminaga (1985) study, mentioned by the authors, was the fact that the interviewer was one of the principal investigators and was not blinded to the study hypotheses or outcomes. This could contribute to a subtle judgment bias, or bias regarding the degree of prompting responses to dietary questions of cases versus controls by a principal investigator with interest in a particular hypothesis.

Selection of Controls

Another critical issue is selection of controls for comparison. The control group should reflect the exposure of interest at the time of interest in the baseline population. In the Tajima and Taminaga (1985) study, controls were selected from the clinic, with exclusion for neoplastic disorders. A large proportion of the controls had gastrointestinal tract disorders, and the authors estimated that 20% of the controls suffered from chronic gastritis. If these conditions affected dietary intakes such that the responses to the food frequency questionnaire reflected changes in intake related to disease onset, or if the intake of cabbage was associated with these diseases, the underlying population would not be correctly represented, making this an inappropriate control group. Most other studies using hospital based controls excluded for gastrointestinal disorders. Another study which used hospital based controls with gastrointestinal problems (subjects admitted for abdominal surgery) was one of the few other studies showing no significant protective effect of cabbage (Miller et al., 1983). Since many medical conditions are related to some dietary factor, use of population based controls is strongly recommended in epidemiologic studies in which diet is being assessed by interview.

The presence of many medical conditions can also result in change in dietary behavior. This is an issue of internal validity and potential bias in the retrospective assessment of diet, and can affect both cases and controls. If each group is affected differently, differential misclassification will result, and either inflation or deflation of the true risk estimate will be the consequence.

Selection of Cases

Most epidemiologic studies of colon cancer take newly diagnosed colon cancer patients as cases. Some take 'prevalent cases'; patients who had been diagnosed within a greater time period, or at a time in the past. The ecologic studies, because of the absence of cancer registries in many countries, most often resort to deaths primarily attributable to colon cancer. Each of these have their biases. Newly diagnosed cases do not represent all incident cases, many individuals with malignant tumors of the colon remain undiagnosed. Controls may have undiagnosed colonic cancer. Retrospective recruitment of cases bias the group towards less sever cases, as the more severe cases may have died in the meanwhile. And the use of mortality as the endpoint limits the generalizability of the findings to individuals who die of the disease; excluding the survivors.

Modern epidemiologists are searching for good early clinical markers of disease and disease risk. In one study, conducted in Norway by Hoff et al. (1986), the case selection was conducted by recruiting a subset of the normal population to colonoscopy and determining the presence or absence of polyps and tumors. This is an improvement which reduces many sources of selection biases. One weaknesses of this are that, in rare diseases, recruitment will be large, costly and inefficient. However, for an outcome such as presence of polyps in men over 50, the prevalence is great enough in western populations to make this feasible. Another is that polyps, as most pre-clinical markers, will not all become carcinomas, and the exposure of interest may have its effect after this stage of carcinogenesis, in which case, no effect of it may be seen on this endpoint. Hoff et al. (1986) did find that the number and severity of polyps was negatively associated with the gram amount of cruciferous vegetable consumed per day as seen in **(Figure 1)**.

DIETARY EXPOSURE ASSESSMENT DIFFERENCES

Unique to nutritional epidemiology of cancer is the dependence on food frequency questionnaires. These are self-administered questionnaires which generally require the subject to determine which of 5 to 7 categories of intake reflects their usual consumption of one or a combination of foods. Absolute amounts, and meal patterns are not usually captured with this tool. Food frequency questions are simplified in their handling at the cost of the subject burden. So, for example, instead of asking someone what they eat per meal, or how often they eat lasagna during lunch, and meatloaf during supper, the food frequency asks: "How often do you eat beef, pork, or lamb as a sandwich or mixed dish, e.g. stew, casserole, lasagne, etc." (Sempos et al., 1992).

This forces the subject to summarized single foods across meals, summarize multiple foods with one another, and to consider what might be meant by etcetera in this case, as compared to its use in another question. In this question, portion sizes are not considered. In other questions in the same questionnaire, subjects are required to adjust frequencies to match their portion sizes. So for example if the given portion size for milk is an 8 oz glass, a women who drinks half a glass daily should report drinking 8 oz glasses of milk 2-4 times per week.

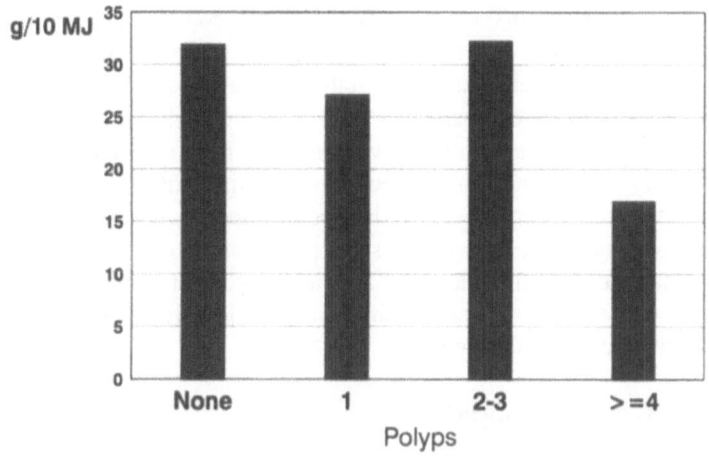

Figure 1. Number of Polyps by cruciferous consumption in Norway (Hoff et al., 1986).

When the tool is designed, decisions are made about the weighing factors used to assign nutrient values to each question. The assignment of a single nutrient value to the question about beef, pork or lamb requires application of a percentage of relative consumption of each. The relative amount of the individual sandwiches and dishes is then applied to all persons using this questionnaire. The assigned levels of protein, fat, thiamine and all other nutrients reflect that set proportion which is assumed to be eaten in this population at large. Individuals who prefer pork will be assigned thiamine values for the beef, pork, lamb. Those who eat lasagna regularly will have ham sandwiches with mayonnaise averaged into their nutrient intake.

This approach to dietary assessment necessarily reduces the variation within a population from three sources: amounts, frequencies and proportions of many specific foods. Amounts, when specified, are at the level of one egg or one tomato, so that individual who deliberately select and consume small eggs or large tomatoes are not differentiated from those who eat the same number of large eggs or cherry tomatoes. One portion of fish is identified as 3-5 oz, one portion of beef at 4-6 oz. Thus, within the same response, there is a 50% difference in quantity. Frequencies of consumption are necessarily truncated, and the number of categories allowed for frequency responses is inversely proportional to the variance lost within categories. Finally, as described above, in the assignment of nutrient values to food frequency questions, assumptions need to be made about the relative frequency of consumption all possible individual foods summed under that question.

All of these issues relate to loss of variance in dietary exposures with this blunt tool. Where dietary intakes vary greatly, and the true variation is captured by the instrument, without differential misclassification of exposures between subjects, this tool may suffice. It remains possible, however, that the inconsistent findings on the importance of individual nutrients, such as dietary fat, in the process of carcinogenesis may well be results of reliance on flawed exposure assessment methods.

An example of this crude measure of exposure showing highly significant associations is one of the earliest epidemiologic studies to report on dietary cabbage exposure levels and colon cancer was conducted in the U.S. by Graham et al. (1978). They used a hospital based case-control study, with 330 rectal and 256 colon cancer cases. As controls, they selected the then popular option of using patients in the hospital for other diseases. Wisely, they chose to exclude anyone with gastrointestinal disease. The method of assessment of prior diet was a food frequency questionnaire which asked separately about the consumption of Brussel sprouts, coleslaw, cabbage, broccoli, and sauerkraut. This was not a crucifera loving population their median frequency of intake was once a month. Thus the distribution of exposure levels was concentrated at the lowest 2 categories of exposure. They did however find that compared to "never" eating cabbage, "sometimes" showed a protective effect, and "once a month" showed an odds ratio of 0.4 which was 60% lower than those "never" consumers. This protective association did not increase at double this low exposure rate. This strong effect at rather modest dosage, which does not increase with much greater levels of intake, is a rather surprising result. Results this extreme, based on measures so simple lead to suspicion that another concurrent behavior may be underlying this effect such that the cabbage eaters differ in other important respects, regarding their colon cancer risk, than non-cabbage eaters.

Food and Food Group Identification

Commonly, in studies based on food frequency questionnaires, the foods being assessed are poorly defined. The tens of thousands of foods available to are to be assigned to a hundred or so groups with many diverse foods grouped into one category. It is often left to the discretion of the subject to place food items in the group they believe most appropriate. It is also possible that foods consumed are recorded twice; in different categories. In the study of Tajima and Taminaga (1985), specific foods included in the Hakusai group are not identified. It is not clear from this report whether the group reported separately as "cabbage" excluded cabbage. The cabbage group did show a protective tendency in this study. Somehow Hakusai and "cabbage" consumption are separated, without the reader knowing whether these were exclusive categories. If so, the cabbage group would rightfully be called non-Hakusai cabbage consumption, which would explain in part the discrepant findings.

A separate issue is the effect of grouping of foods on response behavior. Serdula et al. (1992) found that the more foods grouped together in a category, the greater the proportion of people reporting non-consumption of the items.

Comparing Quantities Consumed

To maximize information distribution across comparison groups, exposures are often categorized by the quintile on tertile of intake of the control group of the population under study. The proportion of cases and controls within quintiles are then compared in the development of odds ratios. Comparison of the magnitude of effect between studies is dependent on the absolute consumption levels which determine group cutoffs, as well as the comparison baseline (either highest or lowest consumption levels). Not only is it unlikely that quintiles will coincide between studies, levels are also sometimes not reported. So, for example, in the study of Tajima and Taminaga, odds ratios are assessed at high, medium and low consumption levels without these ever being defined, making it impossible to compare the exposure frequencies used for analyses with other studies. Another study applies one set of cutoffs to all foods, independent of consumption levels. The case-control study of Peters et al., (1992) conducted in Los Angeles found no association between cabbage consumption and the risk of colon cancer when they entered cabbage intake as a continuous variable in their logistic regression, with the unit of measure being 10 servings per month. In most populations, the mean consumption of cabbage was below 4 servings per month.

Reference Period of Dietary Exposure

The true reference period of dietary exposure is often unknown. Studies resort to one of two strategies to compensate for this: either the subject is asked to recall and report on usual eating behavior a number of years prior to recruitment, or current behavior is taken as a surrogate of long term eating patterns. The assumption is then that, although behavior may have changed, the inter individual differences exceed intra individual variation, leaving relative rankings stable.

In most of the 14 studies on this topic, the reference period of dietary intake is vague. Tajima and Taminaga (1985) reports asking usual diet 1-2 years prior to hospitalization. In other studies current intake was assessed. Only one study was found which treated long term and recent intake differently. Young and Wolf (1988), in a population based study using a cancer registry as the basis of recruitment, and controls drawn from motor vehicle operator lists, attempted to assess diet at three stages of life: before 18, between 18 and 35, and after the age of 35. The subjects were between 35 and 89 at the time of the assessment. A food frequency approach was used. This study did not find time of life of exposure to influence odds rations relating cruciferae or cabbage intake to total risk of colon cancer.

Exposure Amount or Exposure Frequency.

Food frequency questionnaires are designed and applied under the assumption that frequency of consumption is more accurately recalled than usual amounts, and that amounts do not vary substantially between subjects. There is evidence that the first assumption is true (Smith et al., 1991). The use of frequency of intakes, ignores true dose, as the usual portions of these foods differ from individual to individual and from food to food. The four studies on cruciferous vegetables and colon cancer which used quantitative amounts consumed per day provide the strongest and most consistent evidence of a protective relationship across categories of intake. In a large case-control study conducted in Australia, with over 700 subjects with colon and rectal cancer using both community based controls and hospital based controls, a much more extensive method (a Burke diet history) was applied (Kune et al., 1987). In this study, the average consumption of brassica was not low, calculated at 310 g/week. The odds ratios for the extreme quintile comparison was .5 for colon cancer and .64 for rectal cancer. A comparison across all quintiles revealed a significant trend for both men and women. The odds ratios for quintiles of intake, as assessed by diet history questionnaire in Australia (Smith et al., 1991) are presented in **Figure 2**. The results of a second study applying a diet history questionnaire in Belgium (Tuyns et al., 1988) can be seen in **Figure 3**.

Food Preparation

Food preparation methods are seldom captured in the food frequency questionnaires in common use. Depending upon the exposure of interest, valuable information may be lost. The case control study conducted on this topic in Belgium, with over 800 cases, and information from food frequencies, analyzed cooked and raw leafy vegetables separately (Smith et al., 1991). Raw vegetables would be expected to provide greater concentrations of isothiocyanates, as these enzymes needed to cleave the glucosinolate are destroyed by heat. As we might have expected from the biology, the consumption of raw vegetables was even more protective than the consumption of cooked vegetables, even though the amounts consumed raw were only one quarter of those cooked **(Figure 3).**

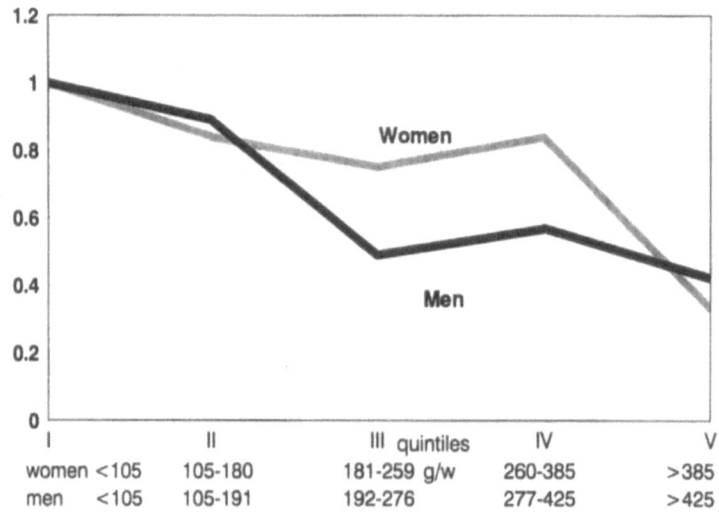

Figure 2. Cruciferous vegetable consumption and colorectal cancer risk in Australia (Kune et al., 1987).

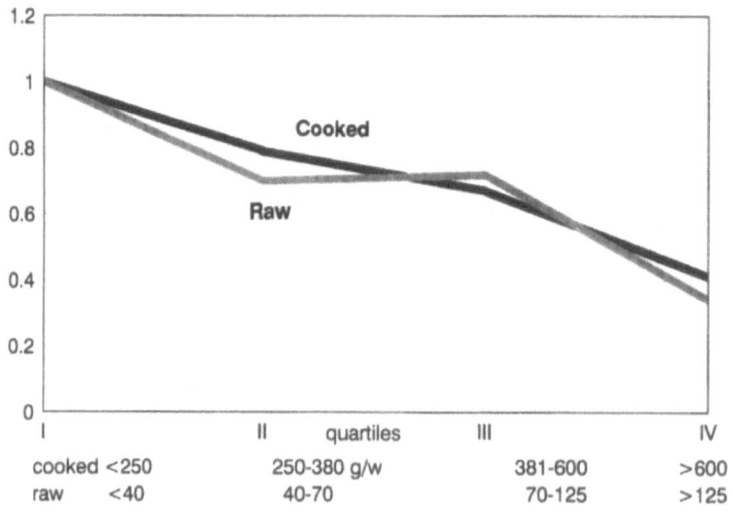

Figure 3. Colon Cancer and "Leafy Vegetable" consumption in Belgium (Tuyns ct al. 1988).

Use of Biomarkers of Dietary Exposure

For all of the reasons mentioned, it is difficult to obtain accurate information on dietary intake in epidemiological studies (Barrett-Connor, 1991; Kohlmeier, 1988). In addition, intake of foods, when assessed by interview or questionnaire, may be recalled differentially between cases and controls. The use of food composition tables may also introduce errors, because of large variations in nutrient content of some foods. Furthermore, when measuring intake, individual variation in absorption and metabolism is not taken into account. Where available, molecular markers of intake often are preferable. Plasma or serum concentrations are the most commonly used biomarkers of exposure. They generally reflect intake during the past few days to weeks. Under the assumption that these measures reflect the internal dose, and under conditions of stable intake levels over a longer period of time, these plasma/serum concentrations may be adequate indicators of exposure. Dietary intake may, however, have been influenced by disease, or the transport proteins may be affected by the cancer. Therefore, markers of accumulated long term exposure offer distinct advantages and may better reflect the exposure levels before change.

For glucosinolates, due to their reactivity and rapid metabolism, no long term markers are known. Antioxidant status can be measured short term by assessment of the oxidative products, such as oxidized nucleotides in urine. However, long term intake assessment is more challenging. Adipose tissue is one of the few media which provide a stable storage site for fat soluble exposures (Kohlmeier and Kohlmeier, in press). It has been used to estimate adipose tissue concentrations of fat soluble antioxidants, such as alpha tocopherol and beta carotene, as reflecting long term dietary intake (Kardinaal et al., 1993). Unfortunately, it is unlikely that biomarkers of previous dietary intake will become readily available for a wide range of nutrients. Also, the need for estimation of energy intake precludes the elimination of subjective dietary assessments in epidemiologic studies in the near future.

CONCLUSIONS

Epidemiologic studies are necessary to quantify the impact of dietary related exposures on carcinogenesis in man. Interventions are then needed to test causality. The current generation of observational epidemiologic studies are limited by the use of blunt instruments to quantify dietary exposures of individuals. Better dietary assessment methodology, which applies quantitative methods for assessing dietary intake, meal by meal show stronger protective effects. These types of methods have until now been used to a limited extent because of the need for well trained interviewers to conduct this extensive interview by hand. The development of the next generation of computer assisted, fully automated dietary assessment methods; which tailor the interview to the subjects' responses, reduce the subject burden, and not just that of the scientist, should enhance our detection power in epidemiologic studies on diet and cancer.

Biomarkers of exposure and internal dose are still poorly developed. Biomarkers may provide good estimates of long term exposures to certain foods and specific nutrients which are stable over time and stored in the body. They are however unlikely to replace assessment of total diet. Stronger, and cognitively smarter tools, which reduce the subject burden of assessing usual diet are urgently needed.

Epidemiologic studies of diet and cancer can also be strengthened by increase study size, improve measurements of diet and by expand the diversity of populations being studies. The latter two approaches are the most cost effective approaches, and should enhance both the internal and external validity of future studies.

REFERENCES

Baker, M.E., 1992, Evolution of regulation of steroidmediated intercellular communication in vertebrates: Insight from flavonoids. signals that mediate plant-rhizobia symbiosis, J. Steroid Biochem. Mol. Biol. 41:301-08.

Barrett-Connor, E., 1991, Nutrition epidemiology: how do we know what they ate? Am. J. Clin. Nutr. 1:54:182S-187S.

Betz, J. and Obermeyer, W., 1993, Effects of processing on the glucosinolate content of broccoli (Brassica oleracea var. botrytis L. subvar cymosa Lam.). FASEB J. 7:A 863.

Block, G., Patterson, B., Subar, A., 1992, Fruit, vegetables, and cancer prevention: A review of the epidemiological evidence, Nutr. Cancer 18:1-29.

Blot, W.J., Li, J.-Y., Taylor, P.R., Guo, W., Dawsey, S., Wang, G.-Q., Yang, C.S., Zheng, S.-F., Gail M, Li G-Y, Yu Y, Liu B-q, Tangrea J, Sun Y-h, Liu F, Fraumeni JF, Jr, Zhang YH, Li B., 1993, Nutrition intervention trials in Linxian, China: Supplementation with specific vitamin/mineral combinations, cancer incidence, and disease-specific mortality in the general population, JNCI 85(18):1483-1492.

Bradlow, H.L., Michnovicz, J.J., Telang, N.T. and Osborne, M.P., 1991, Effects of dietary indole-3-carbinol on estradiol metabolism and spontaneous mammary tumors in mice, Carcinogenesis 12:1571-1574.

Bresnick, E., Birt, D.F., Wolterman, K., Wheeler, M. and Markin, R.S., 1990, Reduction in mammary tumorigenesis in the rat by cabbage and cabbage residue, Carcinogenesis 11:1159-63.

Byers, T. and Perry, G., 1992, Dietary carotenes, vitamin C, and vitamin E as protective antioxidants in human cancers, Ann. Rev. Nutr. 12:139-59.

Castagna, M., Takai, Y., Kaibachi, K., Sano, K., Kikkawa, U. and Nishizuk, Y., 1982, Direct activation of calciumactivated, phospholipid-dependent protein kinase by tumorpromoting phorbol esters, J. Biol. Chem. 257:7847-7851.

Davis, D.L., Bradlow, H.L., Wolff, M., Woodruff, T., Hoel, D.G. and Anton-Culver, H., 1993, Medical hypothesis: xenoestrogens as preventable causes of breast cancer, Environ. Health Perspect. 101:372-377.

Fenwick, G.R., Heaney, R.K. and Mullin, W.J., 1983, Glucosinolates and their breakdown products in food and food plants, CRC Critical Reviews in Food Science and Nutrition 18:123-201.

Fischer S.M., Mills, G.D., Slaga T.J. 1982, Inhibition of mouse skin tumor promotion by several inhibitors of archidonic acid metabolism, Carcionogenesis 3:1243-1245.

Graham, S., Dayal, H., Swanson, M., Mittelman, A. and Wilkinson, G., 1978, Diet in the epidemiology of cancer of the colon and rectum, J. Natl. Cancer Inst. 61:709-14.

Haenszel, W., Locke, F.B. and Segi, M., 1980, A case control study of large bowel cancer in Japan, J. Natl. Cancer Inst. 64(1):17-22.

Hertog, M.G.L., Feskens, E.J.M., Hollman, P.C.H., Katan, M.B. and Kromhout, D., 1993a, Dietary antioxidant flavonoids and risk of coronary heart disease: the Zutphen Elderly Study, Lancet 342:1007-1016.

Hertog, M.G.L., Hollman, P.C.H., Katan, M.B. and Kromhout, D., 1993b, Intake of potentially anticarcinogenic flavonoids and their determinants in adults in The Netherlands, Nutr. Cancer 20:21-29.

Ho, C.-T., 1992, Phenolic Compounds in Food: An Overview, in "Phenolic Compounds in Food and Their Effects on Health II: Antioxidants & Cancer Prevention," M.-T. Huang, C.-T. Ho, and C.Y. Lee, eds., American Chemical Society, Washington, D.C.

Hoff, G., Moen, I.E., Trygg, K., Froelich, W., Sauar, J. Vatn M., Gjoine, E., Larsen, S., 1986, Epidemiology of polyps in the rectumand sigmoid colon: Evaluation of nutritional factors, Scand. J. Gastroenterol. 21:199-204.

Huang, M.T. and Ferraro, T., 1992, Phenolic Compounds in Food and Cancer Prevention, in "Phenolic Compounds in Food and Their Effects on Health II: Antioxidants & Cancer Prevention," M.-T. Huang, C.-T. Ho, and C.Y. Lee, eds., American Chemical Society, Washington, D.C.

Kardinaal, A.F.M., Kok, F.J., Ringstad, J., Gomez-Aracena, J., Mazaev, V.P., Kohlmeier, L., Martin, B.C., Aro, A., Kark, J.D., Delgado-Rodriguez, M., Riemersma, R.A., Veer, P van 't, Huttunen, J.K., Martin-Moreno, J.M., 1993, EURAMIC Study: Antioxidants in adipose tissue and the risk of myocardial infarction, Lancet 342:1379-1384.

Kelly, A., Becker, W. and Helsing, W., 1991, Food balance sheets, in "Food and Health Data: Their Use in Nutrition Policy-Making." W. Becker and E. Helsing, eds., WHO Publications, Copenhagen, Denmark.

Kohlmeier, L., Dortschy, R, 1991, Diet and disease in East and West Germany, Proc. Nutr. Soc. 50:719-727.

Kohlmeier, L., 1993, Health divergence during political division: East and West Germany - 'socioeconomic factors in health', in "Europe Without Frontiers - The Implications for Health," C.E.M. Normand and J.P. Vaughan, eds., John Wiley & Sons, England.

Kohlmeier, L., 1988, Analytical problems in nutritional epidemiology, in: "Epidemiology, Nutrition and Health; Proceedings of the 1st Berlin Meeting on Nutritional Epidemiology," L. Kohlmeier and E. Helsing, eds., Smith-Gordon, London.

Kohlmeier, L. and Dortschy, R, Brassica consumption and colon cancer (submitted).

Kohlmeier, L. Epidemiologic Findings on Colon Cancer and Cruciferae Consumption (submitted).

Kohlmeier, L. and Kohlmeier M., Adipose tissue as a medium for epidemiologic exposure assessment, Environ Health Perspect (in press).

Krinsky, N.I., 1993, Actions of Carotenoids in biological systems, Ann. Rev. Nutr. 13:561-587.

Kune, S., Kune, G.A. and Watson, L.F., 1987, Case-control study of dietary etiological factors: the Melbourne Colorectal Cancer Study, Nutr. Cancer 9:21-42.

Langer, P., 1983, Naturally Occurring Food Toxicants: Goitrogens, in "CRC Handbook of Naturally Occurring Food Toxicants," M. Rechcigl, Jr., ed., CRC Press, Inc., Boca Raton, Florida.

Leighton, T., Ginther, C., Fluss, L., Harter, W.K., Cansado, J., and Notario, V., 1992, Molecular Characterization of Quercetin and Quercetin Glycosides in Allium Vegetables: Their Effects on Malignant Cell Transformation, in "Phenolic Compounds in Food and Their Effects on Health II: Antioxidants & Cancer Prevention," M.T. Huang, C.-T. Ho, and C.Y. Lee, eds., American Chemical Society, Washington, D.C.

Michnovicz, J.J. and Bradlow, H.L., 1991, Altered estrogen metabolism and excretion in humans following consumption of indole-3-carbinol, Nutr. Cancer 16:59-66.

Miller, A.B, Howe, G.R., Jain, M., Craib, K.J.P. and Harrison, L., 1983, Food items and food groups as risk factors in a case-control study of diet and colo-rectal cancer, Int. J. Cancer 32:155-61.

Niedel, J.E., Kuhn, L.J., and Vandenbank, G.R., 1983, Phorbol diester receptor copurifies with protein kinase C, Proc. Natl. Acad. Sci. 80:36-40.

Olson, J.A., 1993, Atwater lecture: The irresistible fascination of carotenoids and vitamin A, Am. J. Clin. Nutr. 57:833-9.

Peters, R.K., Pike, M.C., Garabrant, D., Mack, T.M., 1992, Diet and colon cancer in Los Angeles County, California, Cancer Causes and Control 3:457-73.

Peto, R., Doll, R., Buckley, J.D. and Sporn, M.B., 1981, Can dietary -carotene materially reduce human cancer rates? Nature 290:201-208.

Pfander H., ed., 1987, Key to Carotenoids, 2nd edition, Birkhauser, Basel.

Prochaska, H.J., Santamaria, A.B., and Talalay, P., 1992, Rapid detection of inducers of enzymes that protect against carcinogens, Proc. Natl. Acad. Sci. 89:2394-2398.

Sempos C.T., Briefel R.R., Flegal K.M., Johnson C.L., Murphy R., Woteki C.E., 1992, Factors involved in selecting a dietary survey methodology for national nutrition surveys, Australian J Nutr and Dietetics; 49:96-104.

Serdula, M., Byers, T., Coates, R., Mokdad, A., Simoes, E.J., Eldridge, L., 1992, Assessing consumption of high-fat foods: the effect of grouping foods into single questions. Epidemiology 3:503-8.

Simic, M.G., 1992, Carotenoid free radicals, Methods in Enzymology 213:444-453.

Smith, A.F., Jobe, J.B., Mingay, D.J., 1991, Question-induced cognitive biases in reports of health-related behaviors, Health Psychol. 10:244-251.

Stavric, B., Matula, T.I., Klassen, R., Downie, R.H. and Wood, R.J., 1992, Effect of Flavonoids on Mutagenicity and Bioavailability of Xenobiotics in Foods, in "Phenolic Compounds in Food and Their Effects on Health II: Antioxidants & Cancer Prevention," M.-T. Huang, C.-T. Ho, and C.Y. Lee, eds., American Chemical Society, Washington, D.C.

Stoewsand, G.S., Anderson, J.L. and Munson, L., 1988, Protective effect of dietary Brussels sprouts against mammary carcinogenesis in Sprague-Dawley rats, Cancer Lett. 39:199-207.

Tajima, K. and Tominaga, S., 1985, Dietary habits and gastro-intestinal cancers: a comparative case-control study of stomach and large intestinal cancers in Nagoya, Japan, Jpn. J. Cancer Res. 76:705-16.

Telang, N.T., Suto, S., Wong, G.Y., Osborne, M.P., Bradlow, H.L., 1992, Induction by estrogen metabolite by 16-hydroxyestrone of genotoxic damage and aberrant proliferation in mouse mammary epithelial cells, JNCI 85:634638.

Tuyns, A.J., Kaaks, R. and Haelterman, M., 1988, Colorectal cancer and the consumption of foods: a case-control study in Belgium, Nutr. Cancer 11:189-204.

Wood, A.W., Smith, D.S., Chang, R.L., Huang, M.-T. and Conney, A.H., 1986, Effect of flavonoids on the metabolism of zenobiotics, in "Plant Flavonoids in Biology and Medicine: Biochemical, Pharmacological, and Structure-Activity Relationships," V. Cody, E. Middleton, and J.B. Harborne, eds., Alan R. Liss, Inc., New York.

Young, T.B. and Wolf, D.A., 1988, Case-control study of proximal and distal colon cancer and diet in Wisconsin, Int. J. Cancer 42:167-175.

Zhang, Y., Talalay, P., Cho, C.G. and Posner, G.H., 1992, A major inducer of anticarcinogenic protective enzymes from broccoli: isolation and elucidation of structure, Proc. Natl. Acad. Sci. 89:2399-2403.

DIETARY EFFECTS ON DNA METHYLATION: DO THEY ACCOUNT FOR THE HEPATOCARCINOGENIC PROPERTIES OF LIPOTROPE DEFICIENT DIETS?

Judith K. Christman

Molecular Oncology Program
Michigan Cancer Foundation
Detroit, MI 48201

INTRODUCTION

The first evidence suggesting that dietary deprivation of sources of one-carbon units could influence development of tumors is now almost fifty years old. In 1946, Copeland and Salmon reported that long-term feeding of a choline deficient diet increased the incidence of tumors in the liver and other organs of the rat. Once it was demonstrated that the peanut meal used in the diet was contaminated with aflatoxin (Newberne, 1965), it was assumed that the diet had in some way increased the sensitivity of cells to low levels of carcinogens. This supposition was supported by the demonstration that a variety of different diets deficient in choline and/or other lipotropes (methionine, folate and vitamin B_{12}) increased the tumorigenicity of a wide range of carcinogens in liver and other organs (Rogers, 1975, 1993; Rogers and Newberne, 1980; Rogers et al., 1974; Shinozuka et al., 1978a, b; Lombardi and Shinozuka, 1979). Unequivocal demonstration that deficiency of lipotropes alone was sufficient to cause liver tumors was obtained by feeding of an amino-acid defined (AAD) diet lacking methionine and choline but supplemented with folate, vitamin B_{12} and homocystine (Mikol et al., 1983). In the absence of added carcinogens, this diet produced a high incidence of hepatomas in rats, yet was reported to "afford a slight protection against spontaneous tumor formation in extra-hepatic tissues." These studies raised a number of questions that are yet to be resolved. Why is deficiency of lipotropes carcinogenic? Why is the liver the primary target tissue? Does moderate lipotrope deficiency play a significant role in development of human cancers?

A major difficulty involved in resolving these questions is the central metabolic role played by the lipotropes in maintaining the levels of both folate-derived one carbon groups and the primary methyl donor in transmethylation reactions, S-adenosylmethionine, AdoMet. Because of the interrelatedness of these one-carbon pathways, diets low in methionine and choline can lead to depletion of total folates and severe depletion of folate can lead to decreased levels of AdoMet and ratios of AdoMet/AdoHcy (Eto and Krumdieck, 1986; Horne et al., 1983; Krumdieck, 1983; Kutzbach et al., 1969). In turn, depending on the tissue, decreased levels of folates and/or AdoMet have the potential to interfere with *de novo* synthesis of purines and thymidylate (McMurray, 1983), to affect rates of synthesis of DNA, RNA (McMurray, 1983) and polyamines (Pegg and Hibasami, 1979), the frequency of genetic lesions in DNA (Loeb and Kunkel, 1982; Kuntz, 1982; Sutherland, 1988), the rate and extent of methylation of DNA, RNA, proteins (including histones, p21 ras and other membrane associated proteins and neurohumoral agents (Clarke, 1992; Trautner, 1984; Usdin et al., 1982) as well as the availability of methionine for protein synthesis (Cooper, 1983) and methionine and choline for phospholipid synthesis (Kennedy and Weiss, 1956; Bremer et al., 1960). Several hypotheses to account for the hepatocarcinogenic properties of lipotrope deficient diets emphasize effects of the rapid lipid accumulation observed in livers of rats. It has been proposed that tumor formation is initiated through damage to DNA caused by lipid peroxides (Hinrichsen et al., 1990; Perera et al., 1987; Rushmore et al., 1987). Promotion of initiated cells could then occur through the mitogenic stimulus provided by recurring cycles of hepatocyte death and replacement (Farber and Sarma, 1987; Giambarisi et al., 1982; Goshal et al., 1983). Increases in protein kinase C activity due to diradyl glycerol accumulation in hepatocytes could be important in stimulating cell division and altering gene activity (da Costa et al., 1993). Concurrent interference with DNA methylation due to reduced ratios of AdoMet or decreased levels of AdoMet/AdoHcy (Locker et al., 1986; Shivapurkar, N. and Poirier, L.A., 1983; Wainfan et al., 1989; Wilson et al., 1984) could allow enhanced transcription of genes involved in stimulation of cell growth or expression of a transformed phenotype. In this chapter, the potential role of alterations in DNA methylation as a contributing factor to the carcinogenic effects of dietary lipotrope deficiency will be discussed and evidence supporting this role will be presented.

THE IMPORTANCE OF GENETIC CHANGES IN TUMOR FORMATION

Current investigations on the genetic basis of tumorigenesis suggest that accumulation of multiple genetic defects underlies the progression of initiated cells toward malignancy (Fearon and Vogelstein, 1990). A wide array of gene mutations, rearrangements, deletions, translocations and amplifications have been noted to occur in tumor cell populations (Rowley, 1990). As has been detailed in several hundred reviews published in the last two years, these genomic changes have been associated with altered expression or activity of two classes of growth regulatory genes, proto-oncogenes and suppressor genes. In general, proto-oncogenes are dominant acting genes that encode proteins involved in transduction of positive growth signals (Bishop, 1983). These genes become activated either through mutations that affect the activity or level of their protein products or through mutations and/or translocation or amplification events that cause their normal products to be overproduced. In contrast, suppressor genes are defined as recessive genes whose products constrain unregulated cell growth or promote terminal differentiation. Thus, loss of production or function of the products of these genes, either through mutation of both alleles or through loss of the normal allele in cells that contain one normal and one mutated allele can contribute to tumor cell progression (Cavanee et al., 1983) . Mutations that lead to loss of production of a normal suppressor gene product can occur by gene deletion or by mutations or small deletions that prevent transcription or translation of mRNAs. Suppressor protein function can be lost not only by mutations in critical coding regions of the genes but through overproduction of proteins that bind to and inactivate normal suppressor proteins. These may include mutant homologues of the suppressor protein itself [Dominant-negative mutations (Herskowitz, 1987)].

ROLE OF DNA METHYLATION IN TUMOR DEVELOPMENT

One of the most consistent observations related to DNA methylation and tumor development is an overall decrease in the level of DNA methylation in tumor cells as compared to normal tissue (Gama-Sosa et al., 1983). In extensive studies of genomic changes during development of human colon cancer, it was noted that one of the earliest detectable molecular changes, occurring in small adenomas, involves loss of methyl groups from DNA (Goelz et al., 1985; Fearon and Vogelstein, 1990). However, superimposed on the generalized loss of methylation is an increase in methylation of specific gene regions (Makos et al., 1993) and an increase in the level of DNA methyltransferase (MTase), the enzyme that catalyzes transfer of methyl groups from AdoMet to carbon 5 of cytosine (C) residues in CpG dinucleotides (Christman et al., 1991; el-Diery et al., 1991). Three mechanisms have been proposed to link these alterations in DNA methylation with events occurring in tumor progression. First, 5mC residues in DNA could serve as hot spots for mutation since 5mC is more readily deaminated than C at physiological temperatures (Ehrlich et al., 1990). It has also been reported that a bacterial DNA MTase, HpaII, can deaminate of C residues that it would normally methylate when AdoMet and AdoHcy are absent from the reaction mixture (Shen et al., 1992). Although it remains to be demonstrated that sufficient reduction of AdoMet and AdoHcy levels can achieved in mammalian cells to lead to deamination of C residues by DNA MTase, deamination of C or 5mC in CpG sites could account for the high level C:G to A:T transition mutations observed at CpG sites in some genes (Jones et al., 1991). Second, since inhibition of methylation of C residues in DNA leads to impaired condensation of chromatin, it has been postulated that decreased levels of methylation in DNA may favor genomic instability due to chromosomal nondisjunction (Schmidt et al., 1984). Finally, alterations in DNA methylation may lead to heritable alterations in gene expression. For a number of tissue specific genes, extensive methylation of C residues in regulatory regions is correlated with decreased efficiency of transcription (Cedar, 1988; Christman, 1984; Doerfler, 1983) although there are a few genes where extensive methylation is associated with increased gene activity (Gellerson and Kempf, 1990; Tanaka et al., 1983) Cytosine methylation is thought to affect transcription by altering the interaction of transcription factors with DNA in regulatory regions or by altering the conformation of chromatin in such a way as to interfere with transcription. Methylation of regulatory sequences in DNA can either enhance or prevent binding of transcription factors or other proteins that recognize these sequences (Christman et al., 1992; Comb and Goodman, 1992; Iguchi-Ariga and Schaffner, 1989; Meehan et al., 1989; Watt and Malloy, 1988; Wang et al., 1986) and, depending on whether the proteins have a positive or negative effect on transcription, can either enhance or inhibit RNA production. Obviously, if proto-oncogenes or genes encoding inhibitors of the function of suppressor genes are transcribed more efficiently when they are less extensively methylated or if transcription of suppressor genes is blocked by methylation, stable alterations in methylation could lead to a tumorigenic phenotype in cells that have not suffered mutagenic alterations in the primary structure of DNA.

Despite the plausibility of these hypotheses, experimental evidence that disruption of normal tissue-specific patterns of DNA methylation can lead to development of cancer is indirect. It is known that, in addition to their mutagenic affects, many carcinogens block DNA methylation either by direct alkylation of DNA or by forming adducts with DNA that interfere with its ability to serve as a substrate for DNA MTase (Cox and Irving, 1977; Wilson and Jones, 1983). Hypomethylation of specific sites in proto-oncogenes such as c-myc, c-Ki-ras and c-Ha-ras has been detected in a variety of human and animal tumors and increased expression of one or more of these genes in tumor tissue is a common finding (Bhave et al., 1988; Cheah et al., 1984; Feinberg and Vogelstein, 1983; Nambu et al., 1987; Rao et al., 1989). Finally, feeding of inhibitors of methylation such as 5-azacytidine or L-ethionine induces formation of preneoplastic lesions and tumors in liver and other organs (Carr et al., 1983; Farber, 1963).

DOES LIPOTROPE DEFICIENCY INDUCE STABLE CHANGES IN HEPATO-CYTES?

A useful criterion for identifying changes necessary for oncogenic transformation of normal cells is to determine whether changes induced by treatment with a given carcinogen persist and are present in tumors or preneoplastic lesions that form after cessation of tumor-inducing treatments (Bannasch, 1986). There are several reports suggesting that lipotrope deficiency can cause change(s) in rat liver that are critical to tumor development and that are not reversed when normal levels of lipotropes are restored to the diet. Although the minimum period of lipotrope deficiency required for liver tumor induction in rats has yet to be determined, it has been shown that 15 weeks feeding of an AAD diet supplemented with homocystine but lacking choline, methionine and vitamin B_{12} (MDD) was sufficient to lead to development of neoplastic nodules in the liver which persisted and grew in size during 37 weeks subsequent feeding of an adequate diet (Hoover et al., 1984). Similar persistence of hyperplastic nodules was noted at one year in livers of rats fed a different choline-deficient AAD diet containing L-methionine, vitamin B_{12} and folate for 26 weeks (Nakae et al., 1992) and at 16 months in rats fed a purified choline-devoid diet for 3 months (Chandar and Lombardi, 1988). The results of the latter investigation suggested that 3 months of choline deprivation was sufficient to induce carcinomas in ~10% of the rats.

Table 1 summarizes some of the changes that have been observed in rat liver after short-term feeding methionine and choline deficient AAD diet with (SD) or without (MDD) folic acid and vitamin B12 (Shivapurkar and Poirier, 1983). Many of these effects are common

Table 1. Effects of short-term feeding of an amino-acid defined lipotrope deficient diet on rat liver

Effect	Onset	References
Increased rate of DNA synthesis	(≤1 week)	Wainfan et al., 1989
Increased numbers of mitotic cells	(≤1 week)	Christman et al., 1993b Christman et al., 1993b
	(<6 weeks)	Hoover et al., 1984
Increased activity of DNA MTase	(1-2 weeks)	Christman et al., 1991, 1993a.
Increased activity of tRNA MTase	(1-2 weeks)	Christman et al., 1991, 1993a. Wainfan et al., 1984, 1988
Increased levels of specific mRNAs	(≤1 week)	Dizik et al., 1991
Decreased levels of AdoMet or decreased ratio of AdoMet/AdoHcy	(≤1 week)	Shivapurkar and Poirier 1983
Decreased levels of specific mRNAs	(≤1 week)	Dizik et al., 1991
Decreased methylation of DNA	(1 week)	Wainfan et al., 1989
Decreased methylation of tRNA	(5 days)	Wainfan et al., 1986, 1988, 1989.

Diets are defined in Shivapurkar and Poirier, 1983 and Wainfan et al., 1989.

to lipotrope deficiencies induced by feeding of a variety of other choline or choline and methionine deficient diets (Ghoshal and Farber, 1984; Newberne and Rogers, 1980; Yokoyama et al., 1985; Rogers and MacDonald, 1965). All of these changes persist and increase in magnitude during the first month of lipotrope deficiency. However, with more prolonged dietary lipotrope deficiency (>3 months), some changes in the liver, including elevation in triglyceride content and increased rate of liver cell proliferation, slowly diminish (Chandar and Lombardi, 1988; da Costa et al., 1993). AdoMet/AdoHcy ratios return to normal, although this is primarily due to a continued decrease in the level of AdoHcy rather than an increase in AdoMet levels in the liver (Wilson et al., 1984). In contrast, the 5mC content of liver DNA remains low (Wilson et al., 1984) and the same alterations in the pattern of methylation of genes such as c-*myc* and c-*Ha-ras* that are observed within 1 month persist in the hepatocellular carcinomas that arise many months later (Bhave et al., 1988). This suggests that alterations in DNA methylation could play an important role in hepatocarcinogenesis and has led us to investigate whether alterations in DNA methylation also persist when adequate lipotropes are restored to the diet of lipotrope deficient rats.

WHICH EFFECTS OF LIPOTROPE DEFICIENCY ARE REVERSIBLE?

To determine whether any of the effects we have observed after short-term feeding of MDD persist after restoration of adequate levels of lipotropes to the diet, rats were fed MDD for 4 weeks and then fed a diet adequate in lipotropes [CSD, the same basal AAD diet as MDD but supplemented with 2 g/kg choline chloride, 5.2 g/kg D,L methionine, 5 mg/kg folic acid, and 100 µg/kg vitamin B_{12}]. Livers were examined histologically and biochemically at regular intervals after making this switch. Feeding of CSD causes reversal of many of the effects of lipotrope deficiency within a few weeks, suggesting that simple replenishment of C1 pools is sufficient to restore most normal hepatocyte functions (**Table 2**). Levels of mRNA for all of the genes we have studied to date [c-*myc*, c-*Ha-ras*, c-*fos*

Table 2. Reversibility of the effects of short-term feeding of an amino acid-defined lipotrope-deficient diet

Effect	Time to Reach Control Level*
Increased rate of DNA synthesis	(≤1 week)
Increased numbers of mitotic cells	(≤1 week)
Increased lipid accumulation	(8-10 weeks)
Increased activity of DNA MTase	(1-2 weeks)
Increased activity of tRNA MTase	(1-2 weeks)
Increased (or decreased) levels	(1-3 weeks)
of specific mRNAs	
Decreased methylation of DNA and tRNA	(1-2 weeks)
(Ability to act as a substrate	
for *in vitro* methylation-see text)	
Decreased methylation of CCGG sites in	(>12 weeks-
specific genes	<10 months)

*All parameters are compared with those in livers from age-matched animals fed a lipotrope-sufficient diet. Data are presented in Christman et al., 1993a,b; Chen et al., 1993; Wainfan and Poirier, 1992).

and EGFR] returned to control levels within 1-3 weeks as did levels of active DNA and t-RNA methyltransferases. The rate of DNA synthesis as monitored by BrdUrd accumulation in nuclei of hepatic cells dropped precipitously within 24-48 hours and by 1 week was indistinguishable from that found in livers of age matched rats continually fed CSD. The overall extent of methylation of tRNA in the liver, as measured by ability to accept methyl groups in vitro, was also restored to normal within 1-2 weeks. Similar results were obtained when methyl acceptance of DNAs was measured. It should be noted, however, that mammalian DNA MTase was used to catalyze methyl transfer in this highly sensitive assay for quantitating unmethylated sites in DNA. Since mammalian DNA methyltransferase is at least 40-fold more active in methylating hemi-methylated sites than completely unmethylated sites (Bestor and Ingram, 1983), what is actually measured is the capacity of the liver cells to maintain methylation of DNA at hemi-methylated sites created during DNA replication, not the overall of 5mC content of DNA. Thus, the finding that within one week of restoring lipotropes to the diet (CSD) of rats previously fed MDD the methyl acceptance of liver DNA is indistinguishable from that of liver DNA from rats continually fed CSD indicates that AdoMet and/or the ratio of AdoMet/AdoHcy rapidly rebound to levels that allow efficient maintenance of methylation of newly synthesized DNA.

Only two of the effects of lipotrope deficiency that we have studied persist for more than a month after restoration of adequate levels of lipotropes to the diet. There is persistence of liver cells containing large amounts of lipid and specific hypomethylated CCGG sites in liver cell DNA. Within a few days of restoring lipotropes to the diet, areas of normal appearing hepatocytes are present in the periductular regions of the liver. After 2-4 weeks, these areas of well organized parenchyma have enlarged, but lipid-filled cells (LFC) still remain throughout zone 2 and around the central veins **(Figure 1)**. Unlike the LFC present in all regions of the liver during MDD feeding, these cells fail to incorporate BrdUrd

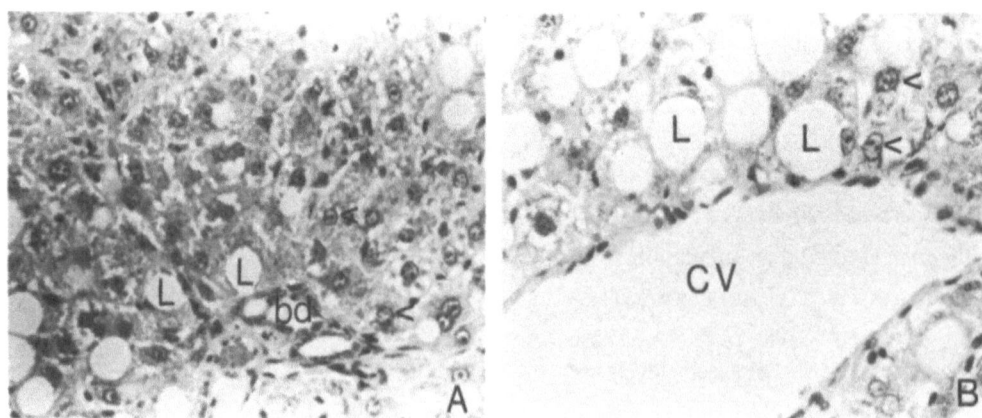

Figure 1. Persistence of lipid filled cells (L) in livers of rats fed a lipotrope deficient diet (MDD) for four weeks followed by feeding of a diet containing adequate levels of lipotropes (CSD).
A. A typical view of a periductular area in the liver two weeks after restoration of lipotropes to the diet. Some lipid filled cells are still present but more than half the field contains parenchymal cells with relatively normal morphology. The bile duct is indicated (bd).
B. A typical area showing the even distribution of lipid filled cells around a central vein (CV). Note the difference in nuclear size in parenchymal cells in A and B (arrows). Magnification in both panels is 400X. Fresh liver samples were fixed in neutral buffered formalin for 24 h. prior to embedding. Sections were prepared for staining with hematoxylin-eosin using standard methods.

into DNA after a 2 h pulse. This suggests that at least some hepatocytes become resistant to the mitogenic effects of lipid accumulation and have lost the capacity to secrete VLDL even when normal lipotrope levels have been restored. (Chen and Christman, unpublished observation). Further investigation will be needed to determine whether the distribution of LFC reflects the pattern of replacement of hepatocytes progressing from the periductular to the central vein areas or a gradient of inter-cellular C1 pools. In either case, after 8-12 weeks of CSD feeding, the percentage of LFC is low and no greater than in age-matched animals fed CDS continuously.

Specific loss of methylation at CCGG sites in liver DNA resulting from feeding of MDD is even more persistent than lipid accumulation and is detectable for at least four months after subsequent feeding of CSD to rats. Southern blot analysis of HpaII sites in the c-*myc* gene is shown as an example (**Figure 2, Panel A**). We have noted similarly persistent loss of methylation at CCGG sites in c-*Ha-ras*, c-*fos* and p*53* genes (Christman et al., 1993a, 1993b, and unpublished). After 10 months, however, DNA from livers of rats previously fed MDD for 4 weeks had essentially the same restriction fragment pattern when

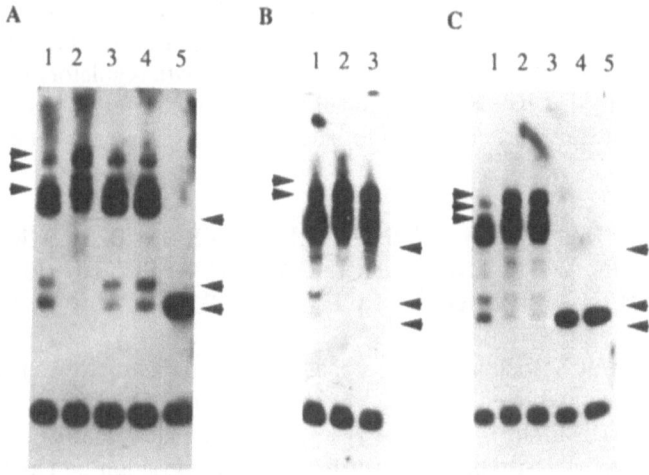

Figure 2. Effect of dietary lipotrope deficiency and subsequent feeding of an adequate diet on methylation of CCGG sites in the c-*myc* gene. Southern blot analysis of HpaII and MspI digested DNAs from rats fed:

Panel A: Lane 1, MDD, 4 weeks; Lane 2, CSD, 10 weeks; Lane 3, MDD, 4 weeks, CSD, 9 weeks; Lane 4,5 MDD 4 weeks, CSD, 12 weeks. DNA in Lanes 1-4 was cut with HpaII which is unable to cleave DNA at CCGG sites when the C residue 5' to G is methylated DNA in Lane 5 was cut with MspI which can cleave at CGG sites regardless of whether this C residue is methylated or not.

Left pointing arrows indicate the position of HpaII fragments whose concentration, relative to those present in HpaII digests of DNA from rats fed CSD (Lane 2), is increased during feeding of MDD and subsequent recovery. Right pointing arrows indicate the position of HpaII fragments whose concentration is decreased during feeding of MDD and subsequent recovery.

Panel B: Lane 1, MDD, 4 weeks; Lane 2, CSD, 11 months; Lane 3, MDD, 4 weeks, CSD 10 months. All DNAs were cut with HpaII. The HpaII cleavage pattern of DNA in Lanes 2,3 is identical suggesting that the normal pattern of liver DNA methylation has been restored by 10 months feeding of CSD after an initial 4 week period of lipotrope deprivation.

Panel C: Lane 1, MDD, 4 weeks; Lane 2, CSD 4 weeks, Lane 3, CSD -folate, 5 weeks. DNA in lanes 1-3 were cut with HpaII; DNA in lanes 4,5 with MspI. The HpaII cleavage pattern of DNA in Lanes 2, 3 is identical and suggests that deprivation of folate alone is not sufficient to cause loss of methylation.

All experimental details are as described in Christman et al., 1993a.

cut with HpaII as DNA from livers of age matched rats fed CSD continually (**Figure 2, Panel B**). We are in the process of determining whether this reversion to the pattern of methylation found in normal liver occurs because *de novo* methylation of completely unmethylated sites in liver DNA occurs at a slow but finite rate or whether "remethylation" is the result of replacement of the majority of parenchymal cells containing hypomethylated DNA by stem cells that have retained or regained a normal pattern of methylation.

Our working hypothesis is that persistent changes resulting from short-term dietary methyl-deficiency are more likely to be neoplastic or preneoplastic in nature than those that are rapidly reversed. Loss of methylation at CCGG sites is persistent and has, as argued above, the potential to lead to alterations in gene expression that could contribute to the promoting and carcinogenic potential of lipotrope deficient diets. Our data is consistent with the proposal that persistent hypomethylation of c-*myc*, c-*fos* and c-*Ha-ras* may be required for the sustained increase in their transcript levels observed in the liver during MDD feeding (Dizik et al., 1991). However, loss of methylation is not sufficient by itself to maintain expression of these genes once lipotropes are restored to the diet and the liver cells become quiescent (Christman et al., 1993a). This suggests that active transcription factors needed for efficient expression of these genes may only be present in proliferating cells but leaves open the question of how persistent hypomethylation in specific genes affects the behavior of hepatocytes. Evidence that hypomethylated growth regulatory genes in the liver are more readily activated by mitogenic stimuli or low doses of carcinogen than the same genes when methylated in patterns normal for adult hepatocytes has not yet been obtained.

DOES THE MECHANISM BY WHICH CHANGES IN DNA METHYLATION OCCUR EXPLAIN WHY THE CARCINOGENIC EFFECTS OF LIPOTROPE DEFICIENCY OCCUR PRIMARILY IN THE LIVER?

A general scheme accounting for how changes in patterns of DNA methylation occur during cell division is presented in **Figure 3**. In the absence of repair-like processes that remove 5mC from DNA (Jost, 1993), it is clear that complete loss of methylation at a specific site in cellular DNA requires that maintenance methylation be blocked through at least two rounds of DNA synthesis and cell division. In the case of lipotrope deficiency, we have shown that levels of active DNA MTase actually increase (Christman et al., 1991; 1993a). Thus, it is most likely that alterations in AdoMet levels or increased ratios of AdoHcy to AdoMet are responsible for inhibiting maintenance methylation. However, for complete loss of methylation at a specific site, it is necessary not only to deplete cells of AdoMet and/or increase the level of AdoHcy relative to AdoMet but to have DNA synthesis and cell division occurring at the same time. In this regard, the liver is unique in its response to lipotrope deficiency. Although lipotrope deficiency does not cause significant changes in AdoMet/AdoHcy ratios or levels in kidney, pancreas or brain (Eloranta, 1977; Shivapukar and Poirier, 1983), it can mediate decreases in AdoMet/AdoHcy in lung and testes that are comparable to those observed in liver (Shivapukar and Poirier, 1983). However, cells in other organs do not accumulate massive amounts of fatty acids nor are they stimulated to divide. The negative effect of folate deficiency on hematopoiesis and DNA replication in lymphocytes is well known (Antony, A., 1990; Babior, 1990; Wickremasinghe and Hoffbrand, 1980). In the course of determining the effects of MDD feeding on DNA synthesis in the liver, we have examined the extent to which BrdUrd is incorporated into nuclei of cells in other tissues and have found that 4 weeks of MDD feeding tends to suppress rather than stimulate normal levels of DNA synthesis in the testes, kidney, lung and colon (Chen and Christman, unpublished). Thus, at least in some tissues, decreased levels of methyl donors may not lead to decreased methylation of DNA because lipotrope deficiency also inhibits DNA synthesis.

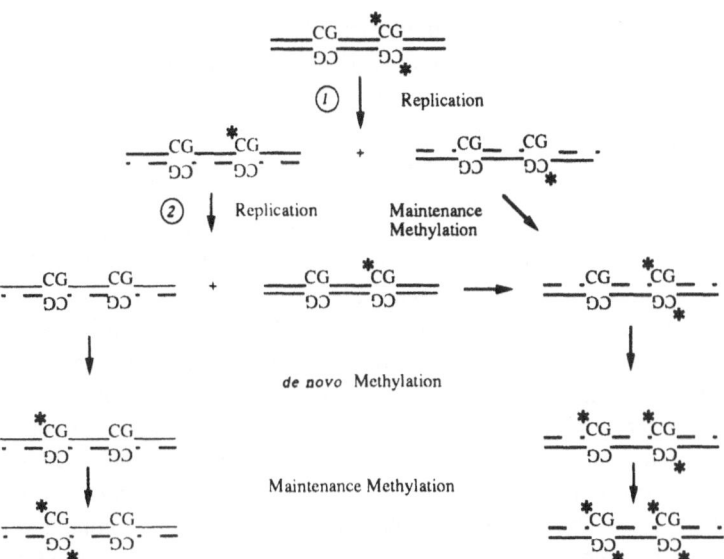

Figure 3. Diagrammatic representation of how DNA methylation patterns can be maintained or altered during DNA replication. Note that two rounds of replication without maintenance methylation are needed to convert a completely methylated site to a completely unmethylated site in 50% of the DNA products. It is not known whether DNA replication is required for *de novo* methylation. *C=5mC

Availability of phosphatidylcholine in the liver regulates secretion of very low density lipoproteins (Yao and Vance, 1988). Since the majority of plasma lipoproteins are synthesized in the liver and phosphatidyl choline accounts for ~75% of the phospholipid in rat lipoproteins (Wu and Windmueller, 1979; Vance and Vance, 1985), it is not surprising that depletion of liver choline leads to rapid and massive accumulation of triglycerides in the liver and not in other tissues. There is a close temporal parallel between triglyceride accumulation and stimulation of DNA synthesis (Chandar et al., 1987; Rogers and MacDonald, 1965) and it has been proposed that cellular necrosis resulting from this accumulation of fat leads to a cycle of persistent liver-cell death and regeneration (Giambarisi et al., 1982; Goshal et al., 1983). Depletion of lipotropes and/or triglyceride accumulation may also have mitogenic effects that are independent of liver damage during the early stages of deficiency, since stimulation of DNA synthesis seems to precede obvious necrosis (Christman et al., 1993b) and the rate of liver cell growth during the first 8 weeks of feeding of a choline deficient diet is greater than needed to replace dying cells (Chandar et al., 1987).

Thus, if changes in DNA methylation are a necessary component in the development of tumors, probability of tumor formation should be highest in liver where increased rates of DNA synthesis and decreased levels of C1 pools occur together. In a like manner, the promotion of carcinogen induced tumors by lipotrope deficiency may depend both on the extent of C1 depletion and the extent to which DNA synthesis is stimulated during repair of carcinogen induced damage or the extent to which initiated cells continue to divide in tissues where normal cells are quiescent.

DOES MODERATE LIPOTROPE DEFICIENCY AFFECT DNA METHYLATION AND INDUCE TUMOR DEVELOPMENT?

Diets as deficient in lipotropes as those employed in experimental studies are unlikely to occur in nature. MDD is so stringent in its effects that rats fed this diet die at ~15 weeks

(Mikol et al., 1983). The diet must be supplemented with folate and vitamin B_{12} if the rats are to survive long enough to develop tumors. However, as noted above, deficiency in any one lipotrope can affect C1 pools. A recent report indicates that rapid hypomethylation of DNA occurs on feeding of MDD supplemented with folate and vitamin B_{12} suggesting that these lipotropes are not sufficient to allow effective maintenance of DNA methylation (Zapisek et al., 1992). Our own studies indicate the effects of 4 weeks feeding of SD and MDD on rat liver are identical with respect to lipid accumulation, mitogenic stimulus of hepatocytes and rate of appearance of hypomethylated CCGG sites in DNA (Chen et al., 1993, ms. in preparation). In contrast, when rats are fed CSD with folate omitted but all other lipotropes present, no change in methylation of CCGG sites can be detected in liver DNA, even after 5 weeks of feeding (**Figure 2, Panel C**). Similar results, i.e., lack of alteration in pattern of methylation of CCGG sites in the c-*myc* gene in liver and colon, have been obtained after omitting folate from an otherwise lipotrope adequate amino acid defined diet for 24 weeks (Kim et al., 1994, ms. in preparation). This diet, which does not contain sulfonamide to prevent utilization of folate produced by gut flora, creates a moderate folate deficiency and enhances development of dimethylhydrazine induced colon cancer but is not sufficient to cause colorectal neoplasia (Cravo et al., 1992). The simplest interpretation of these findings is that changes in DNA methylation are not important for the effects of folate deficiency on tumor formation. However, if loss of methylation at specific sites in liver DNA requires several rounds of DNA replication under conditions of methyl donor deficiency (**Figure 1**), the possibility remains that the number of cells containing DNA with unmethylated sites is too small to detected in DNA extracted from whole liver. Other apparently discrepant results, such as such as the finding that extreme folate deficiency delays rather than enhances formation of neural sheath tumors in transgenic mice (Bills et al., 1992) and the direct correlation found between dietary folate content and incidence of chemically induced mammary tumors in rats (Baggot et al., 1992) may simply reflect a greater effect of folate deficiency on rate of DNA synthesis than on efficiency of maintenance of DNA methylation.

SUMMARY AND PERSPECTIVE

It is improbable that any common human diet could cause a lipotrope deficiency comparable to that induced by the experimental diets we and others have used in animal studies. However, if unmethylated sites in specific genes are produced by exposure to intermittent or moderate dietary methyl deficiency and if remethylation of these sites are as resistant to *de novo* methylation as the unmethylated sites CCGG sites in liver DNA detected in our studies, it is certainly possible that cumulative hypomethylation of genes could contribute to the causation of cancer in humans. In many parts of the world diets that are marginally sufficient in lipotropes are common (Eto and Krumdieck, 1986). In addition, levels of folate may be reduced by excess consumption of alcohol or administration of therapeutic drugs (Porta et al., 1985; Selhub et al., 1991). There is a significantly increased risk for development of leukemia and other cancers in patients recovering from pernicious anemia (Brinton et al., 1989) and deficiency in folate has been epidemiologically linked to development of colorectal, breast, and other dysplasias and cancers in humans (Branda et al., 1991; Butterworth et al., 1982; Giovannucci et al., 1993; Heimberger et al., 1988; Lashner et al., 1989). If persistent changes in DNA methylation are ultimately proven to play a role in development of these cancers, our primary challenges will be to prevent dietary deficiencies in folate through recommendations for consumption of folate rich foods or folate supplements and to develop methods for enhancing re-establishment of normal patterns of DNA methylation in affected tissues.

ACKNOWLEDGMENTS

I thank Dr. Mei-Ling Chen and Dr. Elsie Wainfan for allowing discussion of their unpublished data. Their contributions, along with those Drs. Gholamreza Sheikhnejad, and Mark Dizik and Ms. Susana Abileah, have made this work possible. Our research was supported in part by grants from the American Institute for Cancer Research numbers 92A35 to J.K.C. and 864 to E.W., and the Lloyd and Marilyn Smith Fund.

REFERENCES

Antony, A, 1991, Megaloblastic anemias, in "Hematology: Basic Principles and Practice," Hoffman R, Benz, E.J, Shattil, S.J, Furie, B., Cohen, H.J., eds., Churchill Livingston, New York.

Babior, B.M., 1990, The megaloblastic anemias, in "Hematology (ed 4)," W.J. Williams, E. Beutler, A.J. Erslev, M.A. Lichtman, eds., McGraw-Hill, New York, NY.

Baggott, J.E., Vaughn, W.H., Juliana, M.M. eto, I., Krumdieck, C.L. and Grubbs, C.J., 1992, Effects of folate deficiency and supplementation on methylnitrosourea-induced rat mammary tumors, J. Natl. Cancer Inst. 84:1740.

Bannasch, P., 1986, Preneoplastic lesions as end points in carcinogenicity testing. I. Hepatic preneoplasia, Carcinogenesis 7:689.

Bestor, T.H. and Ingram, V.M., 1983, Two DNA methyltransferases from murine erythroleukemia cells: purification and sequence specificity, Proc. Natl. Acad. Sci. U.S.A. 80:5559.

Bhave, M.R., Wilson, M.F., and Poirier, L.A., 1988, c-H-*ras* and c-K-*ras* gene hypomethylation in the livers and hepatomas of rats fed methyl-deficient, amino acid-defined diets, Carcinogenesis 9:343.

Bills, N.D., Hirichs, S.H., Morgan, R., and Cliff, A.J., 1992, Delayed tumor onset in transgenic mice fed a low-folate diet, J. Natl. Cancer Inst. 84:333.

Bishop, J.M., 1983, Cellular oncogenes and retroviruses, Annu. Rev. Biochem. 52:301.

Branda, R. F., O'Neill, J.P., Sullivan, L. M., and Albertini, R. J., 1991, Factors influencing mutation at the *hprt* locus in T-Lymphocytes: Women treated for breast cancer, Cancer Res. 51: 6603.

Bremer, J., Figard, P.H., and Greenberg, D.M., 1960, The biosynthesis of choline and its relation to phospholipid metabolism, Biochim. Biophys. Acta. 43: 477.

Brinton, L.A., Gridley, G., Hrubec, Z., Hoover, R., and Fraumeni, J.F., 1989, Cancer risk following pernicious anemia, Br. J. Cancer 59:810.

Butterworth, C.E., Hatch, K.D., Gore, H., Meubler, H., and Krumdieck, C.L. 1982, Improvement in cervical displasia with folic acid therapy in users of oral contraceptives, Am. J. Clin. Nutr., 35:73.

Carr, B.I., Reilly, J.G., Smith, S.S., Winberg, C., and Riggs, A., 1984, The tumorigenicity of 5-azacytidine in the male Fischer rat, Carcinogenesis 5:1583.

Cavenee, W.K., Dryja, T.P., Phillips, R.A., Benedict, W.F., Godbout, R., Gallic, B.L., Murphree, A., L., Strong, L.C., and White, R.L., 1983, Expression of recessive alleles by chromosomal mechansims in retinoblastoma, Nature 305:779.

Cedar, H., 1988, DNA methylation and gene activity, Cell 34:5503.

Chandar, N., and Lombardi, B., 1988, Liver cell proliferation and incidence of hepatocellular carcinomas in rats fed consecutively a choline-devoid and a choline-supplemented diet, Carcinogenesis 2:259.

Chandar, N., Amenta, J., Kandala, J.C., and Lombardi, B., 1987, Liver cell turnover in rats fed a choline-devoid diet, Carcinogenesis 8:669.

Cheah, M.S.C., Wallace, C.D., Hoffman, R.M., 1984, Hypomethylation of DNA in human cancer cells: A site-specific change in the c-*myc* oncogene, J.Natl.Cancer Inst. 73:1057.

Christman, J.K., Sheikhnejad, G., Dizik, M., Abileah, S., and Wainfan, E. 1993a Reversibility of changes in nucleic acid methylation and gene expression induced in rat liver by severe dietary methyl deficiency, Carcinogenesis 14:551.

Christman, J.K., Chen, M-L, Sheikhnejad, G., Dizik, M., Abileah, S., and Wainfan, E., 1993b, Methyl deficiency, DNA methylation, and cancer: Studies on the reversibility of the effects of lipotrope-deficient diet, J. Nutr. Biochem. 4:672.

Christman, J.K., Chen, L., Nicholson, R., and Xu, M., 1993, DNA Methylation: A cellular strategy for regulating expression of integrated hepatitis B virus genes? in "Virus Strategies: Molecular Biology and Pathogenesis," W. Doerfler, and P. Bohm, eds., VCH Verlagsgesellschaft, Weinheim, New York.

Christman, J.K., 1984, DNA methylation in Friend Erythroleukemia Cells: The effects of chemically induced differentiation and of treatment with inhibitors of DNA methylation, Current Topics in Microbiol. and Immunol. 108: 49.

Christman, J.K., Chen, L., Sheikhnejad, G., Dizik, M., and Wainfan, E., 1991, Regulation of gene expression by DNA methylation: a link between dietary methyl deficiency and hepatocarcinogenesis, in "The Role of Nutrients in Cancer Treatment," A. Roche, ed., Report of the 9th Ross Conference on Medical Res., Columbus.

Clarke, S., 1992, Protein isoprenylation and methylation at carboxy-terminal cysteine residues, Annu. Rev. Biochem. 61: 131.

Comb, M. and Goodman, H.W. , 1990, CpG methylation inhibits proenkephalin gene expression and binding of the transcription factor AP-2, Nucleic Acids Res.18: 3975.

Cooper, A. J. L., 1983, Biochemistry of sulfur-containing amino acids, Annu. Rev. Biochem. 52:187.

Copeland, D.H., and Salmon, W.D., 1946, The occurrence of neoplasms in the liver, lungs, and other tissues of rats as a resuof prolonged choline deficiency, Am. J. Path. 22: 1059.

Cox, R.D., and Irving, C.C.,1977, Inhibition of DNA synthesis by S-adenosyl-homocysteine with the production of methyl-deficient DNA in regenerating liver, Cancer Res. 37: 222.

Cravo, M.L., Mason, J.B., Dayal, Y., Hutchinson, M. Smith, D., Selhub, J., and Rosenberg, I. H., 1992, Folate deficiency enhances the development of colonic neoplasia in dimethylhydrazine-treated rats, Cancer Res. 52:5002.

da Costa, K, Cochary, E.F., Busztajn, J.K., Garner, S.C., and Zeisel, S. H., 1993, Accumulation of 1,2-*sn*-diradylglycerol with increased membrane-associated protein kinase C may be the mechanism for spontaneous hepatocarcinogenesis in choline deficient rats, J. Biol. Chem. 268: 2100.

Dizik, M., Christman, J.K., and Wainfan, E., 1991, Alterations in expression and methylation of specific genes in livers of rats fed a cancer promoting methyl-deficient diet, Carcinogenesis 12:1307.

Doerfler,W. A., 1983, DNA methylation and gene activity. Annu. Rev. Biochem.52:93 .

el-Deiry, W.S., Nelkin, B.D., Celano, P., Yen, R.W., Falco, J.P., Hamilton, S.R., and Baylin, S.B., 1991, High expression of the DNA methyltransferase gene characterizes human neoplastic cells and progression stages of colon cancer, Proc. Natl. Acad. Sci. U.S.A. 88: 3470.

Erhlich, M., Zhang, X.Y., and Inamdar, N.M., 1990, Spontaneous deamination of cytosine and 5-methylcytosine residues in DNA and replacement of 5-methylcytosine residues and cytosine residues, Mutat. Res. 238:277.

Eto, I. and Krumdieck, C.L.,1986, Role of vitamin B_{12} and folate deficiencies in carcinogenesis, Adv. Exptl. Biol. 206:313.

Eloranta, T.O., 1977, Tissue distribution of S-adenosylmethionine and S-adenosylhomocysteine in the rat, effect of age, sex and methionine administration on the metabolism of S-adenosylmethionine, S-adenosylhomocysteine and polyamines, Biochem. J. 166:521.

Farber, E. and Sarma, D.S.R., 1987, Biology of disease. Hepatocarcinogenesis: A dynamic cellular prospective, Lab. Invest. 56: 4-22.

Farber, E., 1963 ethionine carcinogenesis, Adv. Cancer Res. 7: 380.

Fearon, E.R., and Vogelstein, B. A., 1990, Genetic model for colorectal tumorigenesis, Cell 61:759.

Feinberg, A., and Vogelstein, B., 1983, Hypomethylation of *ras* oncogenes in primary human cancers, Biochem. Biophys. Res. Commun. 111:47.

Gama-Sosa, M.A., Slagle, V.A., Trewyn, R.W., Oxenhandler, R., Kuo, K.C., Gehrke, C.W., and Ehrlich, M., 1983, The 5-methylcytosine content of DNA from human tumors, Nucleic Acids Res. 11: 6883.

Gellerson, B. and Kempf, R., 1990, Human prolactin gene expression: positive correlation between site-specific methylation and gene activity in a set of human lymphoid cell lines, Mol. Endocrinol. 4: 1874 .

Giambarresi, L. K., Katyal, S. L., and Lombardi, B., 1982, Promotion of liver carcinogenesis in the rat by a choline-devoid diet: role of liver cell necrosis and regeneration, Br.J. Cancer 46:825.

Giovannucci, E., Stampfer, M.J. , Golditz, G.A., Rimm, E.B., Trichopoulos, D., Rosner, B.A., Speizer, F.E., and Willett, W.C., 1993, Folate, methionine, and alcohol intake and risk of colorectal adenoma, J. Natl. Can. Inst. 85, 11:875.

Goshal, A.K., Ahluwalia, M., and Farber, E., 1983, The rapid induction of liver cell death in rats fed a choline-deficient methionine-low diet, Am. J. Pathol. 113: 309.

Goshal, A.K., and Farber, E., 1984, The induction of liver cancer by a dietary deficiency of choline and methionine without added carcinogen, Carcinogenesis 5:1367.

Heimburger, D.C., Alexander, B., Birch, R., Butterworth, C.E., Bailey, W.C. and Krumdieck, C. L., 1988, Improvements in bronchial squamous metaplasia in smokers treated with folate and vitamin B12, J.Am. Med. Assoc. 259:1525.

Hinrichsen, L.I., Floyd, R.A., and Sudilovsky, O. 1990 Is 8-hydroxydeoxyguanosine a mediator of carcinogenesis by a choline-devoid diet in the rat liver? Carcinogenesis 11: 1879.

Herskowitz, I., 1987, Functional inactivation of genes by dominant negative mutations, Nature 329:219.

Hoover, K. K., Lynch, P.H., and Poirier, L. A., 1984, Profound postinitiation enhancement by short-term severe methionine, choline, vitamin B_{12} and folate deficiency of hepatocarcinogenesis in F344 rats given a single low-dose diethylnitrosamine injection, J.Natl. Cancer Inst. 73:1327.

Horne, D.E., Cook, R.J., and Wagner, C., 1989, Effect of dietary methyl group deficiency on folate metabolism in rats, J. Nutr. 119: 618.

Iguchi-Ariga, S.M., and Schaffner, W., 1989, CpG methylation of the cAMP-responsive enhancer/promoter sequence TGACGTCA abolishes selective factor binding as well as transcriptional activation, Genes Dev. 3:612 .

Jones, P.A., Buckely, J.D., Henderson, B.E., Ross, R.K,. and Pike, M.C., 1991, From gene to carcinogen: a rapidly evolving field in molecular epidemiology, Cancer Res. 51, 13:3617.

Jost, J-P., 1993, Nuclear extracts of chicken embryos promote an active demethylation of DNA by excision repair of 5-methyldeoxycytidine, Proc. Natl. Acad. Sci. U.S.A. 90:4684.

Kennedy, E.P., and Weiss S.B., 1956, The function of cytidine coenzymes in the biosynthesis of phospholipids, J. Biol. Chem. 222:193.

Kim, Y.I., Christman, J.K., Fleet, J.C., Cravo, M.L., Salomon, R.N., and Mason, J.B., 1994, Is global and gene specific DNA hypomethylation a mechanism by which folate deficiency enhances colorectal carcinogenesis? Abs. Amer. Gastro. Assoc. /Amer. Assoc. for the Study of Liver Diseases. New Orleans, Louisiana.

Krumdieck, C.L., 1983, Role of folate deficiency in carcinogenesis., in "Nutritional Factors in the Induction and Maintenance of Malignancy," Academic Press, New York.

Kuntz, B.A., 1982, Genetic effects of deoxyribonuceotide pool imbalances, Environ. Mutagen. 4: 695.

Kutzbach, C., Galloway, E., and Stokested, E.I., 1969, Influence of vitamin B12 and methionine on levels of folic acid compounds and folate enzymes in rat liver, Proc. Soc. Exp. Biol. Med. 124: 801.

Lashner, B.A., Heidenreich, P.A., Su, G.L., Kane, S.V., and Hanauer, S.B., 1989, Effect of folate supplementation on the incidence of dysplasia and cancer in chronic ulcerative colitis. A case - control study, Gastroenterology 97:255.

Locker, J., Reddy, T.V., and Lombardi, B., 1986, DNA methylation and hepatocarcinogenesis in rats fed a choline-devoid diet. Carcinogenesis 7:1309.

Loeb, L.A., and T.A. Kunkel, 1982, Fidelity of DNA synthesis, Annu. Rev. Biochem. 52: 429.

Lombardi, B., and Shinozuka, H., 1979, Enhancement of 2-acetylaminofluorene liver carcinogenesis in rats fed a choline-devoid diet, Int. J. Cancer 23:565.

Chen, M-L., Abileah, S. Wainfan, E. and Christman, J.K., 1993, Influence of folate and vitamin B_{12} on the effects of dietary lipotrope deficiency, Proc. Amer. Assoc. for Cancer Res. 34: 131.

Makos, M., Nelkin, B.D., Lerman, M.I., Latif, F., Zbar, B., and Baylin, S., 1992, Distinct hypermethylation patterns occur at altered chromosome loci in human lung and colon cancer, Proc. Natl. Acad. Sci.U.S.A., 89: 1929

McMurray, W.C., 1983, "Essentials of Human Metabolism," Harper and Row, Philadelphia.

Meehan, R.R., Lewis, J.D., McKay, S., Kleiner, E.L., and Bird, A.P., 1989, Identification of a mammalian protein that binds specifically to DNA containing methylated CpGs, Cell 58: 499.

Mikol, Y.B, Hoover, K.L., Creasia, D., and Poirier, L.A., 1983, Hepatocarcinogenesis in rats fed methyl-deficient, amino acid-defined diets, Carcinogenesis 4: 1619.

Nakae, D., Yoshiji, H. Mizumoto, Y., Horiguchi, K. Shiraiwa, K. Tamura, K., Denda, A. and Konishi Y. 1992 High incidence of hepatocellular carcinomas induced by a choline deficient L-amino acid defined diet in rats, Cancer Res. 52:5042.

Nambu, S., Inque, K., and Sasaki, H., 1987, Site-specific hypomethylation of the c-*myc* oncogene in human hepatocellular carcinoma, Jpn. J. Cancer (Gann) 78: 695.

Newberne, P.M. and Rogers, A.E., 1980, Labile methyl groups and the promotion of cancer,nju Annu. Rev. Nutr. 6:407.

Newberne, P.M., 1965, Carcinogenicity of aflatoxin-contaminated peanut meal., in "Mycotoxins in Foodstuffs," Massachusetts Institute of Technology, Cambridge, MA.

Pegg, A.E., and Hibasami, H., 1979, The role of S-adenosylmethionine in mammalian polyamine synthesis, in "Transmethylation," Elsevier-North Holland, New York.

Perera, M.I., Betschart, J.M., Virji, M.A., Kaytal, S.L., and Shinozuka, H., 1987, Free radicanjury and liver tumor promotion, Toxicol. Pathol. 15:51.

Porta, E.A., Markell, N. and Dorado, R.D., 1985, Chronic alcoholism enhances hepatocarcinogenicity of diethylnitrosamine in rats fed a marginally methyl-deficient diet, Hepatology 5: 1120.

Rao, P.M., Antony, A., Rajalakshmi, S., and Sarma, D.S.R., 1989, Studies on hypomethylation of liver DNA during early stages of chemical carcinogenesis in rat liver, Carcinogenesis 10:933.

Rogers, A.E., and MacDonald, R.A., 1965, Hepatic vasculature and cell proliferation in experimental cirrhosis, Lab. Invest. 14:1710.

Rogers, A.E., Sanchez, O., Feinsod, F.M., and Newberne, P.M., 1974, Dietary enhancement of nitrosamine carcinogenesis, Cancer Res. 34:96.

Rogers, A.E., 1975, Variable effects of a lipotrope-deficient high-fat diet on chemical carcinogenesis in rats, Cancer Res. 35:2469.

Rogers, A.E., and Newberne, P.M., 1980, Lipotrope deficiency in experimental carcinogenesis, Nutr. Cancer 2:104.

Rogers, A.E., 1993, Chemical carcinogenesis in methyl deficient rat,. J. Nutr. Biochem. 4: 666.

Rowley, J.D., 1990, Molecular cytogenetics: Rosetta stone for understanding cancer - Twenty-ninth G.H.A. Clowes Memorial Award Lecture, Cancer. Res. 50:3816.

Rushmore, T.H., Ghazarian, D.M., Subrahmanyan, V., Farber, E., and Goshal, A.K.,1987, Probable free radical effects on rat liver nuclei during early hepatocarcinogenesis with a choline-devoid low-methionine diet, Cancer Res. 47: 6731.

Schmidt, M., Haaf, T., and Grunert, D., 1990, 5-Azacytidine-induced undercondensations in human chromosomes, Human Genet. 67:257.

Selhub, J., Seyhoum, E., Pomfret, E.A. and Zeisel, S.H., 1991, Effects of choline deficiency and methotrexate treatment upon folate content and distribution, Cancer Res. 51:16.

Shen, J.C., Rideout, W.M,. and Jones, P.A., 1994, High frequency mutagenesis by a DNA methyltransferase, Cell 71:1073.

Shinozuka, H., Kaytal, S.L., and Lombardi, B., 1978a, Azaserine carcinogenesis: Organ susceptibility change in rats fed a diet devoid of choline, Int. J. Cancer 22:36.

Shinozuka, H., Lombardi, B., and Sell, S., 1978b, Enhancement of ethionine liver carcinogenesis in rats fed a choline-devoid diet, J. Natl. Cancer Inst. 61:813.

Shivapurkar, N., and Poirier, L.A., 1983, Tissue levels of S-adenosylmethionine and S-adenosylhomocysteine in rats fed methyl-deficient diets for one to five weeks, Carcinogenesis 4:1052.

Sutherland, G., 1988, The role of nucleotides in human fragile site expression, Mut. Res. 200: 207.

Tanaka, K., Appella, E. , and Jay, G., 1983, Developmental activation of the H-2K gene is correlated with an increase in DNA methylation, Cell 35: 457.

Trautner, T.A., 1984, "Methylation of DNA," Current Topics in Microbiology and Immunology, 108, Springer-Verlag, Berlin.

Usdin, E., Borchardt, R.T., and Creveling, C.R., 1982, "Biochemistry of S-Adenosylmethionine and Related Compounds," Macmillan Press Ltd., London.

Vance, J.E. and Vance, D.E., 1985, The role of phosphatidylcholine biosynthesis in the secretion of lipoproteins from hepatocytes. Can. J. Biochem, Cell Biol. 63: 870.

Wainfan, E., Kilkenny, M. and Dizik, M. 1988 Comparison of methyltransferase activities of pair-fed rats given adequate or methyl-deficient diets, Carcinogenesis, 9:861.

Wainfan, E., Dizik, M., Stender, M. and Christman, J.K., 1989, Rapid appearance of hypomethylated DNA in livers of rats fed cancer-promoting, methyl-deficient diets, Cancer Res. 49:4094.

Wainfan, E., Dizik, M., and Balis, M.E., 1984, Increased activity of rat liver N2-guanine tRNA methyltransferase II in response to liver change, Biochim. Biophys. Acta 799: 288.

Wainfan, E., Dizik, M., Hluboky M., and Balis, M.E, 1986, Altered tRNA methylation in rats and mice fed lipotrope-deficient diets, Carcinogenesis 7:473.

Wang, R.Y., Zhang, X.Y., and Ehrlich, M., 1986, A human DNA binding protein is methylation-specific and sequence-specific, Nucl. Acids Res. 14:1599.

Watt, F., Molloy, P.L. 1988 Cytosine, methylation prevents binding to DNA of a Hela cell transcription factor required for optimal expression of the adenovirus major late promoter, Genes Dev. 2: 1136.

Wickremasinghe R.G. and Hoffbrand A.V., 1980, Reduced rate of DNA replication fork movement in megaloblastic anemia. J Clin. Invest. 65, 1:26.

Wilson, M.J. , Shivapurkar, N. and Poirier, L.A., 1984, Hypomethylation of hepatic nuclear DNA in rats fed with a carcinogenic methyl-deficient diet, Biochem. J. 218:987.

Wilson, V.L. and Jones, P.A., 1983, Inhibition of DNA methylation by chemical carcinogens in vitro, Cell 32:239.

Wu, A.-L., and Windmueller, H. G., 1979, Relative contribution by liver and intestine to individual plasma apolipoproteins in the rat, J. Biol. Chem. 254:7316.

Yao, Z., and Vance, D., 1988, The active synthesis of phophatidycholine is required for very low density lipoprotein secretion from rat hepatocytes, J. Biol. Chem. 263: 2998.

Yokoyama, S., Sells, M.A., Reddy, T.V., and Lombardi, B., 1985, Hepatocarcinogenic and promoting action of a choline-devoid diet in the rat, Cancer Res. 45:2384.

Zapisek, W.F., Cronin, G.M., Lyn-Cook, B.D. and Poirier, L.A., 1992, The onset of oncogene hypomethylation in the livers of rats fed methyl-deficient, amino-acid defined diets, Carcinogenesis 13: 1869.

CHOLESTEROL, CHOLESTEROGENESIS AND CANCER

Marvin D. Siperstein

University of California, San Francisco
and Department of Veterans Affairs Medical Center
4150 Clement Street
San Francisco, CA 94121

INTRODUCTION

The suggestion that deranged cholesterol synthesis may play a role in carcinogenesis is not a new one. With characteristic prescience, Aldous Huxley anticipated the findings of the past few years linking cholesterol metabolism and cancer when he wrote as follows:

"Some of the sterols were definitely poisonous. . . Longbottom had even suggested a connexion between fatty alcohols and neoplasms. In other words, cancer might be regarded, in the final analysis, as a symptom of sterol poisoning" (Huxley, 1939). Over the past few years, however, indeed in the past few months, there has been a remarkable burst of interest and publications suggesting not only that derangements in cholesterogenesis may relate to cancer, but that intervention in this synthetic pathway may specifically inhibit the growth of cancer cells.

FEEDBACK REGULATION OF CHOLESTEROL SYNTHESIS IN CANCER

It has been known since the studies of Gould (1951) and of Tomkins et al.(1953) that feeding cholesterol results in the rapid inhibition of hepatic cholesterol synthesis from acetyl CoA. Our laboratory then demonstrated that the biochemical site of this feedback control of cholesterol synthesis is located specifically at the conversion of β-hydroxy β-methylglutaryl CoA to mevalonate **(Figure 1)**, a reaction catalyzed by the enzyme β-hydroxy β-methylglutaryl reductase (HMG CoA reductase). (Siperstein and Guest, 1966; Siperstein and Fagan, 1966).

As also indicated in **Figure 1**, there are a number of biologically significant isoprene intermediates either on or branching from the pathway of mevalonate to cholesterol. The importance of these post-mevalonate isoprenoid compounds becomes even more significant in view of our observation that this feedback control of the precursor molecule, mevalonic acid, is consistently either impaired or completely deleted in all cancers so far examined **(Figure 2)** (Bricker and Weis, 1972; Elwood and Morris, 1968; Siperstein and Fagan, 1964;

Nutrition and Biotechnology in Heart Disease and Cancer
Edited by J.B. Longenecker *et al.*, Plenum Press, New York, 1995

155

Siperstein et al., 1966; Siperstein, 1970; Sabine et al., 1967. As a result, the synthesis of the numerous compounds derived from mevalonic acid, including farnesyl pyrophosphate, are no longer under feedback control when a cell undergoes malignant change.

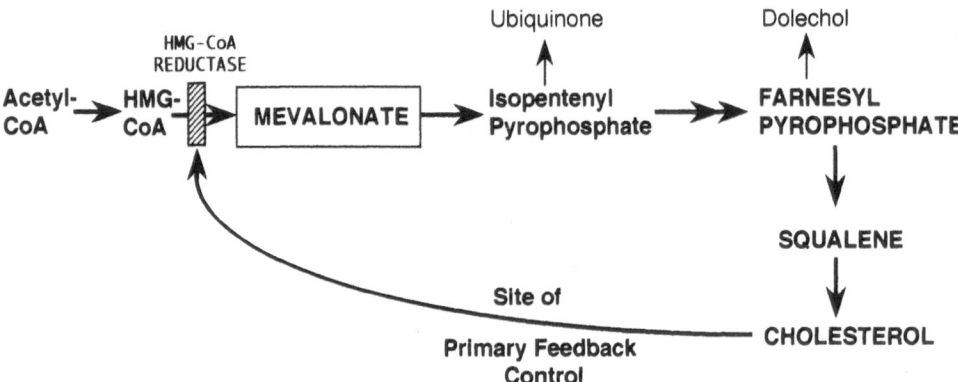

Figure 1. Isoprene synthesis and the site of the cholesterol feedback reaction.

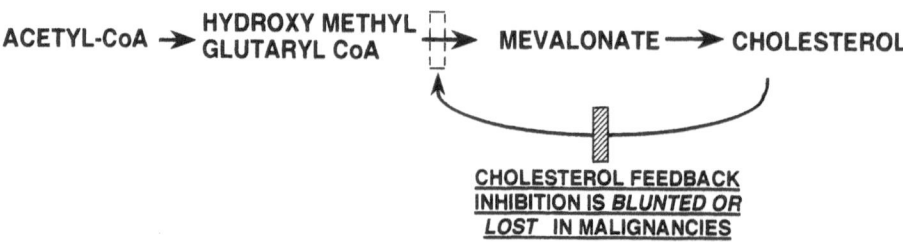

Figure 2. Cholesterol feedback control is impaired or absent in malignant cells.

The absence of cholesterol feedback control of HMG CoA reductase in a typical, highly differentiated liver cancer is illustrated in **Table 1** (Siperstein et al., 1971) A minimal deviation rat hepatoma 9121 was implanted into inbred rats. After cholesterol feeding both the conversion of acetate to cholesterol and the activity of the feedback enzyme, HMG CoA

Table 1. Absence of feedback control of HMG-CoA reductase in HEPATOMA 9121

Tissue	Diet	$(2\text{-}^{14}C)$ Acetate Converted to Cholesterol (nmoles/gm/hr)	HMG CoA Reductase (nmoles/hr)
Liver	Normal	45	0.77 ± 0.11
	5% Cholesterol	96	0.007 ± 0.01
Hepatoma 9121	Normal	46	1.34 ± 0.14
	5% Cholesterol	96	1.69 ± 0.17

reductase in the normal liver were markedly inhibited. By contrast, in the hepatoma tissue cholesterol feeding had no inhibitory effect either on overall cholesterol synthesis or on the activity of HMG CoA reductase. A similar experiment, shown in **Table 2**, demonstrates that in two human hepatomas, in contrast to the case in normal human liver, cholesterol feedback control is minimal or absent (Siperstein and Fagan, 1964; Siperstein et al., 1966). Interestingly, the activation state of HMG CoA reductase in tumor cells is higher than in normal adult, fetal or regenerating tissues (Feingold et al., 1983)), a finding that has been confirmed in at least one other laboratory (Kawata et al., 1990).

Table 2. Absence of cholesterol feedback control in human HEPATOMAS

Tissue	Diet	Cholesterol Synthesis ^{14}C-Acetate Converted to Cholesterol ($\mu\mu$moles/mg tissue)
Normal Liver (5 Subjects)	Low Cholesterol (3 Days)	20
Normal Liver (4 Subjects)	High Cholesterol (3 Days)	5
HEPATOMA 65 Yr. Old Man	Low Cholesterol (3 Days)	76
	High Cholesterol (3 Days	82
HEPATOMA 56 Yr. Old Woman	Low Cholesterol (3 Days)	94
	High Cholesterol (3 Days)	50

CHOLESTEROL FEEDBACK CONTROL IN PREMALIGNANCY

The absence of normal feedback regulation of cholesterol synthesis in a wide variety of malignancies led our laboratory to ask whether the defect in the feedback regulation of isoprene and sterol synthesis might precede malignancy, i.e. is this biochemical lesion present in the premalignant state prior to the development of overt malignancy. To examine this question we took advantage of the ability of the mold toxin, aflatoxin B-1, to cause hepatomas in numerous animal species, including the rat. As indicated by the data in **Table 3**, after only two days of feeding aflatoxin, neither cholesterol nor mevalonic acid synthesis was any longer under normal feedback control.

Furthermore, treatment with aflatoxin caused at least a nine-fold increase in *baseline* mevalonate synthesis (Siperstein, 1966). These results have been comfirmed both in rats (Horton et al., 1972; Siperstein, 1973) and in humans (Bissell and Alpert, 1972).

The finding that feedback control of mevalonate is lost in the premalignant state many months before a hepatoma appears raised the possibility that the overproduction of mevalonate or of one or more of the many isoprene compounds that lie on the pathway of cholesterogenesis could play a role in carcinogenesis.

Table 3. Aflatoxin treatment causes rapid loss of cholesterol feedback regulation in rat liver

Aflatoxin	Dietary Cholesterol	Cholesterol Synthesis mμmoles Acetate-^{14} C	Mevalonate Synthesis mμmoles Acetate-^{14} C
0	0	4.8	25
0	5% on Day 15	0.2	3
1 mg/day on days 1 & 2	0	32	221
1 mg/day on days 1 & 2	5% on Day 15	28	164

Cholesterol and mevalonates synthesis in liver slices on day 16

THE RELATIONSHIP BETWEEN CHOLESTEROGENESIS AND DNA SYNTHESIS

In view of these findings it seemed possible that cholesterol and mevalonate synthesis might regulate cell growth perhaps by influencing DNA synthesis in the normal and malignant cell. To examine this problem Dr. Valeria Quesney-Huneeus in our laboratory measured the activity of HMG CoA reductase and cholesterol synthesis as they relate to DNA synthesis during the various phases of the cell cycle (Huneeus et al., 1979; Quesney-Huneeus et al., 1983). BHK-21 cells were synchronized in tissue culture, following which HMG CoA reductase, cholesterol synthesis, and DNA synthesis from labeled thymidine were measured during each phase of the cell cycle **(Figure 3)**. As expected, the rate of DNA synthesis rose during late G_1, reaching a peak during the S phase of the cycle. Cholesterol synthesis reached a maximum during the growth phase of G_1 when cholesterol is required for cell membrane synthesis. HMG CoA reductase and hence mevalonate synthesis were also very active during the G_1 phase of the cell cycle, presumably to provide the substrate for the concurrent rapid cholesterol synthesis.

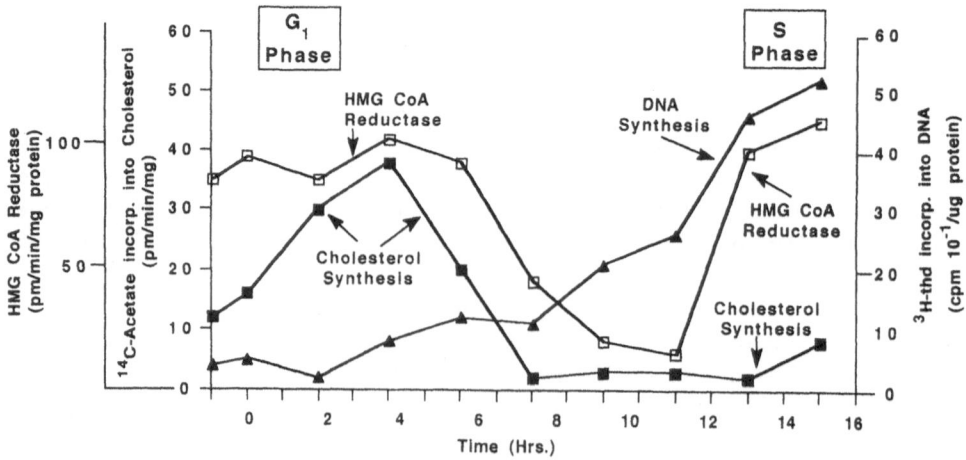

Figure 3. The relationship of cholesterogenesis, β-hydroxy β-methylglutaryl CoA reductase and DNA synthesis during the phases of the cell cycle.

The most important observation of this experiment, however, was that HMG CoA reductase has a second peak of activity that coincides with the rise in DNA synthesis is late G_1 and S-phases. As noted, however, cholesterol synthesis is minimal during these periods of the cell cycle. This observation implies that the synthesis of mevalonate acid may play a role in DNA synthesis that is independent of its requirement for cholesterogenesis.

To examine the possible function of mevalonate in the synthesis of DNA, we made use of compactin, a potent inhibitor of HMG CoA reductase. By adding compactin to synchronized BHK-21 cells in tissue culture, mevalonate synthesis could be almost completely inhibited, and the effect on DNA synthesis of depleting cells of mevalonate during specific periods of the cell cycle determined. As shown in **Figure 4**, inhibition of mevalonate production in fact dramatically reduced DNA synthesis to the baseline levels observed in the G_1 phase and completely prevented the S phase increase in DNA production. (It is likely that the labeled thymidine incorporated into DNA in the presence of compactin during G_1, M and G_2 reflects only DNA repair rather than DNA replication.) It was concluded, on the basis of these experiments, that de novo DNA synthesis during the S phase of the cell cycle is completely dependent upon the presence of mevalonic acid.

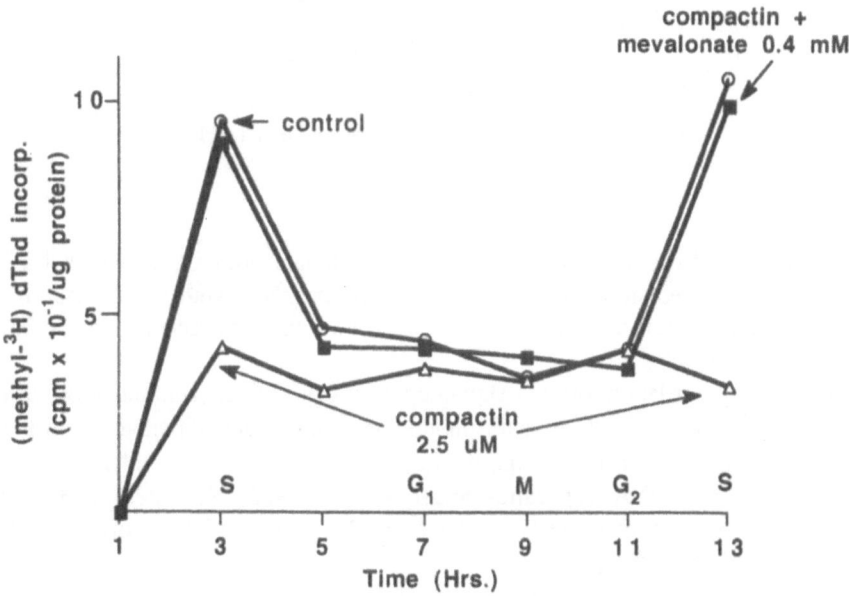

Figure 4. Depriving the cell of mevalonate prevents de novo DNA synthesis; adding back mevalonate restores DNA replication.

A number of earlier studies had suggested that cholesterol itself is required for cell proliferation and DNA synthesis (Chen et al., 1975; Kandutsch and Chen, 1977); however, the results in **Figures 3&4** raised the possibility that mevalonate synthesis in addition to providing an essential precursor for structural cholesterol might directly regulate DNA replication.

To evaluate this hypothesis we examined the effect on DNA synthesis of adding either mevalonate or cholesterol to the compactin treated cells (Huneeus et al., 1979; Qiesmeu-Huneeus et al., 1982). As shown in **Figure 4** the addition of mevalonate to these cells completely reversed the inhibition of DNA synthesis produced by compactin treatment. By contrast, however, supplementing the inhibited cells with cholesterol had no effect upon DNA synthesis (Huneeus et al., 1979; Quesney-Huneeus et al., 1983). These experiments, therefore, provided the first demonstration that mevalonate independent of its role in cholesterol synthesis plays an essential role in initiating DNA replication.

The results of these studies, summarized schematically in **Figure 5**, indicate that mevalonate in fact has a dual role in the cell cycle. First, this acid is required to synthesize the cholesterol needed to form the numerous intracellular and plasma membranes needed for cell growth. Secondly, however, independent of its function in cholesterogenesis, mevalonate is required for DNA replication in both normal and malignant cells.

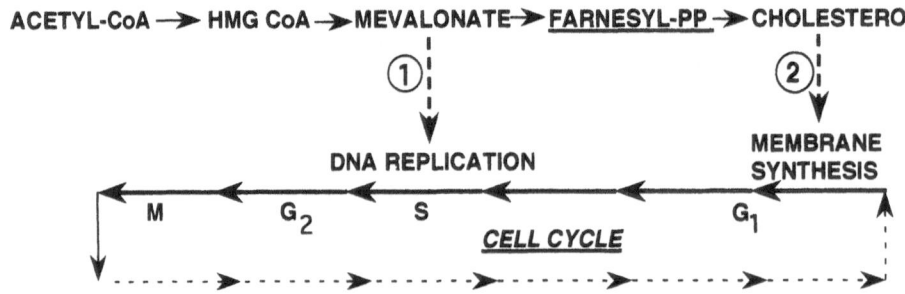

Figure 5. The dual role of mevalonate in the cell cycle.

We have provided some evidence that the mevalonate product, isopentenyl adenine (see **Figure 1**), may mediate the effect of mevalonate on DNA synthesis, at least in some cell systems(Quesney-Huneeus et al., 1980), a suggestion that has recently been supported by Faust and Dice (1991).

More recently our laboratory has shown by means of both specific inhibitors of DNA polymerase α and β as well as with monoclonal antibodies directed against DNA polymerase α, that mevalonate is required for the activity of DNA polymerase α, the enzyme required for DNA replication(Silber et al., 1992; Siperstein and DeVale, unpublished). By contrast, mevalonate has no detectable effect on DNA polymerase β, the polymerase catalyzing DNA repair.

RAS AND CHOLESTEROGENESIS

In 1984, Schmidt et al. (1984) first demonstrated that labeled mevalonate is covalently incorporated into a number of cell proteins. Subsequent studies have shown that the 15 carbon product of mevalonate, farnesyl, is one of two mevalonate derived isoprenes that are incorporated posttranslationally into several cellular proteins (Farnsworth et al., 1989). The mechanism and functions of protein farnesylation have been rapidly elucidated over the past few years (Schafer and Rine, 1992).

These studies have suggested that cholesterogenesis may effect cell proliferation by providing the farnesyl molecule that is incorporated posttranslationally into the normal Ras proteins of the cell (Hancock et al., 1989; Schafer et al., 1990). By way of background, as indicated schematically in **Figure 6**, optimal division and differentiation in many cells requires a functional Ras proto-oncogene, which codes for the synthesis of a normal Ras protein. Under the influence of carcinogens or of oncogenic viruses, the normal Ras proto-oncogene can be transformed into a Ras oncogene, which then codes for the synthesis of a mutant oncogenic Ras protein (see ref. Lowry and Willumsen, 1993; Bokoch and Der, 1993 for recent reviews).

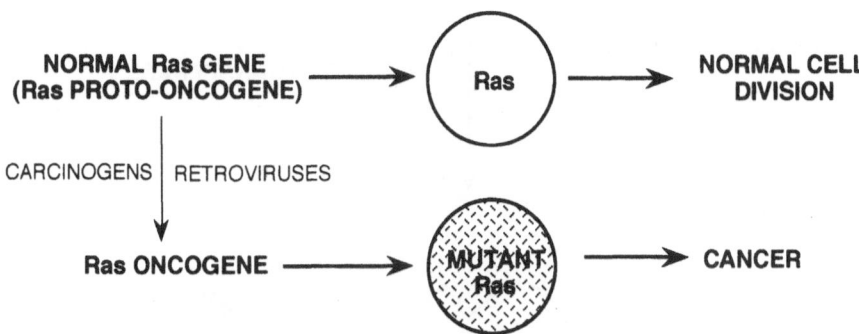

Figure 6. Ras protein normally promotes cell division and differentiation. Mutant Ras can cause malignant transformation.

The Ras oncogene is the commonest oncogene found in malignant cells. This oncogene has in fact been identified in carcinomas of the lung, pancreas, and breast and in a number of types of leukemia. It is present in more than 50% of colon cancers (Bos, 1989).

As indicated in **Figure 7**, it is now well established that farnesylation of this protein is required for the activation of the normal Ras protein, and in the absence of farnesylation the mutant Ras protein will not transform normal into malignant cells (Hancock et al., 1989; Schafer et al., 1990). The enzyme, farnesyl transferase (Schaber et al., 1990; Seabra et al., 1991) catalyzes this reaction in which farnesyl pyrophosphate transfers its farnesyl group to the cysteine that becomes the C-terminal amino acid either of the normal Ras protein or of the oncogene derived mutant Ras protein.

As indicated in **Figure 8**, the farnesyl once bound to Ras serves as a lipophilic link between the Ras protein and the lipid components of the inner surface of the plasma membrane. During this process GTP is added to the Ras protein the result being an active molecule which, through a series of steps now believed to involve translocation of specific protein kinases to the plasma membrane, stimulates cell division (Hall, 1994).

Finally, under normal conditions the Ras bound GTP is converted to GDP by the removal of phosphate through the action of a specific GTPase. The result is an inactive GDP-Ras still bound to the plasma membrane through the farnesyl linkage.

The schematic representation of the comparable reactions for the mutant Ras proteins is indicated in **Figure 9**. In this case farnesyl transferase activates mutant Ras, again by providing the farnesyl that anchors the mutant protein to the plasma membrane. However, in contrast to the normal Ras, the mutant protein can not be inactivated by GTPase, and the mutant Ras is thereby "locked" in an active state, resulting in uncontrolled cell replication and thereby cancer.

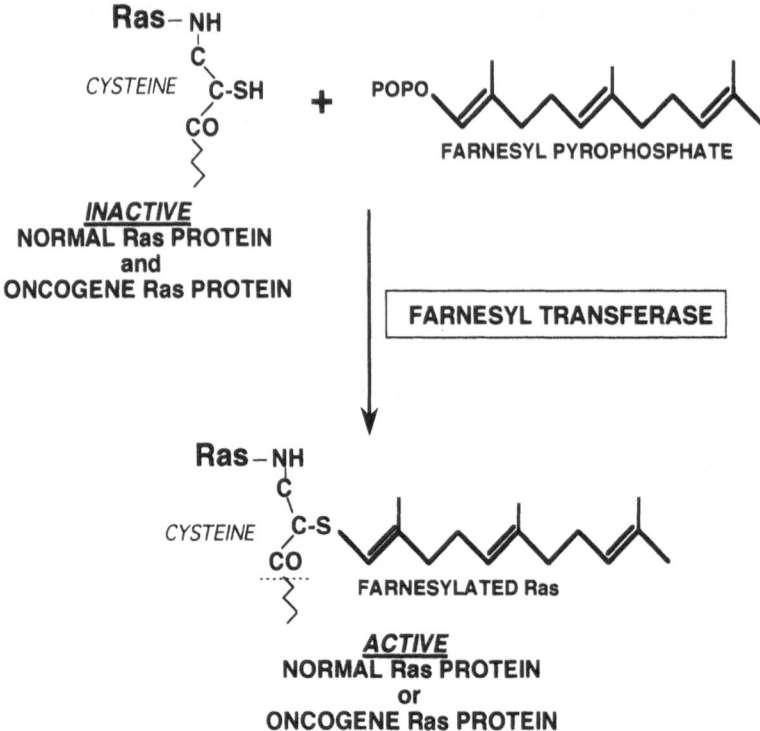

Figure 7. Farnesyl pyrophosphate by transferring its farnesyl group to a cysteine on Ras proteins activates either normal or oncogenic Ras.

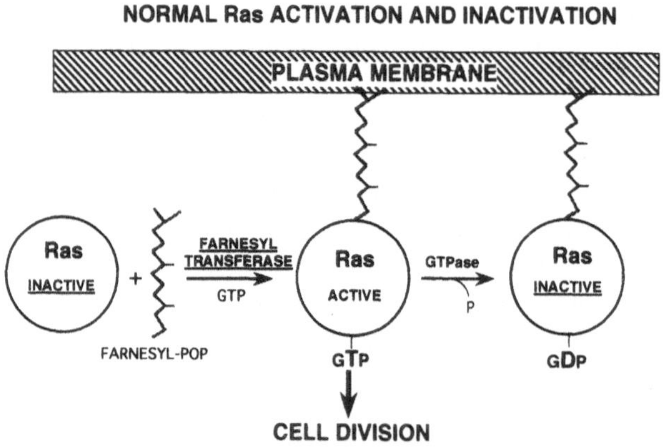

NORMAL Ras ACTIVATION AND INACTIVATION

Figure 8. Farnesylation Ras can bind to the lipids in the plasma membrane, a step required for Ras activation.

IMPLICATIONS FOR CANCER THERAPY

An obvious therapeutic goal based on these findings is to identify a specific inhibitor of farnesyl transferase, which would prevent both the activation of the mutant Ras protein and the uncontrolled cell division that leads to cancer. This formulation is presented

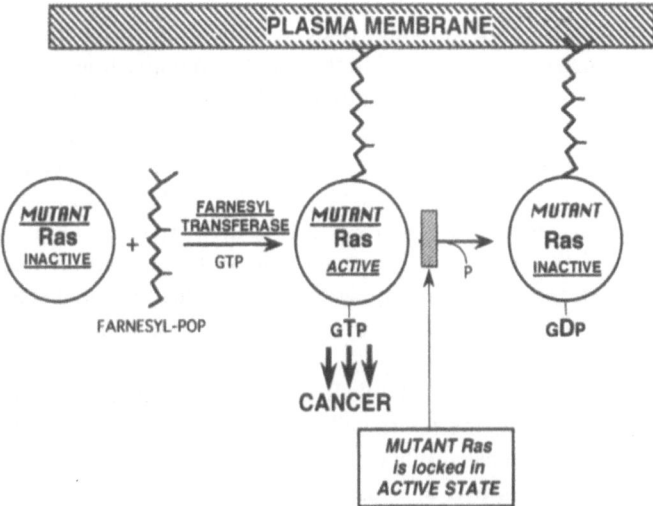

Figure 9. Ras oncogene codes for a mutant Ras that cannot be inactived thereby causing uncontrolled cell division.

schematically in **Figure 10**, in which a specific inhibitor of farnesyl transferase would prevent farnesylation of the mutant Ras protein, thereby preventing its binding to the plasma membrane and rendering this protein inactive. In theory, such a maneuver should neutralize the effect of the mutant Ras protein and prevent the oncogenic Ras from causing malignant transformation.

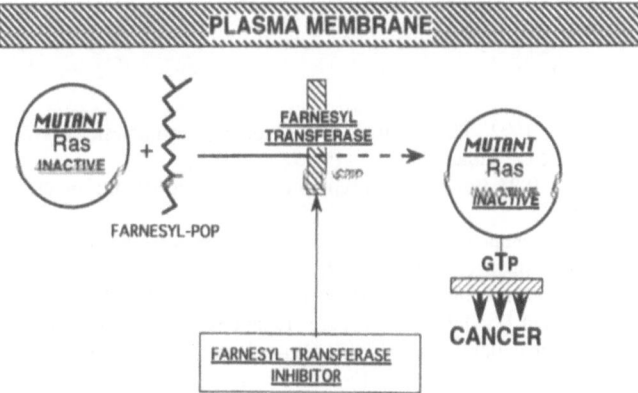

Figure 10. A schematic view of farnesyl transferase inhibitors might inactivate oncogenic Ras proteins.

While it might appear to be extremely difficult to develop an inhibitor that would be specific for farnesyl transferase, in fact, within the past few months at least four laboratories have reported the development of such compounds (Miura et al., 1993; Hara et al., 1993; James et al., 1993; Kohl et al., 1993). In a collaboration between the laboratories of Brown and Goldstein at The University of Texas Southwestern Medical Center and Genentech, Inc., a series of benzodiazepines inhibitors of farnesyl transferase has been synthesized (James et al., 1993). One of these has proved to be extremely effective in inhibiting both farnesyl transferase and the growth of malignant cells in tissue culture. Interestingly this compound did not affect the growth of normal cells and its unanticipated specificity for the malignant cell *in vitro* suggests its possible usefulness as a specific inhibition of cancer growth.

Similarly, Kohl et al. (1993) at Merck Research Laboratories have reported that a tetrapeptide inhibitor of farnesyl transferase will prevent the malignant characteristics of Ras transformation in tissue culture cells. While clearly the use of farnesyl transferase inhibitors in controlling cancer cell growth is at a preliminary stage, the early suggestion of their effectiveness *in vitro* indicates that this or similar approaches to regulating farnesylation may offer promising leads to cancer chemotherapy.

CONCLUSION

Studies of the cholesterol feedback reaction have shown that this regulatory process operates by controlling the synthesis of mevalonate, the precursor of all isoprenes including cholesterol. The further finding that this cholesterol feedback system is deranged in both premalignant and overtly malignant cells, coupled with the demonstration that mevalonate is required for DNA replication, suggested a possible link between derangements of cholesterol synthesis and carcinogenesis. This possibility has been greatly strengthened by the recent recognition of the importance of farnesylation in the activation of the commonest cancer producing gene, the Ras oncogene. Finally, the preliminary evidence that preventing the farnesylation of Ras by the specific inhibition of farnesyl transferase can effectively reverse the malignant phenotype provides a new and exciting approach to the treatment of cancer.

REFERENCES

Bissell, D.M. and Alpert, E., 1972, The feedback control of hepatic cholesterol synthesis in Ugandan patients with liver disease. Cancer Res. 32:149-152.

Bokoch, G.M. and Der, C.J., 1993 Emerging concepts in the Ras superfamily of GTP-binding proteins. FASEB J. 7:750-759.

Bos, J.L.,1989, Ras oncogene in human cancer: A review. Cancer Res. 49:4682-4689.

Bricker, L.A., Weis, H.S. and Siperstein, M.D., 1972, In vivo demonstration of the cholesterol feedback system by means of a desmosterol suppression technique. J. Clin. Invest. 51:197-205.

Chen, H.W., Heninger, H.J. and Kandutsch, A.A.,1975, Relationship between sterol synthesis and DNA synthesis in phytohemagglutinin-stimulated mouse lymphocytes. Proc. Natl. Acad. Sci. USA 72:1950-1954.

Elwood, J.C. and Morris, H.P., 1968, Lack of adaptation in lipogenesis by hepatoma 9121. J. Lipid Res. 9:337-341.

Farnsworth, C.C., Wolda, S.L., Gelb, M.H. and Glomset, J.A.,1989, Human lamin B contains a farnesylated cysteine residue. J. Biol. Chem. 264:20422-20429.

Faust, J.R. and Dice, J.F., 1991, Evidence for isopentenyladenine modification on a cell cycle-regulated protein. J. Biol. Chem. 266:9961-9970.

Feingold, K.R., Wiley, M.H., Moser, A.H. and Siperstein, M.D., 1983, Altered activation state of hydroxymethylglutaryl-coenzyme A reductase in liver tumors. Arch. Biochem. Biophys. 226:231-241.

Gould, R.G., 1951, Lipid metabolism and atherosclerosis. Am. J. Med. 11:209-227.

Hall, A., 1994, A biochemical function for Ras -At last. Science 264:1413-1414.

Hancock, J.F., Magee, A.I., Childs, J.E. and Marshall, C.J., 1989, All Ras proteins are polyisoprenylated but only some are polmitoylated. Cell 57:1167-1177.

Hara, M., Akasaka, K., Akinaga, S., Okabe, M., Nakano, H., Gomez, R., Wood, D., Uh, M. and Tamanoi, F., 1993, Identification of Ras farnesyltransferase inhibitors by microbial screening. Proc. Natl. Acad. Sci. USA 90:2281-2285.

Horton, B.J., Horton, J.D. and Sabine, J.R., 1972, Metabolic controls in precancerous liver. II. Loss of feedbackcontrol of cholesterol synthesis measured repeatedly in vivo during treatment with the carcinogens N-2-fluorenylacetamide and aflatoxin. Eur. J. Cancer 8:437-443.

Horton, B.J., Horton, J.D. and Sabine, J.R., 1973, Metabolic controls in precancerous liver. V. Loss of control of cholesterol synthesis during feeding of the hepatocarcinogen 3'-methyl-4-dimethylaminoazobenzene. Eur. J. Cancer 9:573-576.

Huneeus, V.Q., Wiley, M.H. and Siperstein, M.D., 1979, Essential role for mevalonate synthesis in DNA replication. Proc. Natl. Acad. Sci, USA 76:5056-5060.

Huxley, A. 5, 1939, After Many A Summer Dies The Swan (Chapter 5), Harper and Brothers, New York/London.

James, G.L., Goldstein, J.L., Brown, M.S., Rawson, T.E., Somers, T.C., McDowell, R.S., Crowley, C.W., Lucas, B.K., Levinson, A.D. and Marsters, Jr., J.C., 1993, Benzodiazepine peptidomimetics: Potent inhibitors of Ras farnesylation in animal cells. Science 260:1937-1942.

Kandutsch, A.A. and Chen, H.W., 1977, Consequences of blocked sterol synthesis in cultured cells. J. Biol. Chem. 252:409-415.

Kawata, S., Takaishi, K., Nagase, T., Ito, N., Matsuda, Y., Tamura, S., Matsuzawa, Y. and Tarui, S., 1990, Increase in the active form of 3-hydroxy-3-methylglutaryl coenzyme A reductase in human hepatocellular carcinoma: possible mechanism for alteration of cholesterol biosynthesis. Cancer Res. 50:3270-3273.

Kohl, N.E., Mosser, S.D., deSolms, S.J., Giuliani, E.A., Pompliano, D.L., Graham, S.L., Smith, R.L., Scolnick, E.M., Oliff, A. and Gibbs, J.B., 1993, Selective inhibition of Ras-dependent transformation by farnesyltransferase inhibitor. Science 260:1934-1937.

Lowry, D.R. and Willumsen, B.M., 1993, Function and regulation of RAS. Annu. Rev. Biochem. 62:851-891.

Miura, S., Hasumi, K. and Endo, A., 1993, Inhibition of protein prenylation by patulin. Fed. Euro. Biochem. Soc. 318:88-90.

Quesney-Huneeus, V., Hughes-Wiley, M. and Siperstein, M.D., 1980, Isopentenyladenine as a mediator of mevalonate-regulated DNa replication. Proc. Natl. Acad. Sci. USA 77:5842-5846.

Quesney-Huneeus, V., Galick, H.A. and Siperstein, M.D., 1983, The dual role of mevalonate in the cell cycle. J. Biol. Chem. 258:378-385.

Sabine, J.R., Abraham, S. and Chaikoff, I.L., 1967, Control of lipid metabolism in hepatomas: insensitivity of rate of fatty acid and cholesterol synthesis by mouse hepatoma BW7756 to fasting and to feedback control. Cancer Res. 27:793-799.

Schaber, M.D., O'Hara, M.B., Garsky, V.M., Mosser, S.D., Bergstrom, J.D., Moores, S.L., Marshall, M.S., Friedman, P.A., Dixon, R.A.F. and Gibbs, J.B., 1990, Polyisoprenylation of Ras in vitro by a farnesyl-protein transferase. J. Biol. Chem. 265:14701-14704.

Schafer, W.R., Trueblood, C.E., Yang, C.-C., Mayer, M.P., Rosenberg, S., Poulter, C.D., Kim, S.-H. and Rine, J., 1990, Enzymatic coupling of cholesterol intermediates to a mating pheromone precursor and to the Ras protein. Science 249:1133-1139.

Schafer, W.R. and Rine, J., 1992, Protein prenylation: Genes, enzymes, targets, and functions. Annu. Rev. Genet. 30:209-237.

Schmidt, R.A., Schneider, C.J. and Glomset, J.A., 1984, Evidence for post-translational incorporation of a product of mevalonic acid into Swiss 3T3 cell proteins. J. Biol. Chem. 259:10175-10180.

Seabra, M.D., Reiss, Y., Casey, P.J., Brown, M.S. and Goldstein, J.L., 1991, Protein farnesyltransferase and geranylgeranyltransferase share a common alpha subunit. Cell 429-434.

Silber, J.R., Galick, H., Wu, J.M. and Siperstein, M.D., 1992, The effect of mevalonic acid deprivation on enzymes of DNA replication in cells emerging from quiescence. Biochem. J. 288:883-889.

Siperstein, M.D., 1966, Deletion of the cholesterol negative feedback system in precancerous liver. J. Clin. Invest. 45:1073.

Siperstein, M.D., 1970, Regulation of cholesterol biosynthesis in normal and malignant tissues. Curr. Top. Cell. Regul. 2:65-100.

Siperstein, M.D. and DoVale, H. Unpublished studies.

Siperstein, M.D. and Fagan, V.M., 1964, Deletion of the cholesterol negative feedback system in liver tumors. Cancer Res. 24:1108-1115.

Siperstein, M.D. and Fagan, V.M., 1966, Feedback control of mevalonate synthesis by dietary cholesterol. J. Biol. Chem. 241:602-609.

Siperstein, M.D., Fagan, V.M. and Morris, H.P., 1966, Further studies on the deletion of the cholesterol feedback system in hepatomas. Cancer Res. 26:7-11.

Siperstein, M.D. and Guest, M.J., 1966, Studies on the site of the feedback control of cholesterol synthesis. J. Clin. Invest. 39:642-652.

Siperstein, M.D., Gyde, A.M. and Morris, H.P., 1971, Loss of feedback control of hydroxymethylglutaryl coenzyme A reductase in hepatomas. Proc. Natl. Acad. Sci. USA 68:315-317.

Tomkins, G.M., Sheppard, H. and Chaikoff, I.L., 1953, Cholesterol synthesis by liver. III. Its regulation by ingested cholesterol. J. Biol. Chem. 201:137-141.

Travis, J., 1993, Novel anticancer agents move closer to reality. Science 260:1877-1878.

INHIBITION OF THE INDUCTION OF CANCER BY ANTIOXIDANTS

Thomas J. Slaga

The University of Texas
M.D. Anderson Cancer Center
Science Park - Research Division
Smithville, TX 78757

INTRODUCTION

Free radicals are now known to play an important role in many diseases including aging (Slaga, 1989; Halliwell, 1993). The importance of free radicals in radiation carcinogenesis and free radicals and electrophiles in chemical carcinogenesis is also well recognized. Free radicals and reactive oxygen species are continuously produced *in vivo*. Consequently, organisms have evolved that possess not only antioxidant and electrophile defense systems to protect against them, but also repair systems that prevent the accumulation of oxidatively-damaged molecules (Halliwell, 1993). Since increased free radical levels and levels of electrophilic compounds have been associated with many disease conditions, antioxidants, free radical scavengers, and electrophilic scavengers may be very useful in cancer prevention, cardiovascular disease prevention, immune function augmentation, and increasing the life span of man. Antioxidant enzymes, such as superoxide dismutase, catalase, and glutathione peroxidase, are preventive antioxidants, because they eliminate species involved in the initiation of free radical chain reactions; while small molecule antioxidants, such as ascorbate, the tocopherols, glutathione, and reduced co-enzyme Q_{10} can repair oxidizing radicals directly and therefore are chain-breaking antioxidants. It is well known that ascorbate and tocopherols function synergistically to protect membrane lipids from damage (Buettner, 1993). Other molecules such as metal ion binding proteins, b-carotene, bilirubin, and urate are also very important antioxidants (Halliwell, 1993).

A large number of carcinogens and tumor initiators are known to be metabolized to electrophilic intermediates which bind to DNA and cause mutations. There are specific enzymes which try to detoxify these reactive intermediates directly. Once the genetic material is damaged specific DNA repair enzymes try to reverse the damage (Slaga, 1989). Another important aspect of carcinogenesis, especially the tumor promotional stage, is that carcinogens and tumor promoters give rise directly to free radicals, or indirectly through the generation of free radicals such as superoxide anion and hydroxyl-radical. The production of superoxide anion is a universal phenomenon in all aerobic eucaryotic cells (Fredovich,

Nutrition and Biotechnology in Heart Disease and Cancer
Edited by J.B. Longenecker *et al.*, Plenum Press, New York, 1995

167

1974). Since superoxide anion is a highly reactive radical species and is extremely toxic, most respiring cells possess superoxide dismutase (SOD) enzyme as a defense against oxygen and superoxide anoin toxicity. SOD catalyses a dismutation reaction leading to the formation of hydrogen peroxide and molecular oxygen. The H_2O_2 is then destroyed by cellular catalases in order to prevent the formation of hydroxyl-radical by the Haber and Weiss reaction (Fredovich, 1974). The hydroxyl radical itself is an extraordinarily potent oxidant and could amplify the potential toxicity of oxygen radicals.

CELLULAR ANTIOXIDANT DEFENSE MECHANISMS

Several carcinogens, skin tumor promoters, ultraviolet light, and ionizing radiation have been reported to inhibit the activities of SOD and catalase both in mouse skin and cells in culture (Solanki et al., 1981; Borek and Troll, 1983; Nakamura et al., 1985; Yarosh and Yee, 1990). The treatment of adult mouse skin with a promoting dose of a phorbol ester tumor promoter resulted in a significant decrease in SOD activity (Solanki et al., 1981). This diminutive effect of a tumor promoter on SOD was dose dependent (Solanki et al., 1981). Since SOD is principally a protective enzyme, its lowered activity could lead to adverse effects because superoxide anions may accumulate in the cells. Tumor promoters also caused a greater than 50% reduction in the specific activity of epidermal catalase (Solanki et al., 1981). Several tumor promoters have been shown to have a similar inhibitory effect on epidermal SOD and catalase (Solanki et al., 1981). Similar results have been reported for the effects of tumor promoters, ultraviolet light and ionizing radiation on SOD and catalase in cells in culture (Borek and Troll, 1993; Nakamura et al., 1985). In addition, SOD and catalase added to cells in culture can decrease the damaging effect of radiation and free radical generating chemicals (Borek and Troll, 1983). A SOD mimicing agent, Cu(II) (3,5-diisopropyl salicylate), has been reported to counter the activity of carcinogens and mutagens (Solanki et al., 1984; Kensler et al., 1993).

A large data base exists which suggests a role of SOD in neoplastic growth (Overley and Buettner, 1979). Lowered Cu-Zn SOD activity is a very common characteristic of tumors. There are clear differences in SOD activity in normal and transformed cells. Mn-SOD activity has been found to disappear in a large variety of tumor cells (Amstad et al., 1991). We also reported a low level of SOD in skin tumors which are the result of promotion (Solanki et al., 1981).

The individual roles and the interaction of Cu, Zn-superoxide dismutase (SOD) and catalase in transfectants with human cDNAs of mouse epidermal cells JB6 clone 41 was studied by Amstad and coworkers (1991). Their results suggested that the balance of SOD and catalase plus glutathione peroxidase is more important in preventing oxidant damage than the level of SOD alone. In some cases the SOD overproducing cells were hypersensitive to the formation of DNA single-strand breaks, growth retardation and killing by an extracellular burst of superoxide plus H_2O_2, while the catalase overproducers were protected relative to the parent clone JB6 clone 41. The SOD and catalase overproducers were in general very protected against oxidant damage (Amstad et al., 1991).

There are also many reports that also show chemical carcinogens, tumor promoters and radiation can decrease the activity of DNA repair enzymes in target tissues (Bindahl, 1985) and therefore decrease the effectiveness of the DNA repair enzymes in repairing the damage. Yarosh and Yee (1990) reported that transfected phage T_4 *denV* gene, coding for the pyrimidine dimer specific T_4 endonuclease V or purified T_4 endonuclease V encapsulated in liposomes, was able to increase the repair capacity of both repair-proficient and -deficient human cells. In addition, Kripke and coworkers (1992) reported that the purified T_4 endonuclease V encapsulated in liposomes was capable of decreasing the number of pyrimidine dimers in the epidermis and prevented suppression of both delayed and contact hypersensitivity responses to UV-irradiated mouse skin.

CHEMICAL ANTIOXIDANTS AND RELATED COMPOUNDS

Many different antioxidants have been shown to inhibit the induction of cancer by a wide variety of chemical carcinogens and/or radiation at many target sites in mice, rats, hamsters, and man. **Table 1** summarizes the various carcinogens that antioxidants have been shown to be effective anticarcinogens. The various target organs in different species including man (Li et al., 1993) where antioxidants inhibit the induction of cancer are shown in **Table 2**.

Table 1. Antioxidants protect against the harmful effects of a wide variety of carcinogens[*]

Radiation (Ionizing and Ultraviolet)	Dimethylhydrazine
Polycyclic Aromatic Hydrocarbons	Azodyes
Urethan	Azoxynrethane
Bracken Fern	4-Nitroquinoline oxide
Diethylnitrosamine	Uracil mustard
Aromatic amines	

[*] See Slaga and Digiovanni, 1994 for details.

Table 2. Various target organs in mice, rats, hamsters, and man that antioxidants counteract cancer[*]

Skin	Mammary Gland
Colon	Bone Marrow
Lung	Blood (Leukemias)
Liver	Bladder
Forestomach	Cheek Pouch
Gastrointestinal Tract	Esophagus[**]

[*] See Li et al., 1993; Slaga and DiGiovanni, 1984 for details.
[**] Human esophageal cancer in China is inhibited by a combination of Vitamin E, b-carotene, and selenium.

A wide variety of chemical compounds possessing antioxidant properties have been found to inhibit the formation of chemically induced tumors in experimental animals. A listing of these antioxidants is shown in **Table 3**. The phenolic antioxidants, butylated hydroxytoluene (BHT) and butylated hydroxyanisole (BHA), have been studied extensively, primarily because of their use as food preservatives. However, compounds such as disulfiram (Antabuse), ethoxyquin, and selenium (Se) also have received considerable attention. Although the detailed mechanism(s) of the antagonistic effects of antioxidants are unknown, a number of theories have been advanced to explain their effects. Possible mechanisms include (1) direct interaction of the carcinogen or one of its activated metabolic products with the antioxidant; (2) decreased activities (or alterations) of enzyme pathways responsible for carcinogenic activation; and (3) increased activities of enzyme pathways responsible for detoxifying carcinogens.

Table 3. Antioxidants that are Effective Inhibitors of Carcinogenesis, and Malignant Transformation*

Vitamin C	Nordihydroguaiaretic acid
Vitamin E	Canventol
b-carotene	Oltipraz (Dithiole-thione)
N-acetylcysteine (NAC)	N-acyl-dehydroaniline
Gultathione	Disulfiram
Selenium	Butylated Hydroxyanisole (BHA)
Superoxide Dismutase	3-T-BHA
Catalase	2-T-BHA
Copper (II) bis(diisopropylsalicylate	4-parahydroxyanisole
Epigallocatechin gallate	Butylated Hydroxytoluene
Epicatechin-3-gallate	Ethoxyquin
Epigallocatechin	Flavonoids
Penta-0-galloyl-β-d-glucose	

* See Slaga and DiGiovanni, 1884; Ames et al., 1993; Wang et al., 1991; Komori et al., 1993; Fujiki et al., 1992 for details.

When BHA, BHT, and ethoxyquin were added to the diet, they antagonized the carcinogenic action of 7,12-dimethylbenz(a)anthracene (DMBA) and benzo(a)-pyrene (B[a]P) on the forestomach of mice and mammary gland of rats. In addition, BHA incorporated into the diet protected against pulmonary neoplasms produced by acute exposure to DMBA, B(a)P, urethan, and uracil mustard (Slaga and DiGiovanni, 1984). The carcinogenic effects of other polycyclic aromatic hydrocarbons (PAH) also were inhibited by BHA. The minor isomer of BHA, namely 2-tert-butyl-4-hydroxyanisole (2-t-BHA), was more effective than the major isomer, 3-t-BHA, for antagonizing B(a)P-induced neoplasia of mouse forestomach (Slaga and DiGiovanni, 1984). Other investigations using the two-stage system of mouse skin tumorigenesis have demonstrated that the phenolic antioxidants, BHA and BHT, effectively inhibited tumor initiation by DMBA (Slaga and DiGiovanni, 1984). In this study it was reported that BHA and BHT inhibited the covalent binding of DMBA and B(a)P to epidermal DNA. This effect on PAH activation to binding products could account for the anticarcinogenic activity of these compounds. BHA and BHT also inhibited the carcinogenic action in various animal model systems of several other carcinogens: bracken fern, diethylnitrosamine (DEN), 4-NQO, 2-acetylaminofluorene (AAF), N-OH-AAF, and 1,2-dimethyl-hydrazine (DMH). In addition to the alterations in oxidative metabolism, the phenolic antioxidants have been shown to increase the detoxification pathways for many chemical carcinogens (Slaga and DiGiovanni, 1984).

There is a paucity of data on the mechanism(s) of antagonism produced by the other antioxidants listed in **Table 3**. Disulfirams, Vitamin C, and Vitamin E appear to inhibit chemical carcinogenesis in a similar manner to the phenolic antioxidants by their effect on the metabolism of the carcinogen, their antioxidizing activity, and preventing the formation of carcinogens. The mechanism by which selenium (Se) inhibits chemically induced tumors may be related to its effect on glutathione peroxidase (GSHP) since it is a cofactor for this enzyme.

A number of potent inhibitors of chemical carcinogenesis which appear effective because they either prevent the formation of the ultimate carcinogen and/or they scavenge the reactive ultimate carcinogen is summarized in **Table 4**. The indoles, aromatic isothiocyanates, coumarins, flavonoids, dithiothiones, organosulfides, and glucarates have a potent effect on the metabolism of carcinogens (Wattenberg, 1984). In general, they appear to have a major effect on the detoxification of the carcinogens. Ellagic acid and 2,6-dithiopurine have been shown to be highly potent in scavenging the ultimate (reactive) carcinogenic form of the carcinogen (Wattenberg, 1992; MacLeod et al., 1991). The majority of these chemicals have properties like the phenolic antioxidants such as BHA and BHT. They have both antioxidizing activity and influence the metabolism of carcinogens.

Table 4. Chemicals that inhibit carcinogenesis by preventing the carcinogen from reaching target site *

Indoles	Organosulfides
Aromatic isothiocyanates	Ellagic acid
Coumarins	Glucarate
Flavonoids	2,6-Dithiopurine
Dithiothiones	Phenolic and Polyphenolic antioxidants

* See Slaga and DiGiovanni, 1884; Wattenberg, 1992 for details.

Although the antioxidants have already been discussed in terms of inhibiting complete carcinogenesis and tumor initiation, they also are effective inhibitors of tumor promotion and progression (Slaga and DiGiovanni, 1984; Ames et al., 1993). The majority of antioxidants listed in Table 3 have also been shown to inhibit tumor promotion. For example BHA and BHT inhibited both TPA and benzoyl peroxide promotion in mouse skin. The major isomer, 3-T-BHA, was a very potent inhibitor of TPA promotion in mouse skin. In addition, disulfiram and 4-p-hydroxyanisole were effective inhibitors of TPA promotion (Slaga and DiGiovanni, 1984).

Although the mechanism by which the antioxidants inhibit tumor promotion in mouse skin is not presently known, they may be scavenging radicals generated directly in the case of benzoyl peroxide or indirectly by TPA. The fact that benzoyl peroxide and other free radical-generating compounds, such as lauroyl peroxide and chloroperbenzoic acid, are effective skin tumor promoters suggests that free radicals may be important in tumor promotion (Halliwell, 1993; Kripke et al., 1992). An analogous situation would be the phorbol ester tumor promoters. They can stimulate superoxide unoin production in polymorphonuclear leukocytes, and the antipromoters such as dexamethasone and antioxidants can counteract this effect (Slaga, 1989; Kripke et al., 1992).

Selenium (Se) was found to be an effective inhibitor of skin tumor promotion by croton oil. Se is a necessary cofactor for the enzyme glutathione peroxidase (GSHP) that detoxifies hydrogen peroxide and hydroperoxides within the cell. The possibility exists that Se-dependent GSHP lowers the level of potentially damaging peroxide radicals that are generated from various carcinogenic and promoting chemicals. a-Tocopherol and ascorbic acid have been found to significantly reduce tumor formation induced by DMBA and croton oil. a-Tocopherol also reduced the number of fibrosarcomas induced by 3-methylcholanthrene (MCA) (Halliwell, 1993; Buettner, 1993) and mammary gland adenocarcinomas induced by DMBA. Ascorbic acid was found to inhibit the transformation of C3H/10T 1/2 cells by MCA. The inhibitory effect of ascorbic acid in this study was observed in some cases long after the carcinogen was given. This observation indicates it was active during the promotional stage or possibly progression.

When nordihydroguaiaretic acid (NDGA) was assessed as an antipromoting agent in mouse skin, it was found to be a very potent inhibitor. The possible mechanisms of action may be due to its effect as a scavenger of free radicals and/or its inhibitory effect on arachodonic acid metabolism (Wang et al., 1991; Fischer et al., 1990). NDGA also has an antimutagenic activity which may be related to its inhibitory effect on carcinogen metabolism and DNA-adduct formation (Wang et al., 1991).

Several new antioxidants have recently been found to inhibit chemical carcinogenesis and skin tumor promotion. Caventol was found to inhibit skin tumor promotion through inhibition of tumor necrosis factor a release and protein isoprenylation (Komori et al., 1993). Several polyphenolic antioxidants in green tea have been shown to inhibit chemical carcinogenesis and skin tumor promotion (Fujiki et al., 1992). The main constituent of the green tea polyphenolic antioxidants is (-)-epigallocatechingallate (EGCG) which has been found to be a very potent antipromoter in mouse skin (Fujiki et al., 1992).

A number of antioxidants have been found to inhibit skin tumor promotion and/or progression in mouse skin. Although their mechanism of action is not definitely known, evidence points to several possibilities: (1) they scavenge various radicals generated directly or indirectly by tumor promoters; (2) they increase levels of enzymes that are important in detoxifying cellular radicals; and/or (3) they have other specific functions. Several antioxidants have been shown to have synergistic activities such as vitamins C and E, vitamin E and selenium, and BHA and vitamin E (Buettner, 1993; Warren et al., 1993).

Glutathione, ethyl-ester of glutathione and N-acetylcysteine were also found to inhibit skin tumor promotion and progression. Diethyl maleate, a chemical that reduces glutathione levels, was found to be an effective enhancer of tumor progression. In addition, overexpression of g-glutamyltranspeptidase which leads to a reduction in cellular glutathione levels also enhances tumor progression (Warren et al., 1993). These studies suggest that glutathione is very important in both skin tumor promotion and progression. N-acyldehydroaniline derivatives were found to be very effective inhibitors of skin tumor progression but did not have any significant effect on skin tumor promotion (Vo et al., 1991).

ANTIOXIDANTS INHIBIT ALL STAGES OF CARCINOGENESIS

Evidence is accumulating that suggests that free radicals are important in all stages of chemical carcinogenesis. Several antioxidants have been found to inhibit all stages of carcinogenesis whereas others are more effective against tumor initiation, promotion, or progression. Many of the antioxidants and related compounds such as phenolic antioxidants and vitamins C and E appear to be effective in counteracting the tumor initiating phase of carcinogenesis. This appears related to their antioxidant activity and their effect on carcinogen metabolism. In addition, many antioxidants such as the phenolic and polyphenolic antioxidants are potent inhibitors of the tumor promotion phase of carcinogenesis. Their effect on the free radical defense mechanisms, their antioxidizing activity and their effect on many critical events in tumor promotion, such as arachidonic acid metabolism, possibly explain why the antioxidants are potent inhibitors of tumor promotion. In some cases the antioxidants interact synergistically to inhibit carcinogenesis. A number of antioxidants such as glutathione and some derivatives, cysteine and some derivatives, and N-acyl dehydroaniline derivatives have been shown to inhibit tumor progression. These findings suggest that many antioxidants and related compounds are effective inhibitors of either tumor initiation, promotion, and/or progression. In a number of cases the mechanism(s) of action are related to their abilities to prevent critical carcinogen metabolism, and to increase detoxification pathways for carcinogens and free radicals as well as their antioxidizing activity.

SUMMARY

Many different antioxidants have been shown to inhibit the induction of cancer by a wide variety of chemical carcinogens and/or radiation at many target sites in mice, rats, and hamsters. Evidence is accumulating that suggests that free radicals are important in all stages of chemical carcinogenesis. Both carcinogens and tumor promoters have also been shown to decrease the cellular activity of superoxide dismutase and catalase. A number of antioxidants and related compounds were tested to determine if they would inhibit either skin tumor initiation, promotion, and/or progression. In terms of skin tumor initiation, BHT, vitamin E and C and CuDIPS have been found to inhibit DMBA skin tumor initiation by approximately 50%. The mechanism of action of these compounds appears to be related to their effect on the metabolism of DMBA, since BHT and CuDIPS do not inhibit the initiating activity of BP-diol-epoxide and MNNG. Although several antioxidants do inhibit skin tumor initiation by procarcinogens, antioxidants are in general much more effective inhibitors of skin tumor promotion. BHT, BHA, parahydroxyanisole, disulfiran, and vitamin E and C inhibit skin tumor promotion by TPA and benzoyl peroxide by greater than 90%. We also determined the effect of free radical scavengers on the progression process. Of the agents tested glutathione and N-acyl dehydroalamines were the most effective in reducing carcinoma incidence. Diethyl maleate, a chemical that reduces glutathione levels, was effective in enhancing progression. In addition overexpression of g-glutamylt-ranspeptidase (GGT) which leads to a reduction in cellular glutathione levels also enhances progression. These results suggest that GGT has a functional role in skin tumor progression, and that a number of antioxidants are either effective inhibitors of skin tumor initiation, promotion, and/or progression.

ACKNOWLEDGEMENT

This research was supported by Public Health Service grants CA-34962, CA-34521, CA-34890, and CA-43278 from the National Cancer Institute.

REFERENCES

Ames, B.N., Shigenaga, M.K., and Hagan, T.M., 1993, Oxidants, antioxidants, and the degenerative diseases of aging, Proc. Natl. Acad. Sci. USA 90:7915-7922.

Amstad, P., Peskir, A., Shah, G., Marault, M.E., Moret, R., Zbinden, I., and Cerutti, P., 1991, The balance between Cu,Zn-superoxide dismutase and catalase affects the sensitivity of mouse epidermal cells to oxidative stress, Biochemistry 30:9305-9313.

Bindahl, T., 1985, Carcinogenesis and DNA Repair, Cancer Surveys 4:489-626.

Borek, C., and Troll, W., 1983, Modifiers of free radicals inhibit in vitro the oncogenic actions of x-rays, bleomycin and the tumor promoter 12-0-tetradecanoylphorbol 13-acetate, Proc. Natl. Acad. Sci. USA 80:1304-1307.

Buettner, G.R., 1993, The pecking order of free radicals and antioxidants: Lipid peroxidation, a-locophenol, and ascorbate, Arch. Biochem. Biophys. 300:535.

Fischer, S.M., Patrick, K.E., Patamalai, B., and Slaga, T.J., 1990, Effects of antihistamines on phorbol ester tumor promotion and vascular permeability changes, Carcinogenesis 11:991-996.

Fredovich, I., 1974, Superoxide dismutases, Adv. Enzymol. 41:35-97.

Fujiki, H., Yoshizawa, S., Horiuchi, T., Suganuma, M. Yatsunami, J., Nishiwaki-Matsushima, R., Okuda, T., and Sugimura, T., 1992, Anticarcinogenic effects of (-)-epigallocatechin gallate, Preventive Medicine 21:503-509.

Halliwell, B., 1993, The chemistry of free radicals, in: "Antioxidants, Chemical, Physiological, Nutritional and Toxicological Aspects - Toxicology and Industrial Health," G.M. Williams, ed., Priceton Scientific Publishing Co., Princeton, NJ.

Kensler, T.W., Bush, D.M., and Kozumbo, W.J., 1993, Inhibition of tumor promotion by a biomemetic superoxide dismutase, Science 221:75-77.

Komori, A., Suganuma, M., Okabe, S., Zou, X., Tius, M.A., and Fujiki, H., 1993, Canventol inhibits tumor promotion in CD-1 mouse skin through inhibition of tumor necrosis factor a release and of protein isoprenylation, Cancer Research 53:3462-3464.

Kripke, M.L., Cox, P., Alas, L., and Yarosh, D.B., 1992, Pyrimidine dimers in DNA initiate systemic immunosuppression in UV-irradiated mice, Proc. Natl. Acad. Sci. 89:7516-7520.

Li, J.Y., Taylor, P.R., Li, B., Li, B., Dawsey, S., Wang, G.Q., Ershow, A.G., Guo, W., Liu, S.F., Yang, C.S., Shen, Q., Wang, W., Mark, S.D., Zou, X., Greenwald, P., Wu, Y., and Blot, W.J., 1993, Nutrition intervention trials in Linxian, China: Multiple vitamin/mineral supplemen-tation, cancer incidence, and disease-specific mortality among adults with esophageal dysplasia, J. Natl. Cancer Inst. 85:1492-1498.

MacLeod, M.C., Mann, K.L., Thai, G., Conti, C.J., and Reiners, J.J. Jr., 1991, Inhibition by 2,6-dithiopurine and theopurinol of binding of a benzo(a)pyrene-diol-epoxide to DNA in mouse epidermis and of the initiation phase of two-stage carcinogenesis, Cancer Research 51:4859-4864-4864.

Nakamura, Y., Colburn, N.H., and Gindhart, T.D., 1985, Role of reactive oxygen in tumor promotion: implication of superoxide anion in promotion of neoplastic transformation in JB-6 cells by TPA, Carcinogenesis 6:229-235.

Overley, L.W., and Buettner, G.R., 1979, Role of superoxide dismutase in cancer: A review, Cancer Research 39:1141-1149.

Slaga, T.J., and DiGiovanni, J., 1984, Inhibition of Chemical Carcinogenesis, in: "Chemical Carcinogens," C.E. Searle, ed., ACS Monograph # 182, Washington, D.C.

Slaga, T.J., 1989, Critical events and determinants in multistage skin carcinogenesis, in: "Carcinogenicity and Pesticides," N.N. Ragsdale, and R.E. Menzer, eds., American Chemical Society, Washington, DC.

Solanki, V., Rana, R.S., and Slaga, T.J., 1981, Diminution of mouse epidermal superoxide dismutase and catalase activities by tumor promoters, Carcinogenesis 2:1141-1146.

Solanki, V., Yotti, L., Logani, M.K., and Slaga, T.J., 1984, The reduction of tumor initiating activity and cell mediated mutagenicity of dimethylbenz[a]anthracene by a copper coordination compound, Carcinogenesis 5:129-131.

Vo, T.K.O., Fischer, S.M., and Slaga, T.J., 1991, Effects of N-acyl dehydroalanines on phorbol ester-elicited tumor development and other events in mouse skin, Cancer Letters 60:25-32.

Wang, Z.Y., Agarwal, R., Zhou, Z.C., Bickers, D.R., and Mukhtar, H., 1991, Antimutagenic and antitumorigenic activities of nordihydroguearetic acid, Mutation Research 261:153-162.

Warren, B.S., Naylor, M.F., Winberg, L.D., Yoshimi, N., Volpe, J.P., Gimenez-Conti, I., and Slaga, T.J.,1993, Induction and inhibition of tumor progression, Proc. Soc. Exp. Biol. Med. 202:9-15.

Wattenberg, L.W., 1992, Inhibition of carcinogenesis by minor dietary constituents, Cancer Research (Suppl) 52:2085-2091.

Yarosh, D., and Yee, V., 1990, Skh-1 hairless mice repair UV-induced pyrimidine dimers in epidermal DNA, J. Photochem. Photobiol. 7:173-179.

NUTRIENTS, SIGNAL TRANSDUCTION AND CARCINOGENESIS

Steven H. Zeisel

Department of Nutrition
School of Public Health and School of Medicine
University of North Carolina at Chapel Hill
Chapel Hill, NC 27599-7400

INTRODUCTION

Choline-deficiency is an excellent model system in which to study how a nutrient can influence cell signaling and cause cancer. Choline is universally distributed in all cells. Choline's most important functions are that it is needed to form acetylcholine (a major neurotransmitter), it is needed as a methyl donor, and it is a precursor for the biosynthesis of membrane phospholipids (including phosphatidylcholine, lysophosphatidylcholine, choline plasmalogen, platelet activating factor and sphingomyelin) (Zeisel, 1990).

The importance of choline as a nutrient was accidentally discovered during the pioneering work on insulin (Best and Huntsman, 1932, 1935). Depancreatized dogs developed fatty infiltration of the liver and died. Administration of raw pancreas prevented hepatic damage; the active component was the choline moiety of pancreatic phosphatidylcholine. The term "lipotropic" was coined to describe choline and other substances that prevented deposition of fat in the liver.

Choline is the only nutrient for which dietary deficiency causes development of hepatocarcinomas without exposure to any known carcinogen (Newberne and Rogers, 1986). Choline deficient rats also are markedly sensitized to the effects of administered carcinogens (Newberne and Rogers, 1986). There are several mechanisms suggested for the cancer-enhancing effects of a choline deficient diet. On such a diet, hepatocytes begin to rapidly proliferate but also begin to die more rapidly (Newberne and Rogers, 1986; Chandar et al., 1987; Chandar and Lombardi, 1988). The proliferation-associated increased rate of DNA synthesis could be the cause of greater sensitivity to chemical carcinogens (Ghoshal et al., 1983). Hepatectomy and necrogenic chemicals are examples of stimuli for increased DNA synthesis which result in increased carcinogenesis. Increased DNA synthesis is not the sole cause of carcinogenesis in choline deficiency because the overall rate of liver cell proliferation could be dissociated from the rate at which preneoplastic lesions formed (Shinozuka and Lombardi, 1980). Methylation of DNA is important for the regulation of expression of genetic information. Under-methylation of DNA, observed during choline deficiency (despite

Nutrition and Biotechnology in Heart Disease and Cancer
Edited by J.B. Longenecker *et al.*, Plenum Press, New York, 1995

175

adequate dietary methionine), may be responsible for carcinogenesis (Locker et al., 1986; Dizik et al., 1991). Another proposed mechanism derives from the observation that, when rats eat a choline deficient diet, increased lipid peroxidation occurs within liver (Rushmore et al., 1984). Lipid peroxides in the nucleus could be a source of free radicals that could modify DNA, and cause carcinogenesis. Recently, we have shown that choline deficiency perturbs protein kinase C (PKC) signal transduction, thereby promoting carcinogenesis (daCosta et al., 1993).

CHOLINE DEFICIENCY IN ANIMALS

The demand for choline as a methyl donor is probably the major factor which determines how rapidly a diet deficient in choline will induce pathology. As given in **Figure 1** the pathways of choline and 1-carbon metabolism intersect at the formation of methionine from homocysteine (Zeisel, 1990). Methionine is regenerated from homocysteine in a reaction

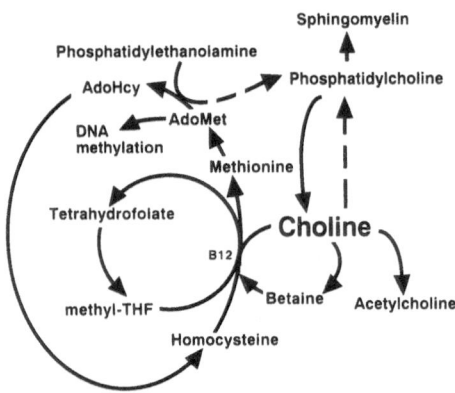

Figure 1. Pathways for choline metabolism. Abbreviations: AdoMet = S-adenosylmethionine; AdoHcy = S-adenosylhomocysteine

catalyzed by betaine:homocysteine methyltransferase, in which betaine, a metabolite of choline, serves as the methyl donor (Finkelstein et al., 1982). Betaine concentrations in livers of choline deficient rats are markedly diminished (Pomfret et al., 1990) as are total folate concentrations (Selhub et al., 1991). The only alternative mechanism for regeneration of methionine is via a reaction catalyzed by 5-methyltetrahydrofolate:homocysteine methyltransferase which uses a methyl group generated *de novo* from the 1-carbon pool (Finkelstein et al., 1988). Methionine is converted to S-adenosylmethionine in a reaction catalyzed by methionine adenosyl transferase. S-adenosylmethionine is the active methylating agent for many enzymatic methylations.

In the human (Zeisel et al., 1991; Buchman et al., 1992, 1993), rat (Lombardi, 1971), hamster (Handler and Bernheim, 1949), guinea pig (Tani et al., 1967), pig (Fairbanks and Krider, 1945; Blair and Newsome, 1985), dog (Hershey and Soskin, 1931; Best and Huntsman, 1932, 1935), and monkey (Hoffbauer and Zaki, 1965) choline deficiency results in liver dysfunction due to accumulation of triacylglycerol within the hepatocyte (Lombardi et al., 1968; Lombardi, 1971; Yao and Vance, 1988; Blusztajn and Zeisel, 1989; Yao and Vance, 1989). Fatty liver begins within hours to days after rats start to eat a choline deficient diet (daCosta et al., 1993). The triacylglycerol produced by liver is delivered to other tissues mainly in the form of very low density lipoprotein (VLDL). Phosphatidylcholine is a required component of the VLDL particle (Yao and Vance, 1988; Yao and Vance, 1989). In choline deficiency, the diminished capacity of liver cells to synthesize new phosphatidylcholine molecules results in the intracellular accumulation of triglycerides because VLDL can not be packaged and secreted. Methionine can substitute for choline, but only as long as phosphatidylethanolamine-N-methyltransferase activity is present (Yao and Vance, 1988). Secretion of high density lipoprotein (HDL) from hepatocytes does not require the synthesis of new phosphatidylcholine molecules (Yao and Vance, 1990).

Choline deficiency also compromises renal function, with abnormal concentrating ability, free water reabsorption, sodium excretion, glomerular filtration rate, renal plasma flow, and gross renal hemorrhage (Griffith and Wade, 1939; Baxter, 1947; Best and Hartroft, 1949; Michael et al., 1975). Diets low in choline content also cause infertility, growth impairment, bony abnormalities, decreased hematopoiesis, and hypertension (Jukes, 1940; Caniggia, 1950; Kratzing and Perry, 1971; Chang and Jensen, 1975).

CHOLINE DEFICIENCY IN HUMANS

Humans require choline for sustaining normal life. Is a dietary source of choline required? Many foods eaten by humans contain significant amounts of choline and esters of choline (Zeisel, 1990). Some of this choline is added during food processing (especially when preparing infant formula (FASEB Life Sciences Research Office, 1975)). Average choline dietary intake (as choline and choline esters) in the adult human (as free choline and the choline in phosphatidylcholine and other choline esters) is more than 7 to 10 mmol/day (FASEB Life Sciences Research Office, 1981; Zeisel, 1981). When we switched humans from a diet of normal foods to a defined diet containing 5 mmol/day we observed that plasma choline and phosphatidylcholine concentrations decreased in most subjects (Zeisel et al., 1991). This suggests that average dietary intake of choline exceeds this level in adults. A kilogram of beef liver contains 50 mmol of choline moiety, thus, it is easy to consume a diet of normal foods (eggs, liver, etc.) that delivers much more choline per day than the calculated average intake (Zeisel et al., 1980). Thus, it is unusual to find humans that consume a choline deficient diet. When healthy humans were fed an experimental choline-deficient diet for three weeks they developed biochemical changes consistent with choline deficiency (Zeisel et al., 1991). These included diminished plasma choline and phosphatidylcholine concentrations, as well as diminished erythrocyte membrane phosphatidylcholine concentrations. Serum alanine transaminase (ALT) activity, a measure of hepatocyte damage, increased significantly during choline deficiency. Choline deficient humans had diminished plasma low density lipoprotein cholesterol (LDL; derived from VLDL) (Zeisel et al., 1991), suggesting that VLDL secretion was impaired. These observations are consistent with the hypothesis that, in humans as in other species, choline is required in the diet.

This requirement may be more apparent in special human populations. The demand for choline in the infant is likely to be larger than the demand for choline in normal adults, as large amounts of choline must be used to make phospholipids in growing organs (Zeisel, 1990). The observed changes that occurred in choline-deficient adult humans might have been greater had we studied growing children. Malnourished humans, in whom stores of choline, methionine and folate have been depleted (Sheard et al., 1986; Chawla et al., 1989), are also likely to need more dietary choline than did our healthy adult subjects. The liver is the primary site for endogenous synthesis of choline. Alcoholics with liver cirrhosis have diminished plasma choline concentration, and have fatty liver which resolves when patients are supplemented with choline (Chawla et al., 1989). Patients fed parenterally become choline deficient and develop hepatic dysfunction; treatment with choline-compounds can reverse this dysfunction (Buchman et al., 1992; Buchman et al., 1993).

OVERVIEW OF SIGNAL TRANSDUCTION

Our understanding of signal transduction within cells has been developed during the last decade. Stimulation of membrane-associated receptors activates neighboring phospholipases, resulting in the formation of breakdown products that are signaling molecules either by themselves (i.e., they stimulate or inhibit the activity of target macromolecules), or after conversion to signaling molecules by specific enzymes (see **Figure 2**). Much of signaling research focuses on phosphatidylinositol derivatives and choline phospholipids, especially phosphatidylcholine and sphingomyelin, as substrates for formation of biologically active molecules that can amplify external signals or that can terminate the signaling process by generating inhibitory second messengers (Zeisel, 1993).

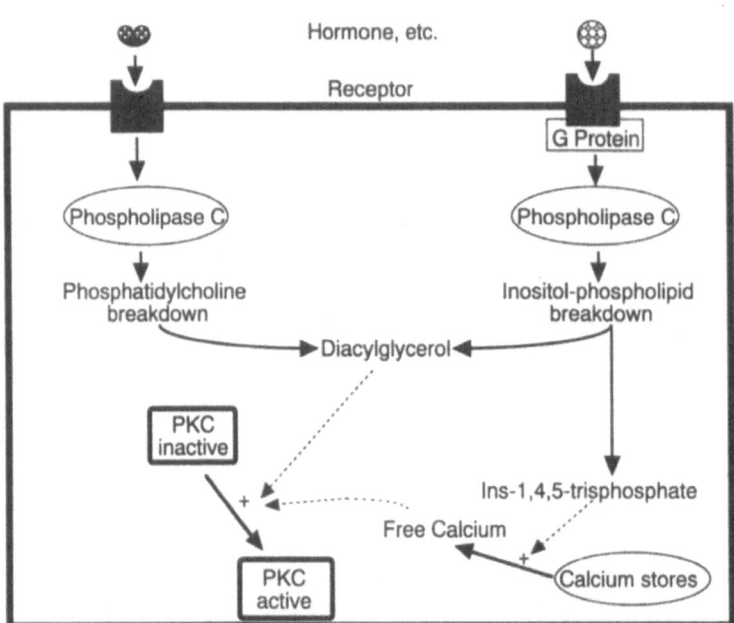

Figure 2. Overview of signal transduction: When cell receptors are activated by a variety of ligands specific phospholipase Cs are activated which hydrolyze phosphatidylinositol-bisphosphate and phosphatidylcholine. This generates inositol-1,4,5-triphosphate (releases calcium from storage pools) and diacylglycerol (binds to protein kinase C and reveals the enzyme's active site).

Signal transduction via receptor mediated hydrolysis of phosphatidylinositol-bis-phosphate has been extensively reviewed elsewhere (Berridge, 1989; Taylor and Marshall, 1992). Briefly, activation of specific receptors leads to altered conformation of the receptor so that it can activate a GTP binding protein (G-protein). These G-proteins are a highly conserved family of membrane associated heterotrimeric proteins. The activation of the G-protein results in the subsequent activation of phospholipase C activity within the plasma membrane. The phospholipase Cs are a family of phosphodiesterases which hydrolyze the glycerophosphate bond of intact phospholipids to generate 1,2-sn-diacylglycerol and an aqueous soluble head group. Numerous phosphatidylinositol-bis-phosphate specific phospholipase Cs exist, and it is believed that specific receptors couple to specific phospholipase C isotypes (Meldrum et al., 1991). In a similar manner, specific receptors appear to be linked to activation of specific phosphatidylcholine-phospholipase Cs (Exton, 1990).

The action of phospholipase C triggers the next event in the signal cascade, which is the activation of PKC. Products generated by phosphatidylinositol-bis-phosphate specific phospholipase Cs include inositol-1,4,5-trisphosphate (Ins-1,4,5-P3) and diacylglycerol. Ins-1,4,5-P3 is a water soluble product, which acts to release calcium from stores in the endoplasmic reticulum. Some PKC isotypes are Ca^{2+} dependent (PKCa, b2, and g). PKC d, e, z, q and h lack a calcium binding-domain (Nishizuka, 1992). Calcium, by increasing membrane occupancy of PKC, places the enzyme in close proximity to phosphatidylserine, a cofactor for PKC activation. Diacylglycerol, the other product of phospholipase C, remains in the plasma membrane, and is both a messenger molecule and an intermediate in the metabolism of lipids. Normally PKC is folded so that an endogenous "pseudosubstrate" region on the protein is bound to the catalytic site, thereby inhibiting activity. The combination of diacylglycerol and Ca^{2+} causes a conformational change in PKC, causing flexing at a hinge region so as to withdraw the pseudosubstrate and unblock the PKC catalytic site. The appearance of diacylglycerol in membranes is usually transient, and therefore PKC is activated only for a short time after a receptor has been stimulated.

The generation of diacylglycerol from membrane phosphatidylcholine amplifies the signal. Phospholipase C and indirectly, phospholipase D, generate this diacylglycerol (Exton, 1990; Qian and Drewes, 1990; Conricode et al., 1992; Liscovitch, 1992). Changes in intracellular calcium may regulate phospholipase D activity (Anthes et al., 1989; Liscovitch, 1992). There is also evidence that receptors with intrinsic tyrosine kinase activity (e.g. epidermal growth factor or platelet derived growth factor receptors) stimulate phosphatidylcholine hydrolysis (Exton, 1990). Sustained activation of PKC is essential for triggering cell differentiation and proliferation (Nishizuka, 1992). Other products of phosphatidylcholine hydrolysis, such as phosphatidic acid, lysophosphatidylcholine and free fatty acids also are second messengers (Besterman et al., 1986; Exton, 1990). Phosphatidic acid can act as a mitogen (Wakelam et al., 1991). Lysophosphatidylcholine stimulates PKC activity (Nishizuka, 1992) but it is a membrane-lytic detergent with potential toxic effects. Lysophosphatidylcholine generation is important in chemotaxis, relaxation of smooth muscle and activation of T-lymphocytes (Nishizuka, 1992). Thus, lysophosphatidylcholine may be a choline-containing putative second messenger generated by phospholipase A2. Recent data indicate that phosphatidylcholine may synergize with diacylglycerol and calcium to activate PKC in lymphocytes (Asaoka et al., 1992) and that modulation of PKC isozymes by phosphatidylcholine may be isoform specific (Sasaki et al., 1993).

The characterization of events that occur down-stream from PKC is just beginning. Serine-threonine kinases and tyrosine kinases catalyze phosphorylation of target proteins distal to PKC. Phosphorylation alters the biochemical properties of these substrates, resulting in a range of cellular responses. These phosphorylation cascades serve to enhance amplification of the original signal. PKC signals impinge on several known intracellular control circuits (Stabel and Parker, 1991). The targets for phosphorylation by PKC include receptors for insulin, epidermal growth factor and many proteins involved in control of gene expression (Nishizuka, 1986; Weinstein, 1990).

CHOLINE DEFICIENCY AND SIGNAL TRANSDUCTION

As discussed earlier, choline deficiency causes fatty liver. We have observed that 1,2-*sn*-diacylglycerol accumulates in this fatty liver (Blusztajn and Zeisel, 1989; daCosta et al., 1993). In plasma membrane from livers of choline deficient rats, diacylglycerol reaches values higher than those occurring after stimulation of a receptor linked to phospholipase C activation (e.g., vasopressin receptor). This results in a stable activation of PKC and/or an increase in the total PKC pool in the cell (daCosta et al., 1993) with changes in several PKC isotypes (at 6 weeks of choline deficiency, amounts of PKC a and d increased 2-fold and 10-fold, respectively).

The accumulation of diacylglycerol and subsequent activation of PKC within liver during choline deficiency may be the critical abnormality that eventually contributes to the development of hepatic cancer in these animals (daCosta et al., 1993). Abnormalities in PKC-mediated signal transduction may trigger carcinogenesis (Weinstein, 1990). Many mitogens activate PKC (Nishizuka, 1992) and the phorbol esters (diacylglycerol analogs) are potent tumor promoters (Nishizuka, 1986). Okadaic acid (an inhibitor of protein phosphatases whose net effect is to increase protein phosphorylation of PKC targets) is also a potent tumor promoter (Hunter, 1991). The expression of oncogenes can perturb PKC-mediated signal transduction in a manner similar to that which we described for choline deficiency. In *erbB*-transformed fibroblasts, diacylglycerol accumulates because diacylglycerol kinase activity does not remove it (Kato et al., 1989). Transformation by *ras* or *src* oncogenes translocates PKC to the plasma membrane (Diaz-Laviada et al., 1990). Diacylglycerol concentrations are elevated *in vivo* in *ras*-transformed liver of neonatal transgenic mice (Wilkison et al., 1989). NIH 3T3 cells transformed with Ha-*ras*, Ki-*ras*, v-*src*, or v-*fms* oncogenes have increased diacylglycerol levels as well as tonic activation and partial down regulation of PKC (Wolfman et al., 1987). Degradation of phosphatidylcholine produces these elevated diacylglycerol levels (Price et al., 1989; Diaz-Laviada et al., 1990). Transfection of fibroblasts so that they overexpress PKC (a, b1 or e) activity causes them to become transformed and tumorigenic (Persons et al., 1988; Krauss et al., 1989; Megidish and Mazurek, 1989; Stabel and Parker, 1991; Cacace et al., 1993). Thus, many observations suggest that choline deficiency causes cancer by increasing diacylglycerol concentration with subsequent sustained activation of PKC.

As noted earlier, choline deficient rats not only have a higher incidence of spontaneous hepatocarcinoma, but that they are markedly sensitized to the effects of a variety of administered carcinogens (Newberne and Rogers, 1986). This has important implications, as choline status may modulate the threshold for sensitivity to environmental carcinogens. In addition, it has direct applicability to parenteral nutrition therapy in humans because many of these patients are also being treated with chemotherapy. When choline deficient male rats were treated with procarbazine, a chemotherapeutic agent, they had 50% higher mammary tumor incidence than did controls treated with procarbazine (Rogers et al., 1990). This could be extremely important if choline deficiency has the same effect in humans. Cancer patients treated with chemotherapeutic agents have increased incidences of secondary primary malignancies compared with patients not so treated (Byrd, 1985). These patients are often fed parenterally and might be choline deficient. Methotrexate, an inhibitor of folate metabolism, exacerbates choline deficiency (Pomfret et al., 1990). This occurs because folate-methyl groups are the only alternative source when choline is not available. The choline depletion secondary to parenteral nutrition in combination with methotrexate treatment might be a very toxic combination.

SUMMARY

Choline phospholipids play major roles in cellular regulation in addition to their essential function as structural components of membranes and lipoproteins. The unique functions of choline phospholipids as hormones (platelet activating factor; 1-alkyl, 2-acetyl-phosphatidylcholine; PAF) and sources (phosphatidylcholine, sphingolipids) of second messengers (sphingosine, diacylglycerol, lysophospholipids, arachidonic acid and its metabolites) may explain how dietary choline influences normal physiological processes as well as a diverse group of pathological processes, including carcinogenesis.

ACKNOWLEDGMENTS

The work described in this review was supported by a grant from the American Institute for Cancer Research.

REFERENCES

Anthes, J. C., Eckel, S., Siegel, M. I., Egan, R. W. and Billah, M. M., 1989, Phospholipase D in homogenates from HL-60 granulocytes: Implications of calcium and G protein control, *Biochem. Biophys. Res. Commun.* **163**, 657-664.

Asaoka, Y., Oka, M., Yoshida, K., Sasaki, Y. and Nishizuka, Y. ,1992, Role of lysophosphatidylcholine in T-lymphocyte activation: involvement of phospholipase A2 in signal transduction through protein kinase C, *Proc. Nat.l Acad. Sci. ,U S A,* **89**, 6447-51.

Baxter, J. H. ,1947, A study of hemorrhagic-kidney syndrome of choline deficiency, *J. Nutr.* **34**, 333.

Berridge, M. J. ,1989, Inositol trisphosphate, calcium, lithium, and cell signaling, *JAMA* **262**, 1834-1841.

Best, C. H. and Hartroft, W. S. ,1949, Symposium on nutrition in preventative medicine: Nutrition, renal lesions and hypertension, *Fed. Proc.* **8**, 610.

Best, C. H. and Huntsman, M. E. ,1932, The effects of the components of lecithin upon the deposition of fat in the liver, *J. Physiol.* **75**, 405-12.

Best, C. H. and Huntsman, M. E. ,1935, Effect of choline on liver fat of rats in various states of nutrition, *J. Physiol.* **83**, 255-274.Besterman, J. M., Duronio, V. and Cuatrecasas, P. ,1986, Rapid formation of diacylglycerol from phosphatidylcholine: a pathway for generation of a second messenger, *Proc. Natl. Acad. Sci. USA* **83**, 6785-9.

Blair, R. and Newsome, F. ,1985, Involvement of water-soluble vitamins in diseases of swine, *J. Animal Sci.* **60**, 1508-17.

Blusztajn, J. K. and Zeisel, S. H. ,1989, 1,2-*sn*-diacylglycerol accumulates in choline-deficient liver, A possible mechanism of hepatic carcinogenesis via alteration in protein kinase C activity? *FEBS Lett.* **243**, 267-270.

Buchman, A. L., Dubin, M., Jenden, D., Moukarzel, A., Roch, M. H., Rice, K., Gornbein, J., Ament, M. E. and Eckhert, C. D. ,1992, Lecithin increases plasma free choline and decreases hepatic steatosis in long-term total parenteral nutrition patients, *Gastroenterology* **102**, 1363-1370.

Buchman, A. L., Moukarzel, A., Jenden, D. J., Roch, M., Rice, K. and Ament, M. E. ,1993, Low plasma free choline is prevalent in patients receiving long term parenteral nutrition and is associated with hepatic aminotransferase abnormalities, *Clin. Nutr.* **12**, 33-37.

Byrd, R. ,1985, Late effects of treatment of cancer in children. *Pediatr. Clin. N. Am.* **32**, 835-856.

Cacace, A. M., Guadagno, S. N., Krauss, R. S., Fabbro, D. and Weinstein, I. B. ,1993, The epsilon isoform of protein kinase C is an oncogene when overexpressed in rat fibroblasts, *Oncogene* **8**, 2095-2104.

Caniggia, A. ,1950, Effect of choline on hemopoiesis, *Haematologica* **34**, 625-627.

Chandar, N., Amenta. J., Kandala. J. C. and Lombardi, B. ,1987, Liver cell turnover in rats fed a choline-devoid diet. *Carcinogenesis* **8**, 669-673.

Chandar, N. and Lombardi. B. ,1988, Liver cell proliferation and incidence of hepatocellular carcinomas in rats fed consecutively a choline-devoid and a choline-supplemented diet, *Carcinogenesis* **9**, 259-263.

Chang, C. H. and Jensen. L. S. ,1975, Inefficacy of carnitine as a substitute for choline for normal reproduction in Japanese quail, *Poultry Sci.* **54**, 1718-20.

Chawla, R. K., Wolf, D. C., Kutner, M. H. and Bonkovsky, H. L. ,1989, Choline may be an essential nutrient in malnourished patients with cirrhosis, *Gastroenterology* **97**, 1514-20.

Conricode, K. M., Brewer, K. A. and Exton. J. H. ,1992, Activation of phospholipase D by protein kinase C. Evidence for a phosphorylation-independent mechanism, *J. Biol. Chem.* **267**, 7199-7202.

daCosta, K., Cochary. E. F., Blusztajn, J. K., Garner, S. C. and Zeisel, S. H. ,1993, Accumulation of 1,2-*sn*-diradylglycerol with increased membrane-associated protein kinase C may be the mechanism for spontaneous hepatocarcinogenesis in choline deficient rats, *J. Biol. Chem.* **268**, 2100-2105.

Diaz-Laviada, I., Larrodera. P., Diaz-Meco. M., Cornet. M. E., Guddal, P. H., Johansen, T. and Moscat, J. ,1990, Evidence for a role of phosphatidylcholine-hydrolysing phospholipase C in the regulation of protein kinase C by *ras* and *src* oncogenes. *Embo J.* **9**, 3907-3912.

Dizik, M., Christman, J. K. and Wainfan, E. ,1991, Alterations in expression and methylation of specific genes in livers of rats fed a cancer promoting methyl-deficient diet, *Carcinogenesis* **12**, 1307-1312.

Exton, J. H. ,1990, Signaling through phosphatidylcholine breakdown. *J. Biol. Chem.* **265**, 1-4.

Fairbanks, B. W. and Krider, J. L. ,1945, Significance of B vitamins in swine nutrition, *N. Am. Vet.* **26**, 18-23.

FASEB Life Sciences Research Office ,1975, Evaluation of the health aspects of choline chloride and choline bitartrate as food ingredients, Report # PB-223 845/9, Bureau of Foods, Food and Drug Administration, Department of Health, Education, and Welfare, Washington DC.

FASEB Life Sciences Research Office ,1981, Effects of Consumption of choline and lecithin on neurological and cardiovascular systems, Report # PB-82-133257, Bureau of Foods, Food and Drug Administration, Department of Health, Education, and Welfare, Washington DC.

Finkelstein, J. D., Martin, J. J. and Harris, B. J. ,1988, Methionine metabolism in mammals. The methionine-sparing effect of cystine, *J. Biol. Chem.* **263**, 11750-4.

Finkelstein, J. D., Martin, J. J., Harris, B. J. and Kyle, W. E. ,1982, Regulation of the betaine content of rat liver, *Arch. Biochem. Biophys.* **218**, 169-73.

Ghoshal, A. K., Ahluwalia, M. and Farber, E. ,1983, The rapid induction of liver cell death in rats fed a choline-deficient methionine-low diet, *Am. J. Pathol.* **113**, 309-314.

Griffith, W. H. and Wade, N. J. ,1939, The occurance and prevention of hemorrhagic degeneration in young rats on a low choline diet, *J. Biol. Chem.* **131**, 567-573.

Handler, P. and Bernheim, F. ,1949, Choline deficiency in the hamster, *Proc. Soc. Exptl. Med.* **72**, 569.

Hershey, J. M. and Soskin, S. ,1931, Substitution of "lecithin" for raw pancreas in a diet of depancreatized dog, *Am. J. Physiol.* **93**, 657-658.

Hoffbauer, F. W. and Zaki, F. G. ,1965, Choline deficiency in the baboon and rat compared, *Arch. Path.* **79**, 364-369.

Hunter, T. ,1991, Cooperation between oncogenes, *Cell* **64**, 249-270.

Jukes, T. H. ,1940, The prevention of perosis by choline, *J.Biol. Chem.* **134**, 789-792.

Kato, M., Kawai, S. and Takenawa, T. ,1989, Defect in phorbol acetate-induced translocation of diacylglycerol kinase in erbB-transformed fibroblast cells, *Febs Lett.* **247**, 247-250.

Kratzing, C. C. and Perry, J. J. ,1971, Hypertension in young rats following choline deficiency in maternal diets, *J. Nutr.* **101**, 1657-61.

Krauss, R., Housey, G., Johnson, M. and Weinstein, I. B. ,1989, Disturbances in growth control and gene expression in a C3H/10T1/2 cell line that stably overproduces protein kinase C, *Oncogene* **4**, 991-998.

Liscovitch, M. ,1992, Crosstalk among multiple signal-activated phospholipases, *Trends in Biochemical Sciences* **17**, 393-399.

Locker, J., Reddy, T. V. and Lombardi, B. ,1986, DNA methylation and hepatocarcinogenesis in rats fed a choline devoid diet, *Carcinogenesis* **7**, 1309-1312.

Lombardi, B. ,1971, Effects of choline deficiency on rat hepatocytes, *Fed. Proc.* **30**, 139-142.

Lombardi, B., Pani, P. and Schlunk, F. F. ,1968, Choline-deficiency fatty liver: impaired release of hepatic triglycerides, *J. Lipid Res.* **9**, 437-46.

Megidish, T. and Mazurek, N. ,1989, A mutant protein kinase C that can transform fibroblasts, *Nature* **342**, 807-811.

Meldrum, E., Parker, P. J. and Carozzi, A. ,1991, The Ptd-Ins-PLC superfamily and signal transduction, *Biochim. Biophys. Acta* **1092**, 49-71.

Michael, U. F., Cookson, S. L., Chavez, R. and Pardo, V. ,1975, Renal function in the choline deficient rat, *Proc. Soc. Exp. Biol. Med.* **150**, 672-76.

Newberne, P. M. and Rogers, A. E. ,1986, Labile methyl groups and the promotion of cancer, *Ann. Rev. Nutr.* **6**, 407-432.

Nishizuka, Y. ,1986, Studies and perspectives of protein kinase C, *Science* **233**, 305-312.

Nishizuka, Y. ,1992, Intracellular signaling by hydrolysis of phospholipids and activation of protein kinase C, *Science* **258**, 607-614.

Persons, D. A., Wilkison, W. O., Bell, R. M. and Finn, O. J. ,1988, Altered growth regulation and enhanced tumorigenicity of NIH 3T3 Fibroblasts transfected with protein kinase C-1 c DNA, *Cell* **52**, 447-458.

Pomfret, E. A., daCosta, K. and Zeisel, S. H. ,1990, Effects of choline deficiency and methotrexate treatment upon rat liver, *J. Nutr. Biochem.* **1**, 533-541.

Price, B. D., Morris, J. D., Marshall, C. J. and Hall, A. ,1989, Stimulation of phosphatidylcholine hydrolysis, diacylglycerol release, and arachidonic acid production by oncogenic ras is a consequence of protein kinase C activation, *J. Biol. Chem.* **264**, 16638-16643.

Qian, Z. and Drewes, L. R. ,1990, A novel mechanism for acetylcholine to generate diacylglycerol in brain, *J. Biol. Chem.* **265**, 3607-10.

Rogers, A. E., Akhtar, R. and Zeisel, S. H. ,1990, Procarbazine Carcinogenicity in methotrexate-treated or lipotrope-deficient male rats, *Carcinogenesis* **11**, 1491-1495.

Rushmore, T., Lim, Y., Farber, E. and Ghoshal, A. ,1984, Rapid lipid peroxidation in the nuclear fraction of rat liver induced by a diet deficient in choline and methionine, *Cancer Lett.* **24**, 251-5.

Sasaki, Y., Asaoka, Y. and Nishizuka, Y. ,1993, Potentiation of diacylglycerol-induced activation of protein kinase C by lysophospholipids: Subspecies difference, *Febs Lett* **320**, 47-51.

Selhub, J., Seyoum, E., Pomfret, E. A. and Zeisel, S. H. ,1991, Effects of choline deficiency and methotrexate treatment upon liver folate content and distribution, *Cancer Res* **51**, 16-21.

Sheard, N. F., Tayek, J. A., Bistrian, B. R., Blackburn, G. L. and Zeisel, S. H. ,1986, Plasma choline concentration in humans fed parenterally, *Am. J. Clin. Nutr.* **43**, 219-24.

Shinozuka, H. and Lombardi, B. ,1980, Synergistic effect of a choline-devoid diet and phenobarbital in promoting the emergence of foci of g-glutamyltranspeptidase-positive hepatocytes in the liver of carcinogen-treated rats, *Cancer Res.* **40**, 3846-3849.

Stabel, S. and Parker, P. J. ,1991, Protein kinase C. *Pharmac. Ther.* **51**, 71-95.

Tani, H., Suzuki, S., Kobayashi, M. and Kotake, Y. ,1967, The physiological role of choline in guinea pigs, *J. Nutr.* **92**, 317-24.

Taylor, C. W. and Marshall, I. ,1992, Calcium and inositol 1,4,5-trisphosphate receptors: A complex relationship, *Trends Biochem Sci* **17**, 403-407.

Wakelam, M. J. O., Cook, S. J., Currie, S., Plamer, S. and Plevin, R. ,1991, Regulation of the hydrolysis of phosphatidylcholine in Swiss 3T3 cells, *Biochem. Soc. Transactions* **19**, 321-324.

Weinstein, I. B. ,1990, The role of protein kinase C in growth control and the concept of carcinogenesis as a progressive disorder in signal transduction, *Adv. Second Messenger Phosphoprotein Res.* **24**, 307-316.

Wilkison, W. O., Sandgren, E. P., Palmiter, R. D., Brinster, R. L. and Bell, R. M. ,1989, Elevation of 1,2-diacylglycerol in ras-transformed neonatal liver and pancreas of transgenic mice, *Oncogene* **4**, 625-628.

Wolfman, A., Wingrove, T. G., Blackshear, P. J. and Macara, I. G. ,1987, Down-regulation of protein kinase C and of an endogenous 80-kDa substrate in transformed fibroblasts, *J. Biol. Chem.* **262**, 16546-16552.

Yao, Z. M. and Vance, D. E. ,1988, The active synthesis of phosphatidylcholine is required for very low density lipoprotein secretion from rat hepatocytes, *J. Biol. Chem.* **263**, 2998-3004.

Yao, Z. M. and Vance, D. E. ,1989, Head group specificity in the requirement of phosphatidylcholine biosynthesis for very low density lipoprotein secretion from cultured hepatocytes, *J. Biol. Chem.* **264**, 11373-11380.

Yao, Z. M. and Vance, D. E. ,1990, Reduction in VLDL, but not HDL, in plasma of rats deficient in choline, *Biochem Cell Biol* **68**, 552-8.

Zeisel, S. H. ,1981, Dietary choline: biochemistry, physiology, and pharmacology, *Ann. Rev. Nutr.* **1**, 95-121.

Zeisel, S. H. ,1990, Biological consequences of choline deficiency in: "Choline Metabolism and Brain Function", "Nutrition and the Brain", Vol. 8, eds., R.V. Wurtman and J.V. Wurtman, Raven Press, New York.

Zeisel, S. H. ,1990, Choline deficiency, *J. Nutr. Biochem.* **1**, 332-349.

Zeisel, S. H. ,1993, Choline phospholipids: signal transduction and carcinogenesis, *FASEB J.* **7**, 551-557.

Zeisel, S. H., DaCosta, K.-A., Franklin, P. D., Alexander, E. A., Lamont, J. T., Sheard, N. F. and Beiser, A. ,1991, Choline, an essential nutrient for humans, *FASEB J* **5**, 2093-2098.

Zeisel, S. H., Growdon, J. H., Wurtman, R. J., Magil, S. G. and Logue, M. ,1980, Normal plasma choline responses to ingested lecithin. *Neurology* **30**, 1226-9.

NUTRITION, IMMUNOLOGY AND CANCER: AN OVERVIEW

Bob G. Sanders[1] and Kimberly Kline[2]

[1]Department of Zoology
[2]Division of Nutritional Sciences
 University of Texas
 Austin, Texas 78712

INTRODUCTION

With the decline in death from infectious diseases in the Western world, cancer has become a leading cause of death, second only to cardiovascular diseases. Projections by the American Cancer Society (1993) are that cancer will develop in one person in three in the United States and that one out of five deaths will be due to cancer. Estimates by the National Cancer Institute that 35% of all cancer deaths may be linked to improper diet points to the belief that nutrition may either help cause or help prevent cancer in humans (Wattenberg, 1985; Kline, 1986; Henderson et al., 1991).

Nutrition, immunology and cancer define three distinct disciplines whose complex interactions are not well understood. Although nutritional effects on immunity and nutritional effects on cancer are documented in the scientific literature, untangling the complex interrelationships among all remains a major challenge. The diet-immune-cancer connection is based on epidemiologic, laboratory, and human intervention research. These studies leave little doubt about the beneficial effects of adequate nutrition, and there is substantial agreement that some cancers are causally related to certain dietary factors, such as high lipid intake (Wattenberg, 1985; Henderson et al., 1991). In an effort to better clarify the connection between foods and the nutrients they contain and specific health conditions, the FDA commissioned the Federation of American Societies for Experimental Biology Life Sciences Research Office (1991) to write up 10 reports on various nutrient-disease relationships. Based on these reports the FDA considers valid the following five claims: calcium and osteoporosis, sodium and hypertension, lipids and cardiovascular disease, lipids and cancer, and folic acid and neural tube defects. The FDA did not find sufficient evidence to support the following claims: zinc and immune function in the elderly, dietary fiber and cancer, dietary fiber and cardiovascular disease, omega-3 fatty acids and heart disease, vitamin A and cancer, vitamin C and cancer and vitamin E and cancer.

Nutrition and Biotechnology in Heart Disease and Cancer
Edited by J.B. Longenecker *et al.*, Plenum Press, New York, 1995

185

NUTRIENT-CANCER CONNECTION

The nutrition-cancer connection was once limited to dietary mutagens and carcinogens that may enhance the carcinogenic process, and dietary inhibitors of carcinogenesis that may protect individuals from cancer. We now know that the role of nutrition in various cancer states is much more complex (Henderson et al., 1991). The development of cancer is a multistep process, driven by genetic and environmental factors (**Figure 1**).

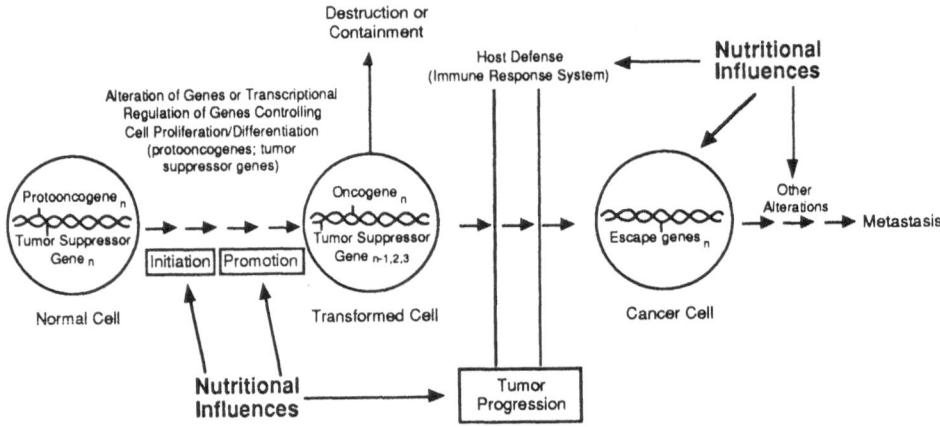

Figure 1. Nutritional influences on the multistep process of carcinogenesis.

Cancer develops when protooncogenes, cellular genes that regulate cell proliferation and differentiation, are genetically altered to oncogenes resulting in cellular escape from normal regulatory controls of cellular proliferation. Cancer can also develop when tumor suppressor genes, normal genes whose products constrain cell proliferation, are inactivated by mutational events. Cancer is the culmination of multiple genetic and epigenetic events resulting in the ability of cells to proliferate in an uncontrolled, invasive manner. One phase of carcinogenesis involves "escape" from the host immune response system. Cancer cells may escape immune destruction and grow undisturbed by suppressing the immune response system or by not expressing appropriate membrane markers required for immune recognition and destruction. Formation of mini-metastases by invasion of secondary sites by cancer cells is a critical event in the evolution of a cancer and one of the most challenging problems facing oncologists today. The expression of several genes are correlated with the ability of the primary cancer cells to migrate or metastasize to secondary sites. Nutritional influences can occur in all stages of carcinogenesis: initiation, promotion, progression, immune escape and metastasis. Important questions that need to be addressed are the mechanisms of action of specific nutrients in this complicated process.

NUTRIENT-IMMUNE-CANCER CONNECTION

Anti-cancer functions of the immune response system and the role that nutrition plays in enhancing cancer immunity are not clear. A general concept of cancer immunology called "immune surveillance theory" implies that cancers arise only if the immune system is somehow impaired or if the tumor cells lose their immunogenicity, thereby allowing them to escape immune detection and destruction. The absence of increased numbers of tumors in immune compromised animals and individuals has led some investigators to question the immune surveillance theory. Nevertheless, studies show that tumor antigens can induce humoral and cell-mediated immune responses which in some cases result in tumor elimination, and that immunostimulants are effective in treating human diseases, including cancers (Hadden, 1993). Key cells in the immune response system with the ability to recognize and eliminate cancer cells are cytotoxic T-killer cells, natural killer cells, macrophages and antibody secreting B cells. Cytotoxic T cells recognize tumor antigens in context of a dual recognition system, and kill tumor cells that express altered self on their cell surface membranes. Nutritional enhancement of numbers of cytotoxic T cells that specifically recognize tumor cells, or nutritional modulation of expression of membrane receptors involved in the dual recognition system of either the cytotoxic T cells or the tumor cells would affect the ability of cytotoxic T cells to destroy tumor cells. Natural killer cells are a population of lymphocytes capable of recognizing and destroying a wide variety of tumor cells. Natural killer cells are not restricted by a membrane recognition system such as that required for cytotoxic T cell killing. Modulation of natural killer cell numbers or function by either positive or negative factors in the diet can impact tumor growth. A third key cell type involved in host defense are macrophages. The antitumor activity of macrophages is thought to be mediated by soluble products. Tumor necrosis factor-alpha is one example of a soluble, potent antitumor factor secreted by macrophages. This factor causes hemorrhaging and necrosis of tumors when injected directly into a tumor mass. The role of antibodies in preventing tumor growth is less clear. activation of the complement system by specific antibodies provides a mechanism for tumor destruction; however, in some cases antibodies have been shown to interfere with cellular destruction of tumor cells by binding to tumor antigens thereby masking them and protecting the tumor cells from cytotoxic T cell or natural killer cell destruction. Nutritional enhancement of the production of antibodies by B lymphocytes that are specific for antigens expressed on tumor cells could therefore have either positive or negative effects on tumor growth. As an example, dietary supplementation or deficiency of retinoic acid can enhance or decrease anti-tumor immunity by modulating receptor expression and functions of T lymphocytes, B lymphocytes and macrophages (reviewed by Watson and Rybski, 1988).

NUTRIENT-IMMUNE-CANCER CYTOKINE NETWORK

The induction of soluble factors that regulate immunity and tumorigenesis may represent an area whereby nutrition is effecting both immunity and carcinogenesis. Soluble factors, called cytokines, secreted by a variety of cells, including immune cells and tumor cells, represent an important mechanism for cell-cell communication. The term cytokine describes a wide variety of protein regulatory molecules that are produced by more than one cell type, that react with distinct receptors, and that can produce markedly overlapping biological effects (Balkwill, 1989; Aaronson, 1991). Most cytokines have multiple biological actions. Cytokine actions can be stimulatory or inhibitory depending on a number of factors including cell type, differentiation state and functional state. For example, cells of a particular lineage are stimulated to proliferate when exposed to a specific cytokine, whereas cells of a different lineage are prevented from growing. Several different cytokines, exhibiting pleiotrophic effects, can be secreted by a single cell type, leading to a cytokine network. Cytokines operate within the cytokine network by stimulating the production of other cytokines, by modulating receptor expression for themselves and other cytokines, or by enhancing or inhibiting the biological activity of other cytokines. For example, the interaction of specific cytokines with specific cellular receptors signal immune cells to proliferate, differentiate and express functional anti-tumor properties. Nutritional influences on cytokine production and secretion and on the expression of specific receptors by immune and tumor cells point to a critical area in which nutrition may impact both the host anti-tumor defense system and the carcinogenic process.

An understanding of the carcinogenesis process must take into consideration the cellular make-up and cytokine network of the microenvironment of the cancer cells. The cellular microenvironment of a solid tumor and the cytokine network, as depicted in **Figure 2** (adapted from Genzyme Biolines, 1993), characterizes the complexity of the interacting factors that affect cancer development in a positive or negative manner.

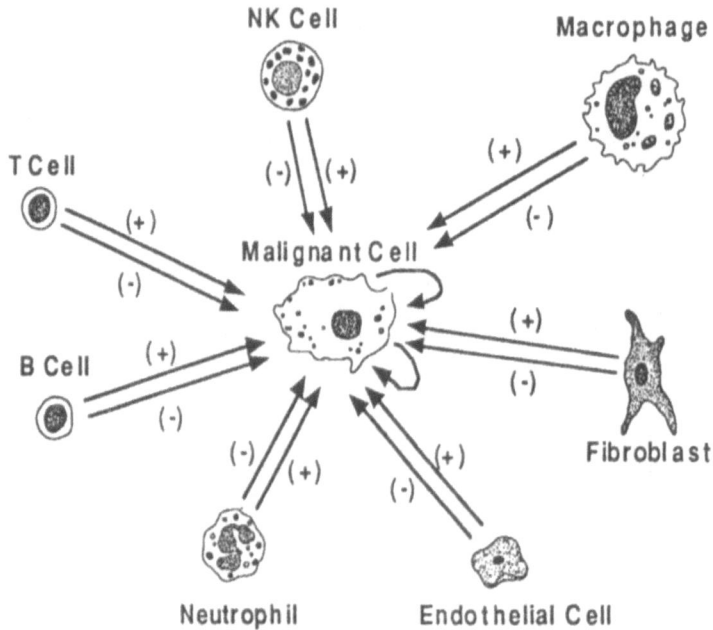

Cells that influence the development of solid cancers includes the cancer cells as well as immune and non-immune cells that surround the cancer mass. We now understand that cytokines, acting in an autocrine or paracine fashion, provide the necessary signals for cell-cell interactions, that either promote or prevent growth of the tumor cells. The model depicted here illustrates a cancer cell as well as immune and non-immune cells that are making and secreting multiple potent soluble cytokines that can directly affect the proliferative capacity of appropriate receptor bearing tumor or non-tumor cells in the microenvironment of a developing cancer.

Nutrient effects on receptor expression, cytokine production, secretion, and activation by tumor, immune, or non-immune cells identify some of the players involved in the complexities of nutrition, immunology and cancer. Studies of nutrient effects on gene expression involving the cytokine network are limited, but such data are necessary before the nutrient-immune-cancer connections can be fully understand. An understanding of the influences of nutrition on the cytokine network and exploiting the production of specific cytokines and receptor expression by normal and tumor cells may lead to a novel method for cancer prevention or treatment.

NUTRIENT-GENE REGULATION

Unraveling the nutrient connection at the gene level has already begun, and data show that nutrients play an active and specific role in governing the expression of specific genes (**Table 1**, Aaronson, 1991; Clarke and Abraham, 1992; Berdanier and Hargrove, 1993). Examples of the regulation of gene transcription by carbohydrates, fatty acids, retinoic acid and iron are provided in a recent review by Clarke and Abraham (1992). Transcriptional regulation of cellular proliferation, differentiation and developmental processes by retinoic acid, a ligand for proteins in the steroid nuclear receptor family which also includes vitamin D, point to a major regulatory role for these two fat soluble vitamins in cell proliferation, differentiation and development (Ross, 1993). Retinoic acid treatment of human HL-60 promyelocytic tumor cells in culture induces the cells to differentiate along the granulocytic pathway (Hall and Watt, 1989). During the past five years remarkable progress has been made in understanding the molecular mechanisms whereby retinoic acid induces cells to differentiate. In 1993 the American Institute of Nutrition hosted a symposium titled "Retinoids: Cellular Metabolism and Activation" that provided an overview of the molecular involvements of retinoids (1993). The nuclei of cells contain retinoic acid receptors which are structurally homologous and functionally analogous to the known receptors for steroid hormones, thyroid hormone and vitamin D. Retinoic acid, functioning as a hormone, regulates the transcriptional activity of a large number of genes (Jump et al., 1993).

Table 1. Nutrient regulation of transcriptional events

	mRNA	
		• transcription
		• processing
		• stability
		• translation
Carbohydrates	↑↓	Glycolytic & Lipogenic enzymes
Fatty Acids	↑↓	Fatty acid synthase
Retinoic Acid	↑↓	Cell Cycle Specific Genes
$1,25(OH)_2vitD_3$	↑↓	Cell Cycle Specific Genes
Divalent Metals	↑↓	Metallothionein synthesis

Nutrient-gene regulation, clearly illustrated by what we know about retinoic acid regulation of genes, constitutes a relatively new scientific frontier that offers great promise for providing basic understanding of how nutrition influences health. The ability of specific nutrients to modulate events regulating gene transcription, message processing, stability and translation illustrate the contributions of nutritional studies to a better understanding of human disorders and identifies an exciting area of nutritional research that will continue to probe the complexities of the relationships between nutrients and health well into the next century. Drs. S. D. Clarke and S. Abraham (1992) wrote that "regulation of gene expression by specific nutrients is now a major frontier for the next generation of nutrition scientists".

VITAMIN E AS AN ANTICANCER AGENT

In our laboratories, we are trying to understand how vitamin E succinate (VES), a derivative of the most biologically active form of vitamin E (RRR-a-tocopherol), prevents the growth of tumor cells *in vitro*. Our studies focus on VES effects on the proliferation of estrogen responsive (ER+) and non-responsive (ER-) human breast cancer cells, and human lymphoid cancer cells. VES has been shown to be a potent inhibitor of both breast cancer and lymphoid cancer cells (**Figure 3**; Bihong Zhao, Kimberly Kline and Bob G. Sanders, unpublished data). VES growth inhibition does not involve cytotoxicity or apoptosis, and VES growth inhibition of ER+ cells is not mitigated by the presence of the positive growth factor, estrogen. VES arrests MDA-MB-435 human breast cancer cell proliferation in the G1 cell cycle phase, and HL-60 promyelocytic tumor cells in a bi-nucleated state.

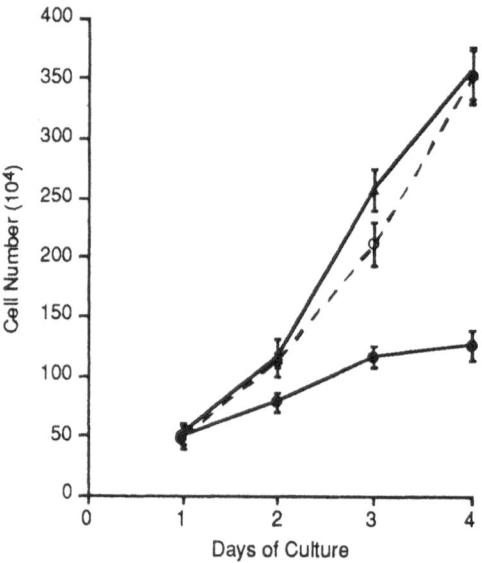

Figure 3. Vitamin E succinate is a potent inhibitor of tumor cell proliferation. Tumor cells were cultured for 1-4 days in the presence of vitamin E. succinate (Solid circles), vehicle (solid triangles) or media only (open circles). Cell proliferation was determined daily by counting cell numbers.

In an effort to understand the mechanisms whereby VES inhibits the proliferation of tumor cells, conditioned media (CM) from VES-growth inhibited breast cancer cells were analyzed and found to contain potent antiproliferative factors. CM from VES-growth inhibited MDA-MB-435 human breast cancer cells exhibits potent antiproliferative activity that is not found in CM from either vehicle-treated or untreated control cells. This antiproliferative activity can function in both an autocrine and paracrine fashion.

Fractionation of CM from VES-treated breast cancer cells, by gel filtration, reveals a major antiproliferative activity of approximately M_r 14,000 (**Figure 4**, Charpentier et al., 1993). The M_r 14,000 antiproliferative factor remains functional following thermal and acid treatments. Antibody neutralization studies show the antiproliferative activity of the M_r 14,000 fraction can be neutralized by an antibody capable of neutralizing the activity of all three mammalian transforming growth factor-beta (TGF-β) subtypes, showing that VES, like retinoic acid, can induce breast cancer cells to secrete the TGF-β cytokine (Wakefield et al., 1990). VES, like retinoic acid, induces biologically active TGF-β, ie 70-75% of the antiproliferative activity in conditioned media from VES-treated human breast cancer cells is "naturally" active, ie does not require heat, acid or protease activation.

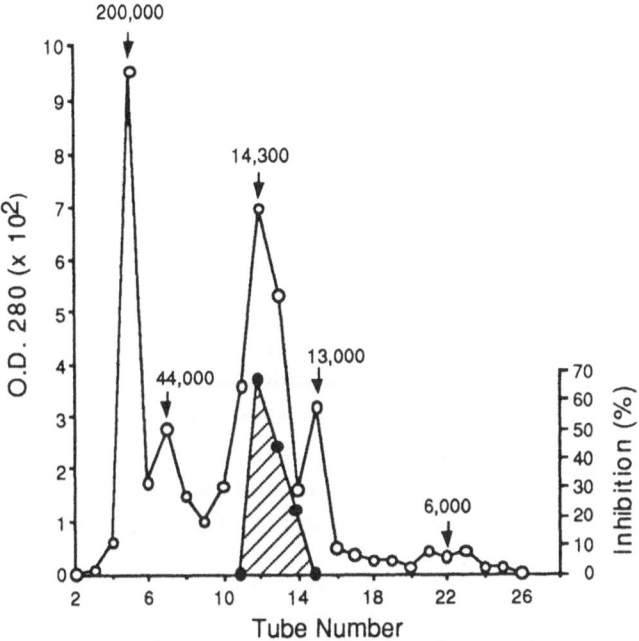

Figure 4. Antiproliferative factor secreted by Vitamin E succinate treated tumor cells. Conditioned media from vitamin E succinate treated tumor cells was fractionated by column chromatography gel filtration. Fractions were assayed for ability to inhibit proliferation of bioassay cells. The conditioned media contained a 14,000 molecular weight antiproliferative factor (closed circle) that was shown to be transforming growth factor-beta (TGF-B). Molecular weights of protein standards are indicated by open circles.

Preliminary studies in our laboratory suggest that VES, like retinoic acid treatment of cancer cells increases the expression of TGF-β receptors, rendering the cells more susceptible to growth inhibition by TGF-β (Ruscetti et al., 1991). These VES studies identify a new class of compounds capable of regulating this important negative growth control factor and its cell surface receptors, placing vitamin E succinate among a select few compounds which have been demonstrated to be capable of modulating the endogenous production of TGF-β, such as the steroids [vitamin D analogs, antiestrogens (tamoxifen) and the synthetic progestin, gestodene] and retinoids (Wakefield et al., 1990).

TGF-β ANTIPROLIFERATIVE FACTORS

A tumor cell mass can be viewed as a heterogeneous group of cells, clonally derived from a particular cell lineage (**Figure 5**). An evolving cancer represents a diverse population of cells that have undergone chromosomal and mutational changes which express different membrane receptors, secrete different soluble factors, exhibit different requirements for growth, and possess different metastatic properties. A successfully growing tumor represents a population of cells that have survived intense selective pressures by expressing cellular phenotypes that mitigate negative growth controls and utilize growth-promoting signals.

An understanding of the TGF-β antiproliferative factors and their receptors will show how cytokines can influence the carcinogenesis process, and will help illustrate possible influences of nutrition on carcinogenesis. Three isoforms or subtypes of TGF-β designated TGF-β 1, β2, and β3 have been demonstrated to be expressed in mammalian tissues. The

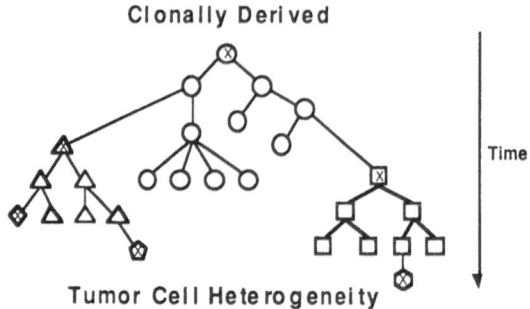

Figure 5. Genetic involvement in the evolution of a clonally derived tumor mass. Clonally derived tumor cells that are capable of responding differently to positive and negative selective factors in their immediate microenvironment. The circles, triangles, squares and octagonals represent genetically distinct mini-tumor cell masses that were derived from a single tumor cell.

TGF-β molecules are first synthesized as large precursor molecules that are processed to - yield a latent molecule consisting of two amino-terminal glycopeptides covalently linked together and non-covalently associated with the mature 25,000 TGF-β polypeptide which consists of two identical 12,500 disulfide-linked monomers (Lyons et al., 1990; Ruscetti et al., 1991). To be biologically active, the 25,000 TGF-β polypeptide must be dissociated from the amino-terminal glycopeptides (Lyons et al., 1990; Massague, 1990). Like other cytokines, the TGF-β subtypes display multiple biological activities and can act via cell surface TGF-β receptors as either positive or negative growth factors. TGF-β receptor molecules exist as three types, referred to as type I, II and III (Massague, 1992). The type I TGF-β receptors are responsible for signalling effects of TGF-β on extracellular matrix, such as induction of fibronectin which is important for metastases, while type II TGF-β receptors mediate induction of growth inhibition and hypophosphorylation of the retinoblastoma gene product. TGF-β receptors I and II have been shown to function as a heterodimer to signal the diverse effects mediated by TGF-β molecules. TGF-β type III receptors which exhibit high affinity TGF-β binding, do not appear to be associated with TGF-β mediated cell responses. However, the type III receptors may influence responses to TGF-β by binding active TGF-β and presenting it to the type II receptors and helping to stabilize the ligand-receptor interactions (Massague, 1992).

Since the regulated growth of cells is dependent on a dynamic balance between stimulatory and inhibitory signals, escape from autocrine or paracine negative growth control mediated by TGF-β is believed to be a critical event in carcinogenesis of many different cell types, including human breast epithelial cells. Several mechanisms can exist whereby cells may lose responsiveness to the inhibitory effects of TGF-β (**Figure 6 I, II, & III**). For example, cells may no longer produce adequate amounts of TGF-β, cells may no longer convert the biologically inactive latent TGF-β form which they secrete to the active form, cells may no longer express TGF-β receptors, or cells may have some alteration in the signal transduction pathway for TGF-β.

I. Absence or alterations in ligand (TGF-β)

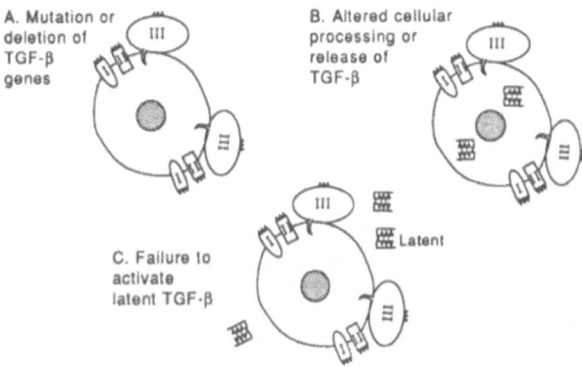

A. Mutation or deletion of TGF-β genes

B. Altered cellular processing or release of TGF-β

C. Failure to activate latent TGF-β

II. Absence or alterations of cell surface receptors for TGF-β

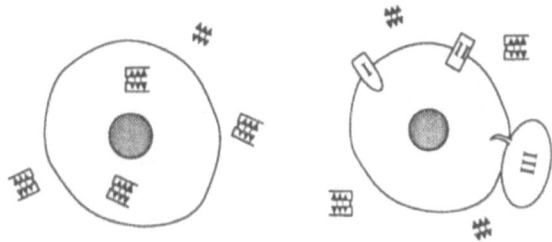

III. Alterations in signaling pathway(s)

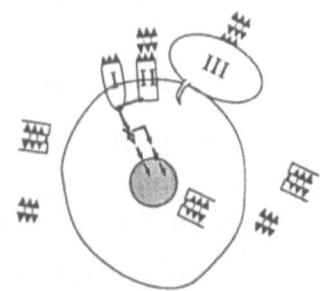

Figure 6. Mechanisms for tumor call escape from the antiproliferative effects of TGF-β. TGF-β is synthesized inside cells and is secreted as biologically inactive latent molecules (triangles with lines) which are converted to biologically active molecules (triangles with lines). TGF-β, reacting with complexes of TGF-β receptors I and II signal tumor cells to cease proliferation. TGF-β receptor receptor III is thought to help stabilize the ligand- receptor complex. Escape from the antiproliferative TGF-β signal is thought to be an important event in various carcinogen processes. This figure depicts three potential mechanisms whereby tumor cells might escape control of proliferation by TGF-β, namely, absence or inactivating alternations in TGF-β molecules (I), absenced or inactivating alterations of TGF-β receptors (II) or blockage of TGF-β induced signaling pathways (III).

Vitamin E succinate and retinoic acid appear to inhibit tumor cell proliferation in culture by up-regulating the production of TGF-β, increasing the conversion of latent to active TGF-b, and by upregulating TGF-β receptor expression. The chemopreventative or chemotherapeutic potential for these compounds remains to be clarified.

193

SUMMARY

There is epidemiological, laboratory and some clinical evidence that certain dietary factors play a role in either promoting or inhibiting cancer development. An understanding of the mechanisms whereby specific nutrients are having effects in the promotion or prevention of cancer is beginning to take shape. Research into nutrient effects on the expression of specific genes, especially cytokine and cytokine receptor expression, will help increase our basic knowledge of cancer biology. Since cytokines can either enhance or suppress both immune defense and cancer growth, increased understanding of nutrient effects on the cytokine network will be beneficial. The regulation of specific gene expression by specific nutrients, indeed, identifies a major frontier for present and future nutritional biologists.

REFERENCES

Aaronson, S. A., 1991, Growth factors and cancer, Science, 254:1146.

American Cancer Society, 1993. Cancer Facts and Figures-1993. American Cancer Society, Inc., Atlanta.

Balkwill, F.R., 1989"Cytokines in Cancer Therapy," Oxford University Press, New York.

Berdanier, C.D. and Hargrove, J.L., 1993, "Nutrition and Gene Expression," CRC Press, Inc., Boca Raton.

Charpentier, A., Groves, S., Simmons-Menchaca, M., Turley, J., Zhao, B., Sanders, B. G., and Kline, K., 1993, RRR-a-tocopheryl succinate inhibits proliferation and enhances secretion of transforming growth factor-beta (TGF-b) by human breast cancer cells, Nutr. Cancer 19:225.

Clarke, S.D. and Abraham, S., 1992, Gene expression: nutrient control of pre- and post-transcriptional events, FASEB J. 6:3146.

Federation of American Societies for Experimental Biology Life Sciences Research Office, 1991, Evaluation of publicly available scientific evidence regarding certain nutrient-disease relationships: articles 1-10, Bethesda.

Genzyme Biolines, 1993. Cytokines: mediators of cell to cell communication in solid human tumors, Genzyme Biolines, 1:1.

Hadden, J.W., 1993, Immunomodulation: immunostimulants, Immunol. Today. 14:275.

Hall, P.A. and Watt, F.M., 1989, Stem cells: the generation and maintenance of cellular diversity, Development 106:619.

Henderson, B.E., Ross, R.K., and Pike, M.C., 1991, Toward the primary prevention of cancer, Science 254:1131.

Jump, D.B., Lepar, G.J., and MacDougald, O.A., 1993, Retinoic acid regulation of gene expression in adipocytes, in: "Nutrition and Gene Expression," C.D. Beardanier and J.L. Hargarove, eds., CRC Press, Inc., Boca Raton.

Kline, K., 1986, Breast cancer: the diet-cancer connection, Nutr. Today (May/June):11.

Lyons, R.M., Gentry, L.E., Purchio, A.F., and Moses, H.L., 1990, Mechanism of activation of latent recombinant transforming growth factor b1 by plasmin, J. Cell Biol. 110:1361.

Massague, J., 1990, The transforming growth factor-b family, Ann. Rev. Cell Biol. 6:597.

Massague, J., 1992, Receptors for the TGF-b family, Cell 69:1067.

Ross, A.C., 1993, Symposium on retinoids: cellular metabolism and activation, J. Nutr. 123:344.

Ruscetti, F.W., Dubois, C., Falk, L.A., Jacobsen, S.E., Sing, G., Longo, D.L., R. H. Wiltrout, R.H., and Keller, J.R., 1991, In vivo and in vitro effects of TGF-b1 on normal and neoplastic haemopoiesis, in: "Clinical applications of TGF-b." G. R. Bock and J. Marsh, eds., John Wiley and Sons, Chichester.

Wakefield, L.S., Kim, J., Glick, A., Winokur, T., Colletta, A., and Sporn, M., 1990, Regulation of transforming growth factor-b subtypes by members of the steroid hormone superfamily, J. Cell. Sci. Suppl. 13:139.

Watson, R.R. and Rybski, J.A., 1988, Immunological response modifiction by vitamin A and other retinoids, in: "Nutrition and Immunology", R.K. Chandra, ed., Alan R Liss, Inc., New York.

Wattenberg, L.W., 1985, Chemoprevention of cancer, Cancer Res. 45:1.

ANEMIA OF MALIGNANCY

Rodger L. Bick

Department of Oncology and Hematology
Presbyterian Hospital of Dallas
Department of Medicine and Pathology
University of Texas Southwestern Medical Center
Dallas, Texas 75231

INTRODUCTION

Anemia of malignancy is a common problem which significantly interferes with the quality of life of patients with cancer and allied diseases. The anemia of malignancy is usually multifaceted, can be because of the malignancy, complications of the malignancy, or resulting from treatment.

NORMAL ERYTHROPOIESIS

A firm understanding of erythroid production and destruction is necessary to understand the basic concepts of anemia of malignancy. We all recall that erythroid hematopoietic elements in the marrow give rise to circulating red cells. Erythroid hematopoietic elements are initially derived from a primitive marrow stem cell called the GEMM (granulocyte, erythroid, monocyte/macrophage and megakaryocyte) stem cell which is committed to non-lymphoid daughter cells, i.e. committed to granulocyte, erythroid, monocyte/macrophage and megakaryocytic precursors which then expand and mature into mature progeny (Lisiewica and Bick, 1993). The earliest cell which can be recognized after derivation from the GEMM stem cell is the proerythroblast which, under the influence of erythropoietin (discussed subsequently) matures through several steps to become a mature erythrocyte. The sequential steps are (1) basophilic erythroblast (2) polychromatophilic erythroblast (3) orthochromatophilic erythroblast (4) nucleated red blood cell (5) reticulocyte and (6) lastly the mature red blood cell (Gulati et al., 1993). Immediately after these red cells are released, a small number are sequestered by the spleen while the majority circulate with a life time of about 120 days. As red cells approach the end of their life span, they are removed from the circulation by the reticuloendothelial system, primarily the liver and spleen wherein they are broken down into hemoglobin and stromal elements with hemoglobin break-down products being excreted in the urine. Marrow

Nutrition and Biotechnology in Heart Disease and Cancer
Edited by J.B. Longenecker *et al.*, Plenum Press, New York, 1995

function can most easily be assessed by use of the reticulocyte count, marrow biopsy to assess cellularity, and in rare instances Fe^{59} uptake (Kjeldsberg et al., 1989). The assessment of red cells in the circulation is easily made by measuring the hemoglobin, hematocrit, or when needed, chromium^{51} survival studies. Additional chromium^{51} survival studies can be done to assess splenic activity in sequestering circulating red cells. Function of the reticuloendothelial system as relates to destruction of red cells can be assessed by haptoglobin, bilirubin, LDH, marrow iron, methemalbumin, and levels of hemoglobinemia. Urinary breakdown of red cell products is best assessed by hematuria, hemoglobinuria, or urinary urobilinogen (Baker, 1993). The cytokine responsible for expansion and maturation of the erythropoietic series has been isolated, purified and cloned and the recombinant form is now generally available for clinical use. Erythropoietin is an 18 Kd protein encoded for on chromosome 17q. Physiologically, erythropoietin is derived from renal cells, and to a lesser extent hepatocytes, in response to hypoxia. Erythropoietin induces expansion/proliferation of erythroid blast forming units (BFU-erythroid), causes maturation through all the stages of erythroid progression, induces globin synthesis in erythroid precursors and, lastly, induces marrow release of reticulocytes into the peripheral blood (Glaspy, 1993).

ANEMIA OF MALIGNANCY

Anemia can significantly interfere with the quality of life with patients with malignancy, and thus is important to recognize and, when possible, treat. Patients with anemia of malignancy suffer some level of weakness, fatigue, light headiness, lassitude, and dyspnea as anemia becomes severe. The particular symptomatology will depend upon how acute or chronic the anemia has developed. The signs of anemia commonly seen in patients with anemia of malignancy are pallor, a systolic flow murmur, and tachycardia. If the anemia has developed acutely, hypotension, especially orthostatic, may be seen. Another problem is that cardiovascular function becomes compromised with a decreased systemic vascular resistance index, decreased venoarterial pressure, decreased arterial pressure, and increased pulse pressure (Bick and Baker, 1990).

The three most common types of anemia seen in malignancy are the hypoproliferative anemias, hemolytic anemias, and blood loss anemia. Uncomplicated hypoproliferative anemia can be viewed as comprised of three defects in descending order of importance as follows: (1) impaired mobilization of reticuloendothelial iron (2) ineffective erythropoiesis and (3) decreased red cell survival. The hypoproliferative anemias are comprised of the typical anemia of chronic malignancy, iron deficiency anemia, marrow metastases, chemotherapy induced suppression of erythropoiesis, radiation therapy induced suppression of erythropoiesis, myelofibrosis, megaloblastic anemia, and pure red cell aplasia; all these may be seen in the patient with malignancy.

Uncomplicated Anemia of Malignancy

The characteristics of uncomplicated anemia of malignancy are that it occurs after several months of the malignancy becoming evident, it usually is associated with normochromic and normocytic red cell indices, as well as a normal MCV, MCH, and MCHC (Bentley, 1982; Cartwright, 1966; Dutcher, 1987). The serum iron is decreased as is the transferrin and the transferrin saturation. Marrow examination reveals normal to increased marrow iron stores, but normal erythroid precursors are usually noted in the marrow. Alternatively, one must always consider occult malignancy when assessing a patient with straight forward anemia of chronic disease, as occult malignancy is one of the most common causes of anemia of chronic disease in the adult. The mechanisms of anemia of malignancy are important. Interleukin 1 (IL-1) is of major importance in the development of anemia in patients with malignancy. Increased levels of Interleukin-1 in malignancy have an indirect affect on suppression of erythropoiesis via it's primary effect on lymphocytes and monocytes. More important, Interleukin-1 causes a lactoferrin release from polymorphonuclear leukocytes. Additionally, Interleukin-1 via it's effect on macrophage activation and pyrexia causes a mild decrease red cell survival; however, it is the lactoferrin release that is the major problem. As IL-1 is released in excess from the monocyte/macrophage system, it induces neutrophils to release lactoferrin into the blood stream. Lactoferrin then competes with transferrin for transferrin bound iron, and shifts the balance to lactoferrin bound iron and free transferrin. Lactoferrin then delivers it's bound iron to macrophages where it remains irreversible and inaccessible for further erythropoiesis. Interferon gamma (IFN-g) and tumor necrosis factor alpha (TNF-a) are also important in the development of anemia of malignancy and they both suppress erythropoiesis; however, the effects of interferon gamma and tumor necrosis factor alpha are somewhat reversible by erythropoietin (Means and Dessyprius, 1993).

The typical uncomplicated anemia of malignancy often responds to recombinant erythropoietin. Many studies have been done documenting this effect in (1) patients with malignancy without chemotherapy, (2) those undergoing multiple types of chemotherapy, and (3) those undergoing nephrotoxic chemotherapy, particularly cisplatinum containing regimens. The use of erythropoietin in uncomplicated anemia of malignancy could clearly decrease the number of transfusions needed and increase the hemoglobin and hematocrit (Case et al., 1993). The response may take quite some time to be noticeable, and often take 4 - 6 weeks and sometimes up to 12 weeks for a response to be noted.

Recombinant Human Erythropoietin (rH-EPO) in Anemia of Malignancy

General guidelines for the use of erythropoietin in patients with uncomplicated anemia of malignancy are as follows: the hematocrit should be less than 33%, and other causes of anemia should be clearly ruled out. The serum erythropoietin levels should be less than 200 units/mL. If these criteria are met, the patients are generally started on erythropoietin at 150 units per kilogram, subcutaneous, three times per week for 8 weeks with assessment of the hematocrit occurring on a weekly basis. After 8 weeks the dose is usually adjusted as follows: if the hematocrit is greater than 40 after the 8 week period erythropoietin is stopped and restarted at 75% of the original dose when the hematocrit, again decreases to 33%. If the hematocrit decreases to greater than 4% within a 2 week period, the erythropoietin is adjusted downward. If the hematocrit increases to 36-40%, then the starting dose is maintained on a P.R.N. basis. If the hematocrit shows no change, or decreases, then an increase of erythropoietin to a maximum of 300 units per kg 3 times a week is suggested. It is important to assess the serum ferritin and supplement the iron if the ferritin is less than 100 NG/mL as otherwise, erythropoiesis cannot be effective.

Hemolytic Anemias Associated with Malignancy

Hemolytic anemias are uncommonly seen in patients with malignancy, and can present as autoimmune hemolytic anemia, drug induced hemolytic anemia, microangiopathic hemolytic anemia, and more rarely hemophagocytic syndrome (Reiner and Spivak, 1988). The autoimmune hemolytic anemias are most commonly seen in chronic lymphocytic leukemia, non-Hodgkin's lymphoma, Hodgkin's disease, angioimmunoblastic lymphadenopathy with dysproteinemia, and ovarian carcinoma. More rarely however, they are seen in a variety of cell types of lung cancer, cervical carcinoma, gastric carcinoma, seminoma, Kaposi's sarcoma, myeloma, thymoma, T-gamma (LGL) leukemia, hairy cell leukemia, and by selected chemotherapy drugs. The chemotherapy induced autoimmune hemolytic anemias are seen with cisplatinum, methotrexate, and 5-Fluorouracil (Doll and

Weiss, 1985; Getaz et al., 1980). These agents must be recalled when seeing a patient on one of these drugs who develops an autoimmune hemolytic anemia.

Microangiopathic Hemolytic Anemias (MAHA's) in Malignancy

The microangiopathic hemolytic anemias are also uncommon, but quite serious when developing in a patient with malignancy. They are most commonly seen with adenocarcinoma and there is some association with mitomycin. Also, the microangiopathic hemolytic anemias can give rise to DIC of malignancy or, alternatively, if the patient with malignancy develops DIC it may trigger a microangiopathic hemolytic anemia syndrome. The most common malignancies associated with microangiopathic hemolytic anemia are gastric carcinoma, breast carcinoma, ovarian carcinoma, and adenocarcinoma of the lung (Sheldon and Slaughter, 1986).

Hemophagocytic Syndrome in Malignancy

The hemophagocytic syndrome is a particularly ominous development whereby marrow macrophages engulf erythroid, as well as other hematopoietic precursors leading to not only severe anemia, but sometimes severe pancytopenia as well. The hemophagocytic syndrome is most commonly seen in patients with gastric carcinoma, lymphoma, and leukemias. The prognosis is usually poor with patients surviving less than a few weeks, despite vigorous therapy, often utilizing cytotoxic agents (Reiner and Spivak, 1988).

Red Cell Aplasia in Malignancy

Paraneoplastic red cell aplasia is another uncommon problem happening in patients with malignancy and can be seen with thymoma, lymphoma, chronic lymphocytic leukemia, T-gamma lymphoproliferative disease (LGL), ALL, CML, breast carcinoma, biliary carcinoma, thyroid carcinoma, small cell lung cancer, and occasionally can be seen with the use of azathioprine (Champlin, 1993).

Myelofibrosis in Malignancy

Some malignancies may induce myelofibrosis, thus giving rise to anemia or pancytopenia. Myelofibrosis most commonly occurs as a complication of prostatic carcinoma, gastric carcinoma, breast carcinoma, all cell types of lung cancer, fibrosarcoma, Hodgkin's disease, and multiple myeloma.

Blood Loss and Fe Deficiency in Malignancy

Blood loss anemia can also be an important part of malignancy, this can come about via several mechanisms including spontaneous bleeding from a tumor, a defect in hemostasis, or post-surgical hemorrhage. When this occurs, it is obviously important to make a diagnosis and identify the bleeding site, as soon as possible (Bick, 1992).

Chemotherapy-Associated Anemia

Chemotherapy agents cause significant marrow suppression, and although megakaryocytes and platelets are more commonly effected than the erythropoietic series one must remember that agents, such as nitrogen mustard, melphalan, busulfan, nitrosoureas, cytarabine, vinblastine, VP-16, and taxol may severely compromise the erythroid, as well as other hematopoietic elements (Rosove and Schwartz, 1990).

Miscellaneous Anemias in Malignancy

Other causes of anemia in malignancy are hypersplenism, and sideroblastic anemias. Several studies have shown up to 30% of patients with solid tumors developing splenic metastases; this is uncommonly appreciated or recognized. When this occurs however, hypersplenism and resultant anemia may ensue. Additionally, hypersplenism as a simple reaction to malignancy can also account for increased splenic sequestration of red cells.

Paraneoplastic Erythrocytosis

Lastly, paraneoplastic erythrocytosis can be seen with some malignancies, and the malignancies which need to be considered when unexplained erythrocytosis is seen include renal cell carcinoma, Wilm's Tumor, sarcoma, hepatoma, cerebellar hemangioblastoma, adrenal tumors, ovarian carcinoma, small cell lung cancer, and thymoma (Messmore and Choudhury, 1993).

LABORATORY EVALUATION

The laboratory evaluation of anemia of malignancy is obviously of importance if a correct diagnosis and thus appropriate therapy is to be delivered. One must evaluate the peripheral blood smear, red cell indices, reticulocyte count, and ferrokinetic parameters including FE/TIBC and free erythrocyte protoporphyrin. In addition, LDH and indirect bilirubin should be performed. If the LDH, reticulocyte count or indirect bilirubin are elevated then a Coombs test should also be performed. Occasionally, bone marrow biopsy and aspirate for assessment of cellularity, myelofibrosis, iron stores, and erythropoiesis is needed, and lastly the patient must be carefully evaluated for potential sites of bleeding.

TREATMENT

Treatment of malignancy is comprised of treating the malignancy itself when possible, and treatment of paraneoplastic problems including defects in hemostasis endocrinopathy's, etc. Also, control of hemorrhage, withdrawal of suspected drugs, immune suppression, the use of packed red blood cells, or erythropoietin may be indicated depending on the laboratory assessment and precise reasons for anemia.

REFERENCES

Baker, W.F., 1993, Clinical evaluation of the patient with anemia, in: "Hematology: Clinical and Laboratory Practice," R.L. Bick, J.M. Bennett and R.K. Brynes, eds., C.V. Mosby, Saint Louis.

Bentley, D.P., 1982, Anemia and chronic disease, Lab. Med. 21:641.

Bick, R.L., and Baker, W.F., 1990, Iron deficiency anemia, Lab. Med. 21:641.

Bick, R.L., 1992, Coagulation disorders in malignancy, Seminars Thrombosis Hemostatis. 18:353.

Cartwright, G.E., 1966, The anemia of chronic disorders, Seminars Hematology. 3:351.

Case, D.C., Bukowski, R.M., Carey, R.W. and Fishkin, E.H., 1993, Recombinant human erythropoietin therapy for anemic cancer patients on combination chemotherapy, J. Natl. Cancer Inst. 85:801.

Champlin, R.E., 1993, Aplastic anemia, in: "Hematology: Clinical and Laboratory Practice," R.L. Bick, J.M. Bennett and R.K. Brynes, eds., C.V. Mosby, Saint Louis.

Doll, D.C. and Weiss, R.B., 1985, Hemolytic anemia associated with antineoplastic agents, Cancer Treatment Reports. 69:777.

Dutcher, J.P., 1987, Hematologic abnormalities in patients with nonhematologic malignancies, Hematology Oncology Clinics North America 1:281.

Gulati, G.L., Ashton, J.K. and Hyun, B.H., 1993, Evaluation of the peripheral blood smear, in: "Hematology: Clinical and Laboratory Practice," R.L., Bick, J.M. Bennett and R.K. Brynes, eds., C.V. Mosby, Saint Louis.

Kjeldsberg, K., Beutler, E., Bell, C., and Hougie, C., 1989, A anemia of chronic disease and normochromic, normocytic nonhemolytic anemias, in: Practical Diagnosis of Hematologic Disorders," K. Kjeldsberg, E. Beutler, C. Bell and C. Hougie, eds., ASCP Press, Chicago.

Lisiewica, J. and Bick, R.L., 1993, Morphology and biochemistry of the myeloid series, in: "Hematology: Clinical and Laboratory Practice," R.L. Bick, J.M. Bennett and R.K. Brynes, eds., C.V. Mosby, Saint Louis.

Means, R.T. and Dessyprius, E.N., 1993, The anemias of chronic disease, renal failure and endocrine disorders, in: "Hematology: Clinical and Laboratory Practice," R.L. Bick, J.M. Bennett and R.K. Brynes, eds., C.V. Mosby, Saint Louis.

Messmore, H.L., and Choudhury, A.M., 1993, Polycythemia, in: "Hematology: Clinical and Laboratory Practice," R.L. Bick, J.M. Bennett and R.K. Brynes, eds., C.V. Mosby, Saint Louis.

Reiner, A.P. and Spivak, J.L., 1988, Hematophagic histiocytosis: a report of 23 new patients and a review of the literature, Medicine 67:369.

Rosove, M.H. and Schwartz, G.E., 1990, Hematologic complications of cancer and cancer treatment, in: "Cancer Treatment (Ed. #3)," C.M. Haskell, ed., W.B. Sanders, Philadelphia.

DIET IN HEART DISEASE AND CANCER

David Kritchevsky

The Wistar Institute
36th and Spruce Street
Philadelphia, PA 19104-4268

INTRODUCTION

It is difficult to do justice to two broad subjects in a short discourse. The principal killer diseases of our time, *coronary disease and cancer*, are both degenerative diseases of multiple etiology. There are defined risk factors for coronary disease - *elevated plasma cholesterol level, elevated blood pressure, cigarette smoking, overweight, male gender*. Cancer, being a more complex and widely physiologically disseminated disease, does not have any particular list of risk factors. Diet has been associated with coronary disease because of its connection with blood lipids and overweight and because the earliest experimental lesions were induced in rabbits by dietary means. The connection of diet with cancer was the subject of heightened attention after Doll and Peto (1981) listed it as a possibly important factor leading to cancer in the U.S.

The following essay will discuss briefly the roles of particular nutrients in both heart disease and cancer. To date, few studies have been reported which attempt to interrelate all of the diet to disease.

ATHEROSCLEROSIS

The interaction among dietary components has not been studied to any extent. We constructed semi purified diets to resemble, in gross terms, the American diets of 1909 and 1972. There are many differences between the diets. Between 1909 and 1972 the ratio of animal to plant protein had risen from 1.07 to 2.16 although the percentage of calories derived from protein had hardly changed (11.6 vs. 12.1%). Percentage of calories from carbohydrate had fallen from 56.1 to 45.1 and the ratio of simple to complex carbohydrate had risen from 0.47 to 1.11. The ratio of animal/plant fat had fallen from 4.77 to 1.61 and percentage of calories derived from fat had risen from 32.2 to 42.2%. If any single one of these changes had been incorporated into an animal diet one might predict drastic effects on lipidemia and cholesterol metabolism. When the two diets were fed to rabbits the 1972 diet was more cholesterolemic and triglyceridemic but not significantly so. Average severity of atherosclerosis in rabbits fed the two diets was 0.71 in rabbits on the 1909 diet and 0.51 in

Nutrition and Biotechnology in Heart Disease and Cancer
Edited by J.B. Longenecker *et al.*, Plenum Press, New York, 1995

201

those fed the 1972 diet. The severity was judged on a 0-4 scale and the values given represent (aortic arch + thoracic aorta) 2. In rats, effects of the diets on serum lipids, cholesterol absorption HMG-CoA reductase activity and cholesterol 7 α hydroxylase activity were virtually the same (Kritchevsky et al., 1983). Hegsted and Ausmann (1988) related the incidence of coronary disease in 18 countries with fat cholesterol and carbohydrate intake. The best correlation with low incidence of coronary disease was with alcohol intake (**Table 1**).

Table 1. Coronary heart disease and fat, cholesterol and alcohol intake

Country	CHD ( )* (per 100,000 males)	Fat ( )* (S/M+P) Kcal**	Cholesterol ( )* (mg/1000 Kcal)	Alcohol ( )* L person $^{-1}$ yr^{-1}
Finland	872 (1)	1.20 (1)	126 (13)	5.90 (16)
USA	793 (2)	0.69 (13)	188 (2)	8.01(10)
Australia	787 (3)	1.01(4)	178 (5)	9.02 (8)
New Zealand	740 (4)	1.19 (2)	206 (1)	7.76 (12)
UK	702 (5)	0.80 (6)	164 (6)	7.53 (13)
Ireland	672 (6)	1.06 (3)	179 (3)	7.02 (15)
Canada	663 (7)	0.84 (5)	179 (3)	8.08 (9)
Denmark	606 (8)	0.74 (9)	145 (9)	9.03 (7)
Sweden	588 (9)	0.76 (7)	152 (8)	5.85 (17)
Norway	581 (10)	0.57 (16)	118 (15)	4.03 (18)
Netherlands	502 (11)	0.67 (14)	129 (12)	7.97 (11)
Belgium	463 (12)	0.75 (8)	157 (7)	9.87 (6)
W. Germany	452 (13)	0.67 (14)	144 (10)	11.59 (2)
Austria	435 (14)	0.72 (12)	124 (14)	9.93 (5)
Italy	309 (15)	0.46 (17)	91 (17)	10.83 (3)
Switzerland	279 (16)	0.73 (11)	136 (11)	10.56 (4)
France	205 (17)	0.74 (9)	104 (16)	14.34 (1)
Japan	15 (18)	0.38 (18)	58 (18)	7.04 (14)

After Hegsted and Ausman (1988)
( ) *Rank
** Saturated FA/Mono + Polyunsaturated Fat

The principal nutritional focus in atherosclerosis has been on dietary fats and cholesterol. Four decades ago saturated fat was shown to be clearly more atherogenic for rabbits than unsaturated fat (Kritchevsky et al., 1954). How, precisely, saturated fatty acids achieve their effects on plasma cholesterol is still undecided. In the 1960's Hegsted and Keys and their colleagues arrived at the following formulas: Δ cholesterol = 2.16 ΔS-1.65 (ΔP) + 6.77 (ΔC)-0.5 (Hegsted et al. 1965) and Δ cholesterol = 1.35 (2 Δ S - Δ P) + 1.5 (ΔZ) where S and P = % of calories from saturated and unsaturated fat, respectively; C= dg/day dietary cholesterol; and Z = (cholesterol/1000 kcal) 1/2 (Keys et al., 1965). Both groups of investigators agreed that stearic acid did not fit into the formula since stearic acid-rich fats such as cocoa butter did not cause the increases in cholesterol predicted by their formulas; Derr et al. (1993) have provided the following formula for calculating changes in total plasma cholesterol: ΔC=2.3 Δ 14:0 + 3.0 Δ 16:0 - 0.8 Δ 18:0-1.0 Δ PUFA. The fatty acids are expressed as energy %. A companion formula for changes in LDL cholesterol has a higher coefficient for 14:0 and lower ones for the other components of the formula. Hayes and Khosla (1992) suggest that the major determinants of plasma cholesterol are linoleic and myristic acids. When the linoleic acid content of the diet falls below 5 en%, myristic acid begins to exert a hypercholesterolemic effect which rises exponentially as energy % from linoleic acid falls. Palmitic and oleic acids exert minor hypercholesterolemic effects when energy % from linoleic acid falls below 3. Hegsted et al. (1993) have reviewed a large number of studies and conclude that saturated fatty acids increase serum cholesterol and are its primary determinants; polyunsaturated fatty acids lower serum cholesterol ; and monounsaturated fatty acids have no independent effect.

Fat

Another determinant of cholesterolemia and certainly of experimental atherosclerosis is triglyceride structure. Peanut oil is unexpectedly atherogenic for monkeys (Vesselinovitch et al., 1974), rats (Gresham and Howard, 1960) and rabbits (Kritchevsky et al., 1971). When peanut oil is randomized (rearranged so that every component fatty acid occurs in each triglyceride position at one-third of its total concentration) it becomes significantly less atherogenic (Kritchevsky et al., 1973). The fatty acid composition of native and randomized peanut oil is identical only triglyceride structure has been altered.

We (Kritchevsky and Tepper, unpublished) have compared the atherogenic effect in rabbits of tallow and lard. Both fats contain about 25% palmitic acid but in lard 90% of the palmitic acid is in the SN2 position and in tallow only about 5% is present in that position. As **Table 2** shows lard is significantly more atherogenic than tallow. Effects of randomized tallow and lard are being tested currently.

Table 2. Atherogenic effect of lard and tallow *

No. of Animals	Lard 16/18	Tallow 16/18
Serum		
Cholesterol, (mg/dl)	662 ± 24	600 ± 48
% HDL-cholesterol	6.3 ± 0.64	9.2 ± 0.78 a
Liver		
Cholesterol,(g/100)g	3.14 ± 0.39	2.02 ± 0.22 b
Atherosclerosis**		
Aortic Arch	1.44 ± 0.19	0.69 ± 0.16 a
Thoracic Aorta	1.06 ± 0.18	0.41 ± 0.13 a

*Kritchevsky and Tepper unpublished
 Rabbits fed semi purified diets containing 2% corn oil and 14% lard or tallow for 2 months.
**Graded on a 0-4 scale
a) $p<0.01$; b) $p<0.02$

Protein

Meeker and Kesten (1940) showed that casein was more atherogenic for rabbits than soy protein. However, a 1:1 mixture of animal and plant protein is no more atherogenic or cholesterolemic for rabbits than a diet containing only plant protein. Hodges et al. (1967) were investigating the effects of fat and simple or complex carbohydrate on cholesterolemia in man. In order to reduce the number of variables the men were switched from mixed protein to plant protein. The fall in cholesterol resulting from this switch was greater than any of the subsequent changes (**Table 3**). Sirtori et al. (1979) tested effects of a prudent diet al..nd of one containing soy protein as virtually the sole protein source. Cholesterol levels fell by 6% in the prudent diet and fell 23% more on the soy protein diet. When the order of diets was reversed cholesterol levels fell by 19% on the soy diet al..nd rose by 9% on the prudent diet.

Table 3. Serum lipids in 6 subjects fed various fats, carbohydrates and protein*

Protein (energy%)	CHO**	% Fat (en%)	Lipids, mg/dl	
			Cholesterol	Triglyceride
Mixed (12)	23/20	45	271 ± 13	159 ± 22
Plant (15)	8/32	45	194 ± 17	122 ± 8
Plant (18)	13/54	15	202 ± 11	109 ± 21
Plant (13)	58/14	15	189 ± 10	161 ± 29
Plant (18)	13/54	15	203 ± 10	119 ± 115
Plant (13)	58/14	15	173 ± 7	203 ± 31
Plant (12)	34/9	45	199 ± 10	135 ± 21
Mixed (12)	23/20	45	267 ± 11	159 ± 14

* After Hodges et al. (1967). Each diet fed for 4 weeks.
**Ratio of simple/complex carbohydrate

Fiber

The role of fiber in the diet has been of great professional and lay interest in the past few years. In general, insoluble fibers such as wheat bran have no effect on cholesterol levels in man. Soluble fibers (pectin, guar gum) will lower total and LDL cholesterol levels. Oat bran has a hypocholesterolemic because of its high content of oat gum or other soluble beta glucans (Pilch, 1987; Schneeman and Lefevre, 1986). In a study of cholesterol levels in Seventh Day Adventists it was shown that the true vegans had significantly lower cholesterol levels than did the lacto-ovo vegetarians or the non vegetarians. The only difference in fiber intake among those groups was in pectin. The vegans ingested 67% more pectin than the lacto-ovo vegetarian and 88% more than the non vegetarian (Kritchevsky et al., 1984).

CANCER

It is probably fair to say that our currently great interest in diet al..nd cancer stems from the paper by Doll and Peto (1981) in which they estimated that diet might be related to 35% of all cancer deaths. Doll and Peto carefully described how they reached their estimate but it was overlooked in the headlong rush to find "anti cancer" foods. The National Research Council (1989), in a discussion of diet al..nd cancer stated, "The data are not sufficient, however, to quantitate the contributions of diet to overall cancer risk, or to determine the quantitative reduction in risk that might be achieved by dietary modifications." The data are provocative however and merit the large amount of effort being expended.

Fiber

Higginson and Oettle (1960) suggested that the rarity of bowel cancer among black Africans could be due to diet, specifically to dietary bulk. Burkitt (1971) proposed that the high fiber diet of Africans populations was responsible for their relative freedom from colon cancer. There have been several reviews of fiber intake and colon cancer risk which provide no unanimity regarding the role of fiber. Pilch (1987) examined data from 18 ecological and 22 case-control studies. A protective effect was found in 67% of the former and 36% of the latter. No effect was observed in 28% of the ecological and 41% of the case control studies. Jacobs (1988) examined 24 ecological studies and found 54% showed protection and 42% showed no effect. In 27 case-control studies he found about equal numbers to be protective (44% or to show no effect (41%). Bingham (1990) reviewed results of 22 case-control studies and found half of them to be protective and 41% to show no effect. Two Australian studies gave conflicting results. Potter and McMichael (1986) in Adelaide found increasing risk of colon cancer with increasing fiber intake whereas Kune et al. (1987). in Melbourne found that fiber intake reduced risk.

Jensen et al. (1982) compared dietary effects on bowel cancer in large and small cities in Finland (Helsinki and Parikkala) and Denmark (Copenhagen and Them). They analyzed 39 variables. A positive correlation with cancer risk was found only for alcohol intake and fecal bile acid concentration. Negative correlations were deduced for saturated fatty acids. starch, carbohydrates, protein, cereals and total fiber. More recently Giovannucci et al. (1992) found fiber intake to be highly negatively correlated with colorectal adenoma in men whereas Sandler et al. (1993) found no significant correlation between fiber intake and colorectal adenoma in either men or women.

One problem is assessing effects of fiber in normal diets is that fiber covaries with both macro-and micronutrients in the diet. Estimates of dietary fiber can be different with and without adjustment of either dietary fat or calories (Freudenheim et al., 1990). A diet high in fiber is also one high in other plant components many of which have been shown to inhibit chemically induced carcinogenesis. Among these components are phenyl isothiocyanate (broccoli, cabbage) and flavonoids (fruits, vegetables) which inhibit covalent DNA binding; beta-carotene (green and yellow vegetables and fruits) and organosulfur compounds (garlic, onions) which inhibit tumor promotion; and indole-3-carbinol (cruciferous vegetables) which induces biotransformation. Dragsted et al. (1993) have written an excellent review of protective factors in fruits and vegetables.

The interactions of nutrients must also be noted. Heilbrun et al. (1989) found a significant negative association between dietary fiber and risk of colon cancer in men consuming less than 61 gms of fat daily but there was no effect of fiber in men ingesting more than 61g of fat/day. Hirayama (1985) found the risks of colon cancer in Japanese who ate meat, but no vegetables to be 18.4/100,000 and vegetables but no meat, to be 13.7/100,000. When both were included in the diet risk fell to 3.9/100,000. Potter and McMichael (1986) suggested that resolution of the fiber-colon cancer question required a multivariate hypothesis.

Fat

Recent reviews of the data relating dietary fat to breast cancer find weak association if any (Goodwin and Boyd, 1987; Byers, 1988; Rogers and Longnecker, 1988). In a study of 89,583 women Willett et al. (1987) found no correlations between the risk of breast cancer and intake of total fat, saturated fat, linoleic acid or cholesterol.

The connection may not be with fat but with calories. Hoffman (1927) laid the blame on caloric excess. The relationship between overweight and cancer incidence and mortality has been demonstrated convincingly (Lew and Garfinkel, 1979; Garfinkel, 1985). Animal studies dating to Moreschi (1909) have shown that general underfeeding or carefully calculated caloric restriction could inhibit growth of spontaneous, transplanted or induced tumors (reviewed by Kritchevsky, 1993a). Lavik and Baumann (1943) showed that a diet high in fat but low in calories led to 48% fewer methylcholanthrene-induced skin tumors in mice than did a diet low in fat but high in calories. Albanes (1987) reviewed the data from 82 studies and found the rations of tumors in mice in whom fat and calories had been varied to be similar to those reported by Lavik and Baumann in 1943. Subsequent studies (Kritchevsky et al., 1984; Klurfeld et al., 1987) have shown that caloric restriction will inhibit chemically-induced mammary or colon cancer even when the restricted diet contains more fat that the control diet .

In one study (Klurfeld et al., 1989) female rats were dosed with 7.12-dimethylbenz (a) anthracene (DMBA) and fed ad libitum on diets in which energy was restricted by 25% but which contained 20 or 26.7% fat. As **Table 4** shows rats fed 20% fat in a diet in which calories were restricted by 25% had fewer and smaller tumors than rats fed 5% fat in an ad-libitum diet.

Table 4. Effects of fat level and 25% caloric restriction on DMBA-induced mammary tumors in rats*

Diet	Tumor Incidence (%)	T/TBR**	Tumor lot (g)	Tumor Burden (g)***
Ad Libitum				
5% Corn Oil	65	1.9+0.3	2.0+0.7	4.2+1.9
15% Corn Oil	85	3.0+0.6	2.3+0.7	6.6+2.7
20% Corn Oil	80	4.1+0.6	2.9+0.5	11.8+3.2
25% Restricted				
20% Corn Oil	60	1.9+0.4	0.8+0.2	1.5+0.5
20% Corn Oil	30	1.5+0.3	1.4+1.0	2.3+1.6
p<	0.005	0.0001	0.0001	0.0001

*   After Klurfeld et al. (1989)
**   Tumors/tumor bearing rat
*** Total tumor weight/rat

The mechanisms by which caloric restriction may exert its tumor inhibiting effect have been reviewed (Kritchevsky, 1993a). They include increasing activity of anti-oxidant enzymes, reducing circulating insulin and mammotrophic hormones, enhanced DNA-repair and reduced oncogene expression. Energy restriction may not inhibit establishment of tumors but it inhibits their growth (Ruggeri et al., 1989) (**Table 5**).

In man there are ample data relating colon cancer risk to caloric intake (Jain et al., 1980; Bristol et al., 1985; Lyon et al., 1987). Lyon et al. (1987) concluded, "Total energy intake must be evaluated before attempting to assign a causal role to any food or nutrient that may be postulated to play a role in colon cancer". There are also data showing that exercise will reduce tumor incidence in humans (Frisch et al., 1985) and rats (Kritchevsky, 1990). Data from several countries also indicate that men who have spent most of their

Table 5. Influence of caloric restriction on incidence and size distribution of DMBA-induced mammary tumors in rats*

Diet	Tumor Incidence (%)	T/TBR**	LP*** tumors (%)	SNP*** tumors (%)	$\dfrac{LP}{SNP}$
Ad libitum	90	5.6+1.7	97	13	6.69
25% Restricted	61	3.4+0.6	77	23	3.35
40% Restricted	20	1.5+0.5	33	66	0.50
p<	0.007	0.05	0.0001	0.0001	---

* After Ruggeri et al. (1989)
** Tumors (tumor bearing rat)
***LP=large palpable tumors; SNP=small, non palpable tumors $\leq$ 100mg

working life at jobs which require considerable energy expenditure are at a lower risk of colon cancer than their more sedentary counterparts (Garabrant et al., 1984; Vena et al., 1985; Ballard-Barbash et al., 1990).

Virtually all dietary guidelines recommend eating a variety of foods and maintenance of ideal weight. The best current advice would seem to be moderation, balance and variety (Kritchevsky, 1993 b).

CONCLUSION

We are accumulating masses of data from experiments in which one macronutrient has been replaced by another-corn oil for coconut oil or sucrose for starch. These experiments lend themselves to simple interpretation but they do not reflect diets as they are eaten. The need for studies for dietary interactions in studies of atherosclerosis has been stressed (Kritchevsky, 1979). The same is necessary in the cancer field. We should be seeking the most beneficial total diet. One which combines as much "safety" as possible with latitude to sample and use all of the foods now available to us. Emphasis on single components leads to the discredited concept of "good" and "bad" foods- we should be seeking healthful dietary patterns.

ACKNOWLEDGMENT

Supported, in part, by a Research Career Award (HL 00734) from the National Institutes of Health.

REFERENCES

Albanes, D., 1987. Total calories, body weight and tumor incidence in mice, Cancer Res. 47: 1987-1992.

Bingham, S.A., 1990. Mechanisms and experimental and epidemiological evidence relating dietary fibre (non starch polysaccharides) and starch to protection against large bowel cancer, Proc. Nutr. Soc. 49: 153-171.

Bristol, J.B., Emmett, P.M., Heaton, K.W. and Williamson, R.C., 1985. Sugar, fat and the risk of colon cancer, Br. Med. J. 291: 1467-1470.

Burkitt, D.P., 1971. Epidemiology of cancer of the colon and rectum, Cancer 28: 3-13.

Byers, T., 1988. Diet al..nd cancer: any progress in the "interim"? Cancer 62:1713-1724.

Derr, J., Kris-Etherton, P.M., Pearson, T.A. and Seligson, F.H., 1993. The role of fatty acid saturation on plasma lipids, lipoproteins and apolipoproteins II. The plasma total and low-density lipoprotein cholesterol response of individual fatty acids, Metabolism 42: 130-134.

Doll, R. and Peto, R., 1981. The causes of cancer: quantitative estimates of avoidable risks of cancer in the United States today, J. Natl. Cancer Inst. 66: 1191-1308.

Dragsted, L.O., Strube, M. and Larsen, J.C., 1993. Cancer-protective factors in fruits and vegetables: Biochemical and Biological Background, Pharmacol. Toxicol. 72: Suppl. 1, S116-S135.

Freudenheim, J.L., Graham, S., Horwath, P.J., Marshall, J., Haughey, B.P.and Wilkinson, G., 1990. Risks associated with source of fiber and fiber components in cancer of the colon and rectum, Cancer Res. 50: 3295-3300.

Frisch, R.E., Wyshak, G., Albright, N.L., Albright, T.E., Schiff, I., Jones, K.P., Witschi, K., Schiang, E. and Marguglio, M., 1985. Lower prevalence of breast cancer and cancers of the reproductive system among former college athletes compared to non-athletes, Br. J. Cancer 52: 889-891.

Garabrant, D.H., Peters, J.M., Mack, T.M., and Bernstein, L., 1984. Job activity and cancer risk, Am. J. Epidemiol. 119: 1005-1014.

Garfinkel, L., 1985. Overweight and cancer, Ann. Int. Med. 103: 1034-1036.

Giovanucci, E., Stampfer, M.J., Colditz, G., Rimm, E.B. and Willett, W.C., 1992. Relationship of diet to risk of colorectal adenoma in men, J. Natl. Cancer Inst. 84: 91-98.

Goodwin, P.J. and Boyd, N.F., 1987. Critical appraisal of the evidence that dietary fat intake is related to breast cancer risk in humans, J. Natl. Cancer Inst. 79. 473-485.

Gresham, G.A. and Howard, A.N., 1960. The independent production of atherosclerosis and thrombosis in the rat, Br. J. Exp. Pathol. 41: 395-402.

Hayes, K.C. and Khosla, P., 1992. Dietary fatty acid thresholds and cholesterolemia, FASEB J. 6: 2600-2607.

Hegsted, D.M., and Ausman, L.M., 1988. Diet, alcohol and coronary heart disease in men, J. Nutr. 118: 1184-1189.

Hegsted, D.M., Ausman, L.M., Johnson, J.A. and Dallal, G.E., 1993. Dietary fat and serum lipids: an evaluation of experimental data, Am. J. Clin. Nutr. 57: 875-883.

Hegsted, D.M., McGandy, R.B., Myers, M.L. and Stare, F.J., 1965. Quantitative effects of dietary fat on serum cholesteorl in man, Am. J. Clin. Nutr. 17: 281-295.

Heilbrun, L.K., Nomura, A., Hankin, J.H. and Stemmermann, G.N., 1989. Diet al..nd colorectal cancer with special reference to fiber intake, Int. J. Cancer 44: 1-6.

Higginson, J. and Oettle, A.G. 1960. Cancer incidence in the Bantu and Cape Coloured races in South Africa: Report of a cancer survey in the Transvaal (1953-1955), J. Natl.Cancer Inst. 24: 589-671.

Hirayama, T., 1985. Mortality in Japanese with LIfe-Styles similar to Seventh Day Adventists: Strategy or risk reduction by life style modification, Natl. Cancer Inst. Monogr. 69: 143-153.

Hodges, R.E., Krehl, W.A., Stone, D.B. and Lopez, A., 1967. Dietary carbohydrates and low cholesterol diets: Effect on serum lipids in man, Am. J. Clin. Nutr. 20: 198-208.

Hoffman, F.L., 1927. Cancer increase and overnutrition, Prudential Ins. Newark, NJ.

Jacobs, L.R., 1988. Fiber and Colon Cancer, Gastroenterol. Clin. North Am. 17: 747-760.

Jain, M., Cook, G.M., Davis, F.G., Grace, M.G., Howe, G.R. and Miller, A.B., 1980. A case-control study of diet al..nd colorectal cancer, Int. J. Cancer 26: 757-768.

Jensen, O.M., MacLennon, R. and Wahrendorf, J., 1982. Diet, bowel function, fecal characteristics and large bowel cancer in Denmark and Finland, Nutr. Cancer. 4: 5-19.

Keys, A., Anderson, J.T. and Grande, F., 1965. Serum cholesterol response to changes in the diet.. IV. Particular saturated fatty acids in the diet, Metabolism. 14: 776-787.

Klurfeld, D.M., Weber, M.M. and Kritchevsky, D., 1987. Inhibition of chemically-induced mammary and colon tumor promotion by caloric restriciton in rats fed increased dietary fat, Cancer Res. 47: 2759-2762.

Klurfeld, D.M., Welch, C.B., Lloyd, L.M. and Kritchevsky, D. 1989. Inhibition of DMBA-induced mammary tumorigenesis by caloric restriction in rats fed high fat diets, Int. J. Cancer 43: 922-925.

Kritchevsky, D., 1979. Dietary interactions in "Nutrition, Lipids and Coronary Heart Disease. A Global View", R.I. Levy, B.M. Rifkind, B.H. Dennis and N.D. Ernst eds. Raven Press, New York. 229-246.

Kritchevsky, D., 1985. Nutrition and cardiovascular disease in "Nutritional Pathology", H. Sidransky, ed. Marcel Dekker Inc. New York pp. 127-160.

Kritchevsky, D., 1990. Influence of caloric restriction and exercise on tumorigenesis in rats, Proc. Soc. Exp. Biol. Med. 193: 35-38.

Kritchevsky, D., 1993a. The Quartercentenary Lecture. Undernutrition and chronic disease: cancer, Proc. Nutr. Soc. 52: 39-47.

Kritchevsky, D., 1993b. Dietary guidelines. The rationale for intervention,Cancer 72: 1011-1014.

Kritchevsky, D., Moyer, A.W., Tesar, W.C., Logan., J.B., Brown, R.A., Davies, M.C. and Cox, H.R., 1954. Effect of cholesterol vehicle in experimental atherosclerosis, Am. J. Physiol. 178: 30-32.

Kritchevsky, D., Tepper, S.A., Czarnecki, S.K., Klurfeld, D.M. and Story, J.A., 1981. Experimental atherosclerosis in rabbits fed cholesterol-free diets. 9. Beef protein and textured vegetable protein, Atheroscler. 39: 169-175.

Kritchevsky, D., Tepper, S.A. and Goodman, G., 1984a. Diet, nutrition intake and metabolism in populations at high and low risk of colon cancer: Relationship of diet to serum lipids, Am. J. Nutr. 40: 921-926.

Kritchevsky, D., Tepper, S.A., Morrissey, R.B., Klurfeld, D.M. and Story, J.A., 1983. Comparison of diets approximating American intake of 1909 and 1972: Effects on lipid metabolism in rabbits and rats, Nutr. Rep. Int. 28: 1-8.

Kritchevsky, D., Tepper, S.A., Vesselinovitch, D. and Wissler, R.W., 1971.Cholesterol Vehicle in Experimental Atherosclerosis XI. Peanut Oil, Atheroscler. 14: 53-64.

Kritchevsky, D., Tepper, S.A., Vesselinovitch, D. and Wissler, R.W., 1973. Cholesterol Vehicle in Experimental Atherosclerosis, 13. Randomized Peanut Oil. Atheroscler. 17: 225-243.

Kritchevsky, D., Weber, M.M. amd Klurfeld, D.M. 1984b. Dietary fat versus caloric content in initiation and promotion of 7.12-dimethylbenz (a) anthracene-induced mammary tumorigenesis in rats, Cancer Res. 44: 3174-3177.

Kune, S., Kune, G.A. and Watson, L.F., 1987. Case-control study of dietary etiological factors: the Melbourne Colorectal Cancer Study, Nutr. Cancer 9: 21-42.

Lavik, P.S. and Baumann, C.A., 1943. Further studies on the tumor-promoting action of fat, Cancer Res. 3: 749-756.

Lee, H.P., Gourley, L., Duffy, S.W., Esteve, J., Lee, J. and Day, N.G., 1989. Colorectal cancer and diet in an Asian population-a case-control study among Singapore Chinese, Int. J. Cancer. 43: 1007-.

Lew., E.A., and Garfinkel, L., 1979. Variations in mortality by weight in 750,000 men and women, J. Chronic Dis. 32: 563-576.

Lyon, J.L., Mahoney, A.W., West, D.W., Gardner, J.W., Smith, K.R., Sorenson, A.W. and Stanish, W., 1987. Energy intake: its relation to colon cancer, J. Natl. Cancer. Inst. 78: 853-861.

Meeker, D.R. and Kesten, H.D., 1941. Effect of high protein diets on experimental atherosclerosis of rabbits, Arch. Path. 31: 147-162.

Moreschi, C., 1909. Beziehungen zwischen Ernahrung und Tumorwachstum, Z. Immunitatsforsch. 2: 651-675.

National Research Council 1989. Diet Health: Implications for reducing chronic disease risk, Nat. Academy Press, Washington, DC p. 701.

Pilch, S.M., (ed) 1987. Review of physiological effects and health consequences of dietary fiber, FASEB, Bethesda, MD.

Potter, J.D. and McMichael, A.J., 1986. Diet al..nd cancer of the colon and rectum: a case-control study, J. Natl. Cancer Inst. 76: 557-569.

Rogers A.E., Longnecker, M.P. 1988. Biology of disease: dietary and nutritional influences on cancer. A review of epidemiological and experimental data, Lab Invest. 59: 729-759.

Ruggeri, B.A., Klurfeld, D.M., Kritchevsky, D. and Furlanetto, R.W., 1989. Growth factor binding to 7.12-dimethylbenz (a) authracene-induced mammary tumors from rats subject to chronic caloric restriction, Cancer Res. 49: 4135-4141.

Sandler, R.S., Lyles, C.M., Peipins, L.A., McAuliffe, C.A., Woosley, J.T. and Kupper, L.L., 1993. Diet al..nd risk of colorectal adenomas: Macronutrients, cholesterol and fiber, J. Natl. Cancer Inst. 85: 884-891.

Schneeman, B.O. and Lefevre, M., 1986. Effects of fiber on plasma lipoprotein composition In "Dietary Fiber: Basic and Clinical Aspects" G.V. Vahouny and D. Kritchevsky eds. Plenum Press, New York, pp. 309-321.

Sirtori, C.R., Gatti, E., Mantero, O. Conti, F., Agradi, E., Tremoli, E., Sirtori, M., Fraterrigo, L., Tavazzi, L. and Kritchevsky, D., 1979. Clinical experience with the soybean protein diet in the treatment of hypercholesterolemia, Am. J. Clin. Nutr. 32: 1645-1658.

Vena, J.E., Graham, S., Zielezny, M., Swanson, M.K., Barnes, R.E. and Nolan, J., 1985. Lifetime occupational exercise and colon cancer, Am. J. Epidemiology 122: 357-365.

Vesselinovitch, D., Getz, G.S., Hughes, R.H. and Wissler, R.W., 1974. Atherosclerosis in the rhesus monkey fed three food fats, Atheroscler. 20: 303-321.

Willett , W., Stampfer, M.J., Colditz, B.A., Rosner, B.A., Hennekens, C.H., Speizer, F.E., 1987. Dietary fat and the risk of breast cancer, N. Engl. J. Med. 316: 22-28.

CARCINOGENS IN FOODS: HETEROCYCLIC

AMINES AND CANCER AND HEART DISEASE

Richard H. Adamson and Unnur P. Thorgeirsson

Division of Cancer Etiology
National Cancer Institute
Bethesda, MD 20892

INTRODUCTION

Laboratory and epidemiologic studies continue to accumulate evidence that dietary factors contribute to both the cause and prevention of chronic diseases found in the Western societies, especially cancer and cardiovascular disease. A number of naturally occurring carcinogens are found in foods and these compounds might contribute to the etiology of both these diseases. Data presented in this paper demonstrate that some of the heterocyclic amines (HCAs) formed as a result of cooking are both carcinogenic and cardiotoxic in animals, and it is possible that the HCAs may play a role in the etiology of both these diseases in humans.

NATURALLY OCCURRING CARCINOGENS IN FOODS

A number of naturally occurring carcinogens are found in foods due to numerous factors, including production by the plant(s), contamination by fungal toxins, or formation during the processing or cooking of foods. **Table 1** lists eight different carcinogens that are found in foods, but this list is by no means comprehensive. Ethyl carbamate or urethane is likely formed by the interaction of carbamyl phosphate present naturally in yeast with ethanol formed by the process of fermentation. Thus, ale, beer, bread, wine, yogurt and other foods contain this carcinogen (Ough, 1976). Nitrosamines are formed in various foods , especially foods having a high amine content, prior to consumption or in the acidic environment of the stomach as a result of nitrosation. Bacon and cured meats that contain added nitrite to prevent botulism have been of major concern (Preussmann and Eisenbrand, 1984). However, two factors have lessened this concern: first, reduction of the amount of added nitrite and second, addition of sodium ascorbate or sodium erythorbate, which inhibits nitrosation. Various carcinogenic hydrazines are found in both wild and cultivated edible mushrooms (Toth, 1984). The aflatoxins are a group of four fungal metabolites produced by the fungus *Aspergillus flavus*; of this group, aflatoxin B_1, is the most toxic and carcinogenic. This compound has been identified in commercial samples of

Nutrition and Biotechnology in Heart Disease and Cancer
Edited by J.B. Longenecker *et al.*, Plenum Press, New York, 1995

211

peanuts, peanut meal, corn and other foods in the United States (Stoloff, 1976), but other foods are also contaminated in other countries (Wogan and Busby, Jr., 1980). Benzene is an important industrial solvent and is found not only in gasoline and cigarette smoke, but it also is present in a number of foods, including cooked meat, eggs and various fruit (IARC, 1982). The average dietary intake may be around 250 μg/day. Benzo(a)pyrene and other polycyclic aromatic hydrocarbons (PAHs) have been found in meat following smoking or broiling or grilling (Lijinsky and Shubik, 1964; Grasso, 1984). Also, during the cooking of meat, poultry and fish, a number of highly mutagenic and carcinogenic heterocyclic amines (HCA) have been found.

Table 1. Naturally occurring carcinogens in foods

Carcinogen	Food
Ethyl carbamate	Ale, beer, bread, wine, yogurt
N-nitrosamines	Cured meats, fish, cheese
Hydrazines	Mushrooms
Aflatoxins	Peanuts, peanut meal, corn
Fusariums	Corn, barley, wheat
Benzene	Meat, eggs, various fruit
Benzo(a)pyrene and other PAHs	Meat
Heterocyclic amines (HCA)	Meat, fish, poultry

HETEROCYCLIC AMINES IN COOKED FOODS

The production of mutagenic HCAs was first reported by Sugimura and coworkers in 1977 (Sugimura et al., 1977). Since that time, using initially the Ames test and then various analytical techniques, , more than 20 different HCAs have been isolated and identified from different foods, especially cooked beef, pork, poultry and fish and additional HCAs have been detected in other foods but not yet identified (Adamson, 1990; Felton and Knize, 1990; Eisenbrand and Tang, 1993). The HCAs found in cooked foods are one of two types: those having an amino group attached to the 2 position of an imidazo ring (imidazo type) and those having an amino group attached to the pyridine ring (nonimidazole type). These two types can be separated by treatment with nitrite under acid conditions since the nonimidazole type is deaminated and does not demonstrate mutagenicity following treatment, whereas the imidazole type still demonstrates mutagenicity. An extract from cooked beef and sardines demonstrated that 75 to 88% mutagenicity remained following treatment with acid nitrite, suggesting that the imidazole type mutagens are those present in Western foods (Tsuda et al., 1980; 1985). The high mutagenicity of various HCAs in two of the Ames tester strains, TA 98 and TA 100, is summarized in **Table 2.** Both the nonimidazole HCAs (Trp-P-1 and Trp-P-2) and the imidazole types (IQ, MeIQ, MeIQx, PhIP, and 4,8 DiMeIQx) are mutagenic. In general, the imidazole type is more mutagenic in both of the tester strains. In contrast, three other known carcinogens, aflatoxin B_1, benzo(a)pyrene and N,N-diethylnitrosamine, are much less mutagenic (Sugimura and Wakabayashi, 1990).

The HCAs in cooked meat are formed primarily from the reaction of amino acids with creatinine at high temperatures, although in some instances the presence of sugar may also facilitate the reaction (Jagerstad et al., 1984;Felton and Knize, 1990; Shioya et al., 1987). More recently, Knize et al., 1994, reported that arginine and 1-methylguanidine, a fragment of arginine, plus amino acids produced mutagenic compounds. Both the cooking temperature and time are important variables in HCA formation; however, temperature appears to be the more important factor (Pariza et al., 1979; Bjeldanes et al., 1983).

Table 2. Mutagenicity of various heterocyclic amines and carcinogens in *Salmonella typhimurium* TA 98 and TA 100[1]

Compound	Revertants/μg	
	TA 98	TA 100
Trp-P-1	39,000	1,700
Trp-P-2	104,200	1,800
IQ	433,000	7,000
MeIQ	661,000	30,000
MeIQx	145,000	14,000
PhIP	1,800	120
4,8 DiMeIQx	183,000	8,000
Aflatoxin B_1	6,000	28,000
Benzo(a)pyrene	320	660
N,N-Diethylnitrosamine	0.02	0.15

[1] After Sugimura and Wakabayashi, 1990

CARCINOGENICITY OF THE HETEROCYCLIC AMINES

Ten of the HCAs have been synthesized in enough quantity for bioassay in rodents and all ten were found to be carcinogenic. Six of the HCAs tested were of the nonimidazole class and in mice the target organs for this class of HCAs was generally that of the liver and blood vessels. In rats, two of these compounds, Glu-P-1 and Glu-P-2, also induced tumors in the small and large intestine. All four of the imidazole type HCAs tested were carcinogenic, inducing a variety of tumors in rodents (Ohgaki et al., 1991). We selected three HCAs, IQ (2-amino-3-methylimidazo[4,5-*f*]quinoline), 8-MeIQx (2-amino-3,4,8-dimethylimidazo[4,5-*f*]quinoxaline), and PhIP (2-amino-1-methyl-6-phenylimidazo[4,5,*b*]pyridine) for evaluation of carcinogenicity and other toxic effects in nonhuman primates **(Figure 1)**. The criteria for selecting the compounds for study included structure, amount in food, the mutagenic activity, the spectrum of tumors induced in rodents and the availability of the compound **(Tables 3** and **4)** (Adamson et al., 1991)

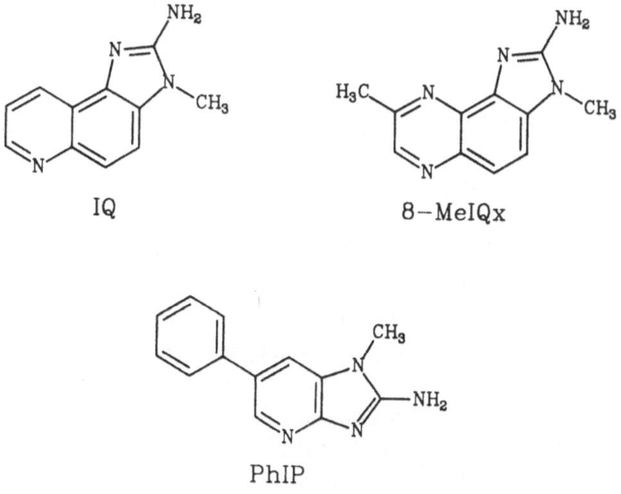

Figure 1. The chemical structures of IQ, 8-McIQx and PhIP.

Table 3. Carcinogenicity of three imidazole type HCAs in rodents

HCA	Species	Route of Administration and Dose	Site of Tumors
IQ	Mouse	Diet 0.03%	Liver, lung, forestomach
	Rat	Diet 0.03%	Small and large intestine, liver, skin Zymbal gland
8-MeIQx	Mouse	Diet 0.06%	Liver, lung lymphoma, leukemia
	Rat	Diet 0.04%	Liver, Zymbal gland, clitoral gland, skin
PhIP	Mouse	Diet 0.04%	Lymphoma
	Rat	Diet 0.04%	Colon, breast

Summarized from Ohgaki et al., 1991

Table 4. Amount of HCAs in cooked foods and extracts

HCA	Food Type	Amount (ng/g)
IQ	Fried ground beef	0.02 - 0.6
	Broiled sardine	4.9 - 20
	Broiled salmon	0.3 - 1.8
	Bact. grade beef extract	41 - 142
8-MeIQx	Fried ground beef	0.1 - 2.4
	Smoked dried mackerel	0.8
	Food grade beef extract	3.1 - 28
	Bact. grade beef extract	58.7 - 527
PhIP	Fried ground beef	15
	Fried codfish	69.2
	Fried bacon	53

Summarized from Felton and Knize, 1990

The compounds were prepared daily as a suspension in hydroxypropyl cellulose (HPC) and administered to cynomolgus monkeys by gavage five times a week (Monday -Friday) at a dose of 10 or 20 mg/kg body weight. Concurrent HPC-treated monkeys and untreated monkeys served as controls. Additional parameters and information on this nonhuman primate colony have been published (Adamson et al., 1994; Thorgeirsson et al., 1994). Thus far, neither 8-MeIQx nor PhIP have induced any tumors following a time period as long as 60 months and 42 months, respectively. Also, based on studies of activation of these carcinogens and DNA adduct studies in this species, we speculate that 8-MeIQx may not induce tumors in cynomolgus monkeys but it is likely that PhIP will eventually induce tumors in this species (Adamson et al., 1991; Davis et al., 1993). In contrast, IQ induced primary hepatocellular carcinoma (HCC) in 55% of the monkeys at the 10 mg/kg dose and in 95% of the animals at the 20 mg/kg dose **(Table 5)** (Adamson et al., 1994). The average total dose given to the monkeys at the 10 mg/kg level was 39.7 grams and 45.1 grams at the 20 mg/kg level. The average latent period was 59.6 months at the lower dose and for the 20 mg/kg dose, 43.4 months. The involvement of IQ-induced HCC ranged from a single tumor nodule to multiple nodules occupying a large proportion of the liver. Tumor development was not associated with liver fibrosis or cirrhosis and, in the majority of the cases, few well- defined nodules were present at the time of necropsy **(Figure 2)**. The most common microscopic appearance of the HCC was that of a trabecular pattern **(Figure 3)**. Different histological patterns were frequently seen among individual tumor nodules in the same liver and even within the same nodule.

Table 5. Induction of hepatocellular carcinoma in nonhuman primates by IQ

Dose	No. Treated	No. with Tumors	% with Tumors	Average Total Dose[a]	Average Latent Dose[a]
10 mg/kg	20	11	55	39.7 grams	59.6 months
20 mg/kg[a]	20	19	95	45.1 grams	43.4 months

[a]Figures are for 18 cynomolgus monkeys; in addition two rhesus monkeys were treated with IQ and tumors were also induced in this species.

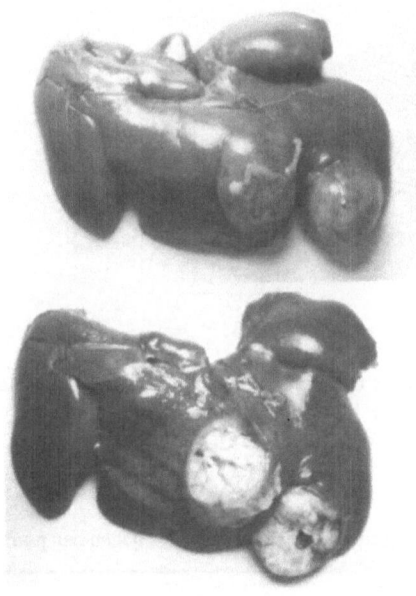

Figure 2. Gross appearance of the liver of a cynomolgus male monkey dosed with a total of 73.5 g of IQ over a period of 69 months. This monkey had four separate nodules of hepatocellular carcinoma, two of which are apparent on the photograph protruding from the anterior aspect of the liver (top). A cross section of the liver (bottom) depicts the cut surface of the two well-defined tumor nodules which measured 3 cm and 3.5 cm in the largest diameter.

CARDIAC DAMAGE BY HETEROCYCLIC AMINES

Studies of IQ-DNA adducts by the ^{32}P-postlabeling method in nonhuman primates showed high levels of the adducts, especially the major C-8 guanine adduct in the liver, kidney and heart (Adamson et al., 1991). Therefore, a systematic study of cardiac pathologic changes associated with chronic administration of IQ to cynomolgus monkeys was undertaken (Thorgeirsson et al., 1994b). At necropsy, the heart from IQ-treated monkeys was removed and perfused with 4% formaldehyde and 1% glutaraldehyde for 60 minutes at 80-100 mm Hg. Following pressure perfusion fixation, sections were taken, prepared for electron microscopy, processed for light microscopy and stained with hematoxylin-eosin (H and E) and Masson trichrome. **Table 6** summarizes some of the pathology seen by light microscopy. These included interstitial fibrosis with myocyte hypertrophy and atrophy, myocyte necrosis associated with inflammation (lymphocytes and macrophages) **(Figure 4A)**, inflammation without myocyte necrosis, especially myocyte hypertrophy and vasculitis (infiltration of vascular walls) (Thorgeirsson et al., 1994b).

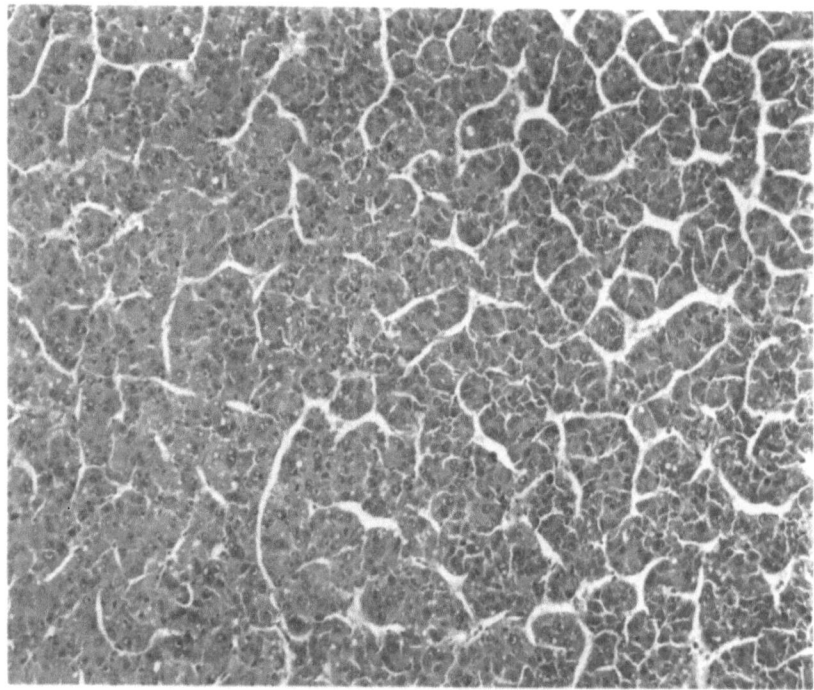

Figure 3. Light microscopic appearance of a hepatocellular carcinoma, trabecular pattern. The histological patterns often varied between different IQ-induced tumor nodules and even within the same tumor. A trabecular pattern was most commonly seen (H&E, x125).

Table 6. Myocardial changes in IQ-treated nonhuman primates[a]

Treatment and No.		Fibrosis	MN	I/N	I	MH	Vasculitis
Controls	(2)	0	0	0	0	0	0
IQ	(10)	4/10	5/10	4/10	4/10	6/10	2/10

[a]As observed by light microscopy
 Abbreviations are: MN, myocyte necrosis; I/N, inflammation associated with myocyte necrosis;
 I, inflammation without myocyte necrosis; MH, myocyte hypertrophy.

Ultrastructural examination of the myocardium from IQ-treated animals revealed lesions in 9 of the 10 hearts examined **(Table 7)**. Myofibrillar loss and disorganization of the parallel alignment of sarcomas occurred in 7 cases **(Figure 4B)** and in 4 of these cases, necrotic myocytes were also present. Mitochondrial abnormalities ranging from mild to moderate mitochondrial swelling, clearing of normal matrix densities, and disruption of mitochondrial cristae occurred in 7 of the 10 hearts from IQ-treated macaques **(Figure 4C)**. Normal ultrastructural architecture was present in one of 10 IQ-treated animals and in the 2 controls. These histopathological findings are consistent with the initial stages of toxic cardiomyopathy. However, it remains to be clarified whether IQ-DNA adducts play a role in the development of the cardiac lesions. Although gross cardiac abnormalities were not observed in the animals dosed with IQ, the possibility must be considered that long-term exposure to HCAs may eventually lead to compromised cardiac function.

Recently similar findings have been reported in rats dosed with IQ or PhIP. Also, in primary cultures of fetal rat myocytes exposed to the activated forms of the carcinogens, N-OH-IQ and N-OH-PhIP, lactate dehydrogenase (LDH) leakage increased in proportion to the dose, but was greater in cells exposed to N-OH-IQ than to the activated form of PhIP (Davis et al., 1994).

Table 7. Electron microscopic findings in hearts from IQ-treated nonhuman primates

Treatment No.		Lipid	ML	MN	MA
Controls	(2)	0	0	0	0
Treatment	(10)	2/10	7/10	4/10	7/10

Abbreviations are ML, myofibrillar loss; MN, myocyte necrosis; MA, mitochondrial abnormalities.

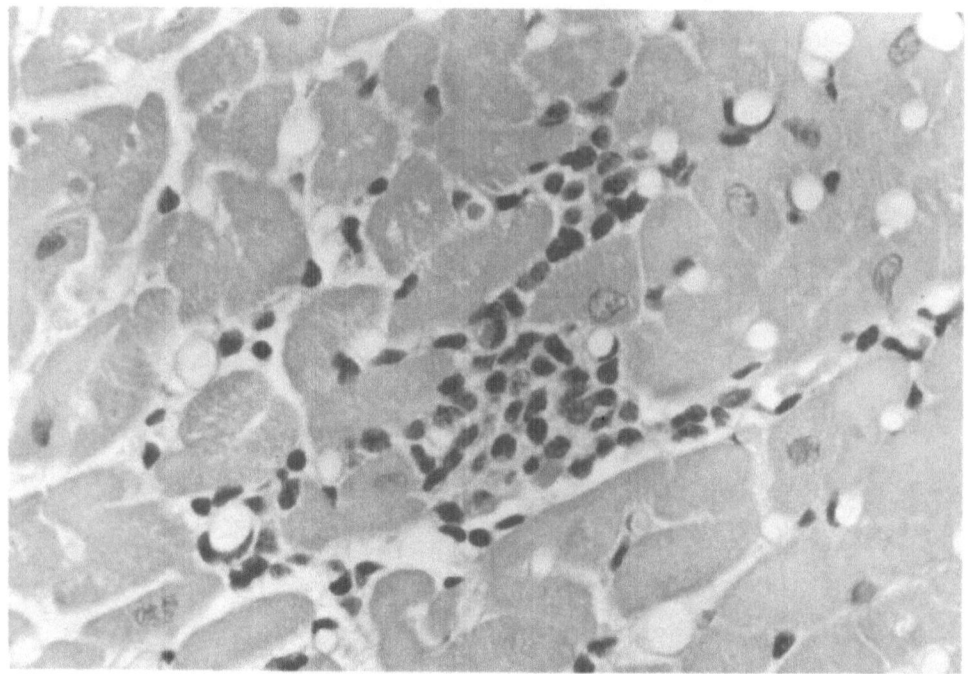

Figure 4A. Light microscopic appearance of focal chronic (lymphocytes and macrophages) inflammatory infiltrate with myocyte dropout, myocardium from the left ventricle of a cynomolgus monkey dosed with IQ for 57 months (H&E x320).

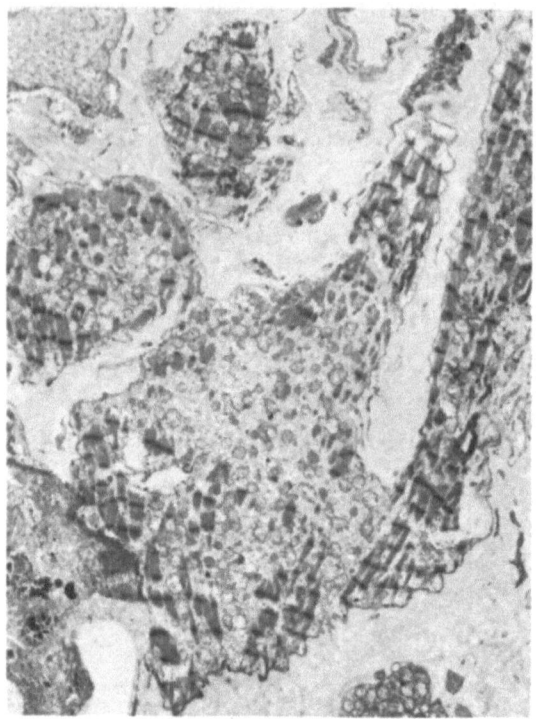

Figure 4B. Electron micrograph of myocardium from an animal dosed with IQ for 90 months. Myocytes show myofibrillar loss (x6000).

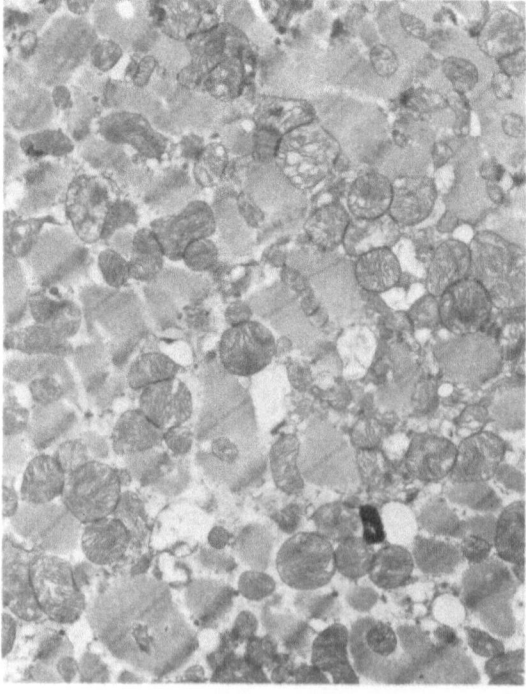

Figure 4C. Electron micrograph of myocardium from an animal dosed with IQ for 73 months. Some of the mitochondria show irregularities and dissolution of the cristae and disruption of sarcomeres (x18000).

REDUCTION OF INTAKE OF HCAs

Quantitation of HCA consumption and identification of all the HCAs in foods remains to be determined. However, individuals can take prudent precautions (Adamson, 1990) to lessen the intake of HCAs: vary the method of cooking meat; eat beef that is cooked medium instead of well done; microwave fish, poultry and bacon more often; if barbecuing, microwave first and discard the juice, especially for poultry and shortribs; stew and poach, where applicable; and avoid making gravy from dry drippings.

SUMMARY

Carcinogens occur naturally in the foods we eat, including a number of HCAs that have been identified in foods (beef, pork, poultry and fish) as a result of cooking. These compounds are formed during the normal cooking process by the reaction of creatine with various amino acids. The HCAs have been identified as a result of their high mutagenic activity in the Ames test. The HCAs can be separated into two types, the nonimidazole and the imidazole type, the latter of which is the predominant type present in Western foods. Both types of HCAs have been found to be carcinogenic in rodent bioassays. Of the three imidazole compounds presently under evaluation in nonhuman primates, IQ has been found to be a potent cacinogen, inducing hepatocellular carcinoma in a majority of the animals in approximately one-seventh of their life span. In addition, a high proportion of the nonhuman primates also had focal IQ-induced myocardial lesions as observed by both light and electron microscopic findings. This information, along with other toxicology data on the HCAs, much of which is cited in this paper, allows the inference to be made that HCAs may be a risk factor for both cancer and cardiovascular disease in humans.

REFERENCES

Adamson, R.H., 1990, Mutagens and carcinogens formed during cooking of foods and methods to minimize their formation, Cancer Prev. November 1-7.

Adamson, R.H., Snyderwine, E.G., Thorgeirsson, U.P., Schut, H.A.J., Turesky, R.J., Thorgeirsson, S.S., Takayama, S., and Sugimura, T., 1991, Metabolic processing and carcinogenicity of heterocyclic amines in nonhuman primates, in "Xenobiotics and Cancer," L. Ernster, H. Esumi, Y. Fujii, H.V. Gelboin, R. Kato, and T. Sugimura, eds., Japan Scientific Societies Press, Tokyo.

Adamson, R.H., Takayama, S., Sugimura, T., and Thorgeirsson, U.P., 1994, Induction of hepatocellular carcinoma in nonhuman primates by the food mutagen 2-amino-3-methylimidazo[4,5-ƒ]quinoline, Environ. Health Perspect. 102:190-193.

Bjeldanes, L.F., Moris, M.M., Timourian, H., and Hatch, F.T., 1983, Effects of meat composition and cooking conditions on mutagenicity of fried ground beef, J. Agric. Food Chem. 31:18-21.

Davis, C.D., Schut, H.A.J., Adamson, R.H., Thorgeirsson, U.P., Thorgeirsson, S.S., and Snyderwine, E.G., 1993, Mutagenic activation of IQ, PhIP and MeIQx by hepatic microsomes from rat, monkey and man: low mutagenic activation of MeIQx in cynomolgus monkeys in vitro reflects low DNA adduct levels in vivo, Carcinogenesis 14:61-65.

Davis, C.D., Farb A., Thorgeirsson, S.S., Virmani, R., and Synderwine, E.G., 1994, Cadiotoxicity of heterocyclic amine food mutagens in cultured myocytes and in rats, Toxicol. Appl. Pharmacol. 124:201-211.

Eisenbrand, G., and Tang, W., 1993, Food-borne heterocyclic amines. Chemistry, formation, occurrence and biological activities. A literature review, Toxicology 84:1-82.

Felton, J.S., and Knize, M.G., 1990, Heterocyclic-amine mutagens/carcinogens in foods, in "Handbook of Experimental Pharmacology," C.S. Cooper, and P.L. Grover, eds., Springer-Verlag, Berlin/Heidelberg, vol. 94/I.

Grasso, P., 1984, Carcinogens in food, in "Chemical Carcinogenes," 2nd ed., C.E. Searle, ed., ACS Monograph 182, American Chemical Society, Washington, DC.

International Agency for Research on Cancer, 1982, IARC Monographs on the Evaluation of the Carcinogenic Risk of Chemicals to Humans, vol. 29, Some Industrial Chemicals and Dyestuffs, IARC, Lyon.

Jagerstad, M., Olsson, K., Grivas, S., Negishi, C., Wakabayashi, K., Tsuda, M., Sato, S., and Sugimura, T., 1984, Formation of 2-amino-3,8-dimethylimidazo[4,5-*f*]quinoxaline in a model system by heating creatinine, glycine and glucose, Mutat. Res. 126:239-244.

Knize, M.G., Cunningham, P.L., Avila, J.R., Jones, A.L., Griffin, E.A., Jr., and Felton, J.S., 1994, Formation of mutagenic activity from amino acids heated at cooking termperatures. Food Chem. Toxicol. 32:55-60.

Lijinsky, W., and Shubik, P., 1964, Benzo(a)pyrene and other polynuclear hydrocarbons in charcoal-broiled meat, Science 145:53-55.

Ohgaki, H., Takayama, S., and Sugimura, T., 1991, Carcinogenicities of heterocyclic amines in cooked food, Mutat. Res. 259:399-410.

Ough, C.S., 1976, Ethylcarbamate in fermented beverages and foods. I. Naturally occurring ethylcarbamate, J. Agric. Food Chem. 24:323-328.

Pariza, M.W., Ashoor, S.H., Chu, F.S., and Lund, D.B., 1979, Effects of temperature and time on mutagen formation in pan-fried hamburger, Cancer Lett. 7:63-69.

Preussman, R., and Eisenbrand, G., 1984, N-Nitroso carcinogens in the environment, in "Chemical Carcinogens," 2nd. ed., C.E. Searle, ed., ACS Monograph 182, American Chemical Society, Washington, DC.

Shioya, M., Wakabayashi, K., Sato, S., Nagao, M., and Sugimura, T., 1987, Formation of a mutagen, 2-amino-1-methyl-6-phenylimidazo[4,5,*b*]pyridine (PhIP) in cooked beef, by heating a mixture containing creatinine, phenylalanine and glucose, Mutat. Res. 191:133-138.

Stoloff, L., 1976, Occurrence of mycotoxins in foods and feeds, in "Mycotoxins and Other Fungal Related Food Problems," J.V. Rodricks, ed., American Chemical Society, Washington, DC.

Sugimura, T., Nagao, M., Kawachi, T., Honda, M., Yahagi, T., Seino, Y., Sato, S., Matsukura, N., Matsushima, T., Shirai, A., Sawamura, M., and Matsumoto, H., 1977, Mutagens-carcinogens in foods with special reference to highly mutagenic pyrolytic products in broiled foods, in "Origins of Human Cancer," H.H. Hiatt, J.D. Watson, J.A. Winsten, eds., Cold Spring Harbor Laboratory Press, Cold Spring Harbor.

Sugimura, T., and Wakabayashi, K., 1990, Mutagens and carcinogens in food, in "Mutagens and Carcinogens in the Diet," M.W. Pariza, H.V. Aeschbacher, J.S. Felton, and S. Sato, eds., Wiley-Liss, Inc., New York.

Thorgeirsson, U.P., Dalgard, D.W., Reeves, J., and Adamson, R.H., 1994, Tumor incidence in a chemical carcinogenesis study of nonhuman primates, Reg. Toxicol. Pharmacol. 19:130-151.

Thorgerisson, U.P., Farb, A., Virmani, R., and Adamson, R.H., 1994b, Cardiac damage induced by 2-amino-3-methylimidazo[4,5-*f*]quinoline in nonhuman primates. Environ. Health Perspect. 102:194-199

Toth, B., 1984, Synthetic and naturally occurring hydrazines and cancer, J. Environ. Sci. Health, C2:51-102.

Tsuda, M., Negishi, C., Makino, R., Sato, S., Yamaizumi, Z., Hirayama, T., and Sugimura, T., 1985, Use of nitrite and hypochlorite treatments in the determination of the contributions of IQ-type and non-IQ type heterocyclic amines to the mutagenicities in crude pyrolized materials, Mutat. Res. 147:335-341.

Tsuda, M., Takahashi, Y., Nagao, M., Hirayama, T., and Sugimura, T., 1980, Inactivation of mutagens from pyrolysates of tryptophan and glutamic acid by nitrite in acid solution, Mutat. Res. 78:331-339.

Wogan, G.N., and Busby, W.F., Jr., 1980, Naturally occurring carcinogens, in "Toxic Constituents of Plant Foodstuffs," 2nd ed., I.E. Liener, ed., Academic Press, New York.

GENETIC ENGINEERING OF FOODS TO REDUCE THE RISK OF HEART DISEASE AND CANCER

Vic C. Knauf and Daniel Facciotti

Calgene, Inc.
1920 Fifth Street
Davis, California 95616

INTRODUCTION

That good nutrition promotes good health is obvious. More recently, with some visibility in the popular press, the role of bad nutrition and poor diet habits has been linked to increased chances of heart disease and cancer. Despite general acceptance of the theory, practice in the form of the American diet falls short of implementing these simple principles.

Aside from vitamin supplements, surprisingly little effort has been applied to improving the nutritional bases of foods in the common diet. Of course, novel processing and harvesting technology now allow year-round availability of many "fresh" fruits and vegetables. But examples of tailoring source food compositions are relatively few. Plant breeding succeeded in removing erucic acid from rapeseed oil: dietary erucic acid had been implicated in smooth muscle lesions. Plant breeders have also selected varieties of corn with altered sugar-starch ratios as an example of altering chemical composition.

A technology complementary to existing plant breeding methods is the now established process of the genetic engineering of plants and animals. Although examples of animal engineering include current experiments with "low fat, lean pigs," we will focus our discussion on specific targeted changes in the nutritional base of plant-derived foodstuffs. We acknowledge our inability to identify or prioritize such possible changes as to their direct health benefits, but offer this paper with hope that interdisciplinary efforts can do so.

STATUS OF THE TECHNOLOGY

There are a number of methods by which laboratory-modified genes can be introduced into plants - a process which results in transgenic plants (Corbin and Klee, 1991). Most methods reliably result in the introduced genetic material stably integrated in a plant chromosome. Thus subsequent progeny from a transgenic plant will also be transgenic. For a crop like canola, one transgenic seed can become 600 transgenic seeds in six months (one generation), and then can become 360000 transgenic seeds in two generations (one year). The

Nutrition and Biotechnology in Heart Disease and Cancer
Edited by J.B. Longenecker *et al.*, Plenum Press, New York, 1995

221

plants generally look identical to the original starting plant; an exception might be deliberately modified flower color (Napoli et al., 1990) - or seedless fruit. Of transgenic lines now scaled up and in field trials, more common are plants (tobacco, tomato, potato, canola, soybean, corn, etc.) that are herbicide-resistant or that resist insect feeding (Knauf, 1991).

These applications of genetic engineering have helped accelerate an assortment of associated technologies. For instance, genes from viruses, bacteria, fungi, and animals have all been adapted for function in transgenic plants - as well as genes from a range of plant sources. For example, Stark et al. (1992) have used a gene derived from a bacterial gene to redirect the amounts of starch produced in potatoes. The enzyme coded by this gene is not regulated by the typical control mechanisms effective on the native potato enzyme; thus the introduced gene has dramatic effects on plant metabolism. Foreign genes can also be programmed with tissue specificity; i.e., an introduced gene will only express in seed (Radke et al., 1988) or perhaps only when some external stimulus is applied. There appear to be at least two means by which natural plant genes can be "turned down" by introducing either antisense (Kramer et al., 1989) or co-suppression (Napoli et al., 1990) gene constructs. We'll consider three examples which show some of these diverse elements.

BXN Cotton

The chemical herbicide bromoxynil has the advantageous trait of killing many of the major weed pests in cultivated cotton. From an environmental standpoint, it is a preferred agrichemical because it breaks down quickly in soil and its application rates are much lower than for herbicides currently used with only partial success in cotton. However, bromoxynil has a problem: it kills cotton.

Calgene scientists McBride and Stalker (Stalker et al., 1988) succeeded in isolating a strain of the soil bacterium *Klebsiella* which degrades bromoxynil into safe and natural metabolites. This degradation process relies on a single nitrilase type enzyme encoded by a single gene. The gene was cloned, adapted to express in plant tissue, and transferred into cotton by a process called co-cultivation (Corbin and Klee, 1991). The transgenic cotton plants thus express a nitrilase. When a bromoxynil-based herbicide is applied to this transgenic cotton, complete resistance is observed. We use this example to show how an undesired chemical constituent (in this example, a herbicide) can be removed. Another approach can be used to prevent the accumulation of certain plant components, and/or to produce more of a given component.

FLAVRSAVR™ Tomato

Genes that encode enzymes are transcribed into messenger RNAs which are subsequently translated into proteins. It has been observed that a plant gene can be configured in the lab so that upon reintroduction into a transgenic plant, it will generate an "antisense" mRNA so that there may be binding between the antisense mRNA and naturally occurring sense mRNAs derived from related genes. Although the mechanism is not well-established, it's clear that this approach can result in less native protein being produced as a result of the presence of an antisense gene.

In the case of the FlavrSavr™ tomato (Kramer et al., 1989), it had been suggested that the process by which a tomato fruit naturally goes soft is mediated by an enzyme called polygalacturonase (PG) among other factors. The gene encoding this key enzyme was cloned and engineered so that in a transgenic tomato, mRNA partially complementary to the normal PG mRNA was produced. Reduced PG enzyme was observed with concomitant changes in the rates at which a ripe tomato fruit turns soft. Thus these transgenic tomatoes resemble other tomatoes but can be allowed to ripen on the vine and still be firm enough to ship to grocery stores.

In another application of this technology, Jean Kridl and her colleagues (Knutzon et al., 1992) used an antisense gene construct based on the stearoyl-ACP desaturase gene from rapeseed. Transgenic canola with down-modulated stearoyl-ACP desaturase have less unsaturated fatty acids and impressively high levels of stearate fatty acids in seed oils (**Figure 1**). By decreasing a specific enzyme, more of a specific component was accumulated in the seeds of transgenic canola. Interestingly, this appears to be a case where seed-specific expression limiting the phenotypic change to storage lipids is important in that stearoyl-ACP desaturase is an essential factor in leaf metabolism.

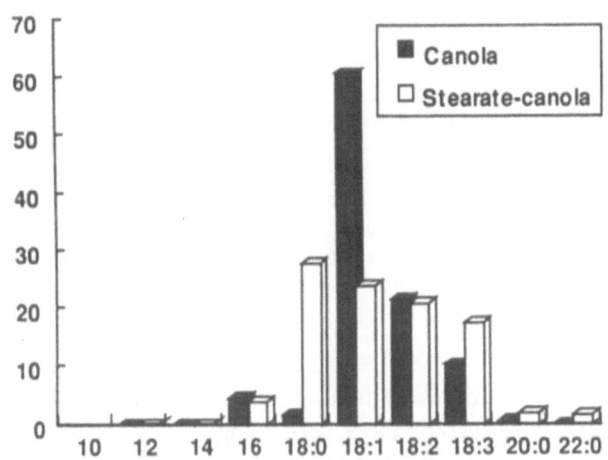

Figure 1. Fatty acid composition in seed triglycerides of normal canola and high stearate transgenic canola

Co-suppression is another method which has been used with some success to lower levels of specific enzymes in transgenic plants (Napoli et al., 1990). Either antisense or co-suppression can be considered as ways to block the accumulation of undesirable compounds by interfering with enzymes necessary for synthesis, or to redirect a plant pathway to accumulate more of a given compound by preventing its further conversion to less desired components.

High Laurate Rapeseed

Most seed oils contain the same kinds of fatty acids that are found in the structural lipids of plant tissues: 16:0 (palmitic), 18:0 (stearic), 18:1 (oleic), 18:2 (linoleic), and 18:3 (linolenic). However, some plants contain different kinds of fatty acids in seed oils and for the most part these more unusual fatty acids are not found in structural lipids or in tissues other than the seed. For example, both the coconut seed endosperm and the oil palm kernel contain vegetable oils containing approximately 50% C12:0 (lauric) fatty acids as opposed to other chain lengths. These vegetable oils are used in large volumes as a source for lauric acid which is a raw material for the soap and detergent industries.

In an effort to develop a domestic source of this raw material, Maelor Davies and his colleagues at Calgene (Pollard et al., 1991) studied the mechanism by which the California bay tree accumulates as part of its seed oil a lauric acid content of 60%. Fatty acids are

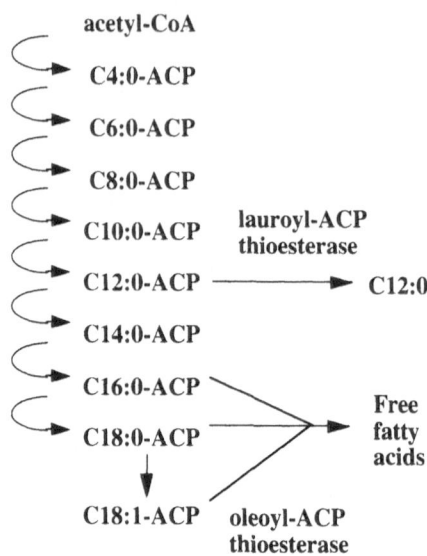

Figure 2. Synthesis of free fatty acids in chloroplasts of higher plants. Elongation of fatty acids from the 4-carbon butyryl-ACP occurs in two-carbon increments while carried as a thioester of acyl carrier protein (ACP). Acyl-ACP thioesterases cleave the acyl group from ACP to produce free fatty acids. Desaturation of C18:0-ACP to C18:1-ACP is catalyzed by stearoyl-ACP desaturase.

initially synthesized by repeated two-carbon additions (**Figure 2**) to a growing acyl-chain attached via a covalent thioester bond to a scaffolding protein called acyl carrier protein (ACP). When the growing acyl chain is at the C18:0-ACP stage, the above-mentioned stearoyl-ACP desaturase can convert C18:0-ACP to C18:1-ACP. There is also an enzyme known as acyl-ACP thioesterase that can cleave the thioester bond between ACP and any of C16:0, C18:0, and C18:1 fatty acids esterified to ACP. The free fatty acid produced can then be used to synthesize glycerol lipids including seed storage triacylglycerols (TAG). Thus, plants produce C12:0-ACP as an intermediate for longer chain fatty acids, but do not produce C12:0 as a free fatty acid to incorporate into TAG. The Calgene group identified a C12:0-ACP thioesterase activity in seeds of the California bay tree which can intercept C12:0-ACP and release C12:0 for TAG synthesis. The enzyme was purified, a corresponding cDNA was cloned, and the gene was adapted for transfer into canola plants (Voelker et al., 1992).

The transgenic canola plants appear identical to normal canola but seeds contain up to 45% C12:0 fatty acids incorporated into TAG (**Figure 3**). Moreover, the C12:0 fatty acids were efficiently esterified to the number one and/or three positions of the TAG molecules but not at the number two position. Coconut oil TAG contains C12:0 equally distributed over all three positions. This may have some health relevance since C12:0 and other saturated fats are associated with hypercholesteremia; however, recent data indicate that it may be the saturated fat content at the number two position that is the critical factor (Redgrave et al., 1988). The transgenic plants yield well and the seeds germinate well.

Transfer of the California bay tree TE gene into canola thus demonstrates the principle of redirecting a plant metabolic pathway by introducing a gene from a different organism. The TE gene not only converts the chemical composition of a foodstuff in a directed and predicted manner, but also in this case for the TAG structure, has led to a novel food oil previously unavailable except in small synthetic laboratory samples.

In summary, the BXN cotton example shows that plants can be "trained" to remove undesired components; the antisense PG tomato shows that the synthesis of certain components can be prevented; the antisense stearoyl-ACP desaturase further shows that the accumulation patterns for natural components can be rearranged; and the bay TE example shows that plants can be reprogrammed to produce novel compounds via introduction of genes from other living organisms.

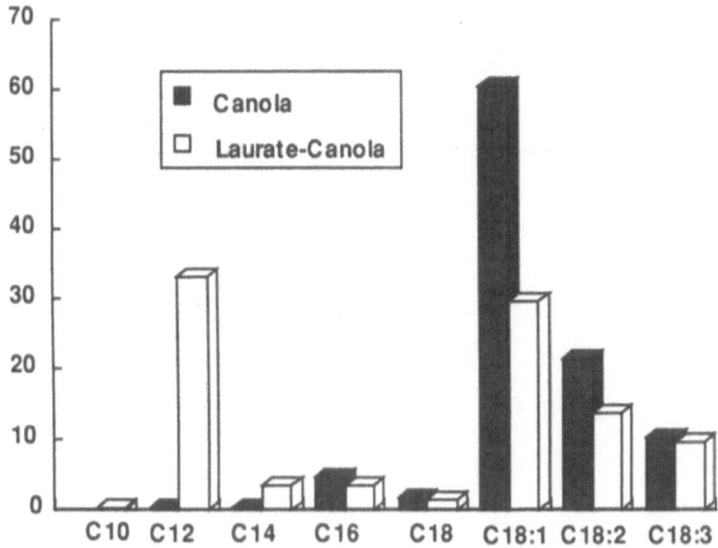

Figure 3. Fatty acid composition of seed oil from transgenic rapeseed expressing the *Umbellularia californica* lauroyl-ACP thioesterase gene.

DIRECTIONS

As some plant seeds contain vegetable oils rich in C12:0 fatty acids, others store C8:0 and C10:0 fatty acids in triacylglycerol form. None of the plants naturally rich in medium-chain-triglycerides (MCTs) are in commercial production. MCTs are commercially available, however, as synthetic products using the small fraction of C8:0 and C10:0 fatty acids "left over" from splitting coconut oil to produce C12:0 fatty acids for the soap and detergent industry. Thus the amount of coconut C12:0 used for other purposes determines the quantity of synthetic MCTs that can be made, and then at relatively high costs. Most MCTs until recently have been used in clinical and medical settings.

MCTs are believed easier to digest, to contain fewer calories on a molar basis, and to be catabolised into energy more readily than anabolised into stored fat. The health impact of widely available and widely consumed MCTs in the American diet is not known and has been in fact moot since there are not sufficient or economical supplies of MCTs. Just as Calgene scientists found a gene encoding C12:0-ACP TE from a C12:0-rich oil containing plant seed, we have also found a similar gene from a seed with oils rich in C8:0 and C10:0 fatty acids (K. Dehesh, personal communication). Thus we anticipate a transgenic oilseed that will make MCTs available in amounts and a price that will allow its inclusion in many foods. Will such use patterns lower cholesterol levels in consumers who are at-heart-disease-risk? Over time, will the availability of such a fat in processed foods help alleviate cancer rates attributed to high fat diets?

Other data suggest that long chain polyunsaturated fatty acids such as eicosapentaenoic acid (EPA, C20:5) and docosahexenoic acid (DHA, C22:6) in the diet help prevent heart disease. Animals including humans and fish, fungi, and algae all synthesize these compounds, with some algae and fungi notably in the TAG form. Higher plants are not known to produce EPA or DHA. Although plants do synthesize significant amounts of the precursor C18:3 α-linolenic, the remaining steps of elongation and further desaturation may be very complex and mediated by a extensive set of enzymes (**Figure 4**). Essential co-factors for enzymatic

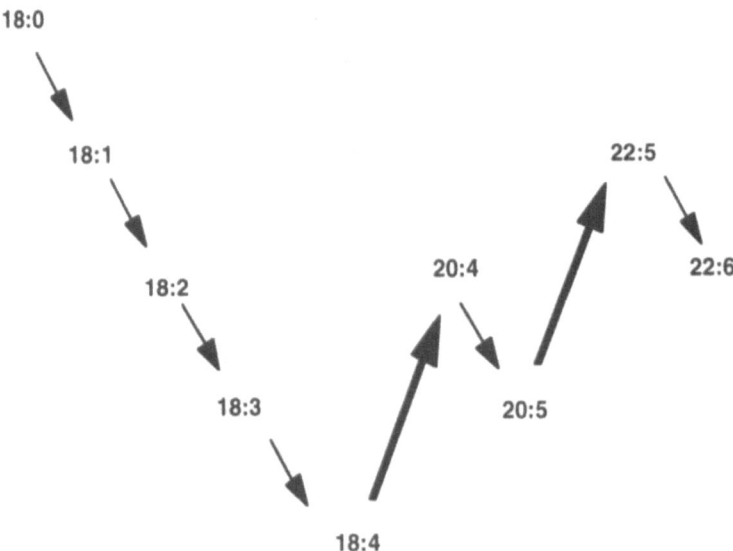

Figure 4. Synthetic pathway for docosahexenoic acid (DHA). Addition of double bonds by a D6 desaturation (18:3 to 18:4) and a D5 desaturation (20:4 to 20:5) and finally, a D4 desaturation (22:5 to 22:6) presumably each require an electron carrier and a membrane-bound terminal desaturase enzyme. Elongation from 18:4 to 20:4 and from 20:4 to 22:5 are believed to require malonyl-CoA and at least four distinct enzyme activities (ketoacyl synthase, ketoacyl reductase, enoyl dehydrase, enoyl reductase). The nature of the substrate and/ or carrier proteins is speculative.

reactions may not be present in plants or perhaps not in the right subcompartment of the plant cell. Similarly, the elongation and desaturation substrates may be lipid molecule intermediates that plant cells do not produce or recognize. EPA and DHA may not be stable in a plant oil or extract. The possible incorporation of EPA or DHA (or intermediates in their synthesis) into structural lipids may have negative effects on viability. Nonetheless, the limited commercial supply of EPA and DHA, the unpalatibility of the fishhead source products, the cost, and the increasing data base for the beneficial efficacy of these compounds all underscore the need for an economical and plentiful supply that crop plants might provide. As more becomes known about the genetic base for EPA and DHA synthesis in other organisms, perhaps it will be possible to embark on a project to produce them in higher plants. Two more natural compounds, α-carotene and β-carotene, also enjoy a good reputation as antioxidants of possible importance in preventing cancer. The pathway which leads to their synthesis in plants is known as the isoprenoid pathway or the carotenoid pathway. It's a very important pathway that also leads to natural dyes, to vitamin E, even to medicines like digitoxin in some plants. Is this a pathway which can be reprogrammed to produce more α-carotene and β-carotene? For those who don't like carrots, perhaps higher levels could be engineered in tomatoes? In basic grain foodstuffs like wheat, corn, rice?

Does the caffeine content of coffee and tea represent a dietary risk for health? Should there be an initiative to genetically block caffeine synthesis in the source plant? If indole glucosinolates in broccoli have anticancer properties (see Kohlmeier, this volume), is it worth considering transferring that trait into apples or some medium a little more popular than broccoli? Can the nature of good dietary fiber become better understood so that fiber modifications might enhance the benefit?

SUMMARY

Gene manipulation techniques can be used to increase, decrease, or add specific proteins to the edible parts of transgenic crop plants. With some basic understanding of plant biosynthetic pathways, then, the targeting of genes encoding specific enzymes allows the direct modification of the biochemical composition of foods. At Calgene, we have engineered the chemical composition of canola vegetable oils. Transgenic canola which are otherwise exactly like regular canola plants produce seed with oils a) that are modified in average fatty acid carbon chain length, b) that are modified in content of saturated fatty acids (both lower and higher), or c) that contain structured lipids. In principle, although the gene target may not be obvious, the relative amount of lipids compared to other nutrients can be decreased or increased in foods like peanut or soybean. The oil content of some food products might be modified to enhance levels of medium chain triglycerides or to contain "fish oil" fatty acids - without the cost or olfactory disadvantages. On a broader scope, the amino acid composition of proteins in basic grains is being pursued by several groups. Specific vitamin contents such as Vitamin A or E might be enhanced in basic foodstuffs; type and content of fiber may eventually be manipulated. Specific components such as caffeine or phytic acid conceivably can be eliminated in the source plant, negating the need for processing steps that add cost and that lessen flavor and nutrition. Clearly, the biochemical compositions of foods is complex and varied. Moreover, tailoring foods to better meet our needs is an expensive and lengthy process. The best possible understanding of how nutrition promotes good health is necessary to direct and prioritize our efforts to improve plants as food sources.

Custom tailoring the chemical composition of lipids and other food components is demonstrated. The range of possible applications is wide. How best do we use these tools to help prevent heart disease and cancer? Prevention is obviously the most cost-effective treatment. Incorporating a rational, science-based approach to disease prevention in the basic diet is ideally a very cost-effective means to reach a susceptible population with comparatively painless and effortless compliance. In fact, it appears clear that disease conditions other than heart disease and cancer may be specifically addressed via diet. These conditions include cystic fibrosis and other malabsorption problems, diabetes, and obesity.

Currently, however, the organizations with the skills to execute "designer foods" typically lack the nutritional expertise necessary to evaluate the health-related opportunities. Nutrition literature often appears controversial, sometimes characterized as "fads." We suggest that this situation calls for an integrated interdisciplinary approach to define targets, to execute proposed development projects and to evaluate the results in terms of real benefits.

REFERENCES

Corbin, D. R., and Klee, H. J., 1991, *Agrobacterium tumefaciens* - mediated plant transformation systems, Curr. Opinon in Biotechnol. 2:147.

Knauf, V.C., 1991, Agricultural progress: engineered crop species, field trial results and commercialization issues, Curr. Opinion in Biotechnol. 2: 199.

Knutzon, D.S., Thompson, G.A., Radke, S.E., Johnson, W.B., Knauf, V.C., and Kridl, J. C., 1992, Modification of *Brassica* seed oil by antisense expression of a stearoyl-ACP desaturase gene, Proc. Natl. Acad. Sci. USA 89:2624.

Kramer, M., Sheehy, R. E., and Hiatt, W. R., 1989, Progress towards the genetic engineering of tomato fruit softening, Trends Biotechnol., 7:191.

Napoli, C., Lemieux, C., and Jorgensen, R., 1990, Introduction of chimeric chalcone synthase gene into *Petunia* results in reversible co- suppression of homologous genes *in trans*, Plant Cell 2:279.

Pollard, M. R., Anderson, L., Fan, C., Hawkins, D. J., and Davies, H. M., 1991, A specific acyl-ACP thioesterase implicated in medium chain fatty acid production in immature cotyledons of *Umbellularia californica*, Arch. Biocem. Biophys., 284:306.

Radke, S. E., Andrews, B. M., Moloney, M. M., Crouch, M.L., Kridl, J.C., and Knauf, V. C., 1988, Transformation of *Brassica napus* L. using *Agrobacterium umefaciens:* developmentally regulated expression of a reintroduced napin gene, Theor. Appl. Genet., 75:685.

Redgrave, T.G., Kodali, D.R., and Small, D. M., 1988, The effect of triacyl-*sn*-glycerol structure on the metabolism of chylomicrons and triacylglycerol-rich emulsions in the rat, J. Biol. Chem., 263:5118.

Stalker, D. M., McBride, K. E., and Malyj, L. D., 1988, Herbicide resistance in transgenic plants expressing a bacterial detoxification gene, Science 242:419.

Stark, D. M., Timmerman, K. P., Barry, G. F., Preiss, J., and Kishore, G. M., 1992, Regulation of the amount of starch in plant tissues by ADP glucose pyrophos-phorylase, Science 258:287.

Voelker, T. A., Worrell, A. C., Anderson, L., Bleibaum, J., Fan, C., Hawkins, D.J., Radke, S. E., and Davies, H. M., 1992, Fatty acid biosynthesis redirected to medium chains in transgenic oilseed plants, Science 257:72.

NEW DIRECTIONS IN DIETARY STUDIES IN CANCER:

THE NATIONAL CANCER INSTITUTE

Peter Greenwald, Carolyn Clifford, Susan Pilch
Jerianne Heimendinger, and Gary Kelloff

Division of Cancer Prevention and Control
National Cancer Institute
9000 Rockville Pike
Building 31, Room 10A52
Bethesda, Maryland 20892

INTRODUCTION

Epidemiologic and laboratory research conducted during the past two decades offers compelling evidence that diet modifies cancer risk. Dietary factors may account for a greater proportion of all cancers occurring in Western societies than does any other category of environmental exposure, except perhaps smoking. Although the exact proportion is unknown, it has been crudely estimated that 35 percent of overall cancer deaths are attributable to diet (Doll and Peto, 1981). Thus, research on diet, nutrition, and cancer is receiving strong emphasis at the National Cancer Institute (NCI).

The NCI sponsors and performs a broad range of research comprising basic nutritional, epidemiologic, clinical/metabolic, and laboratory studies on the relationship between dietary factors and cancer risk, as well as randomized, controlled intervention trials to evaluate the effects of dietary modification or nutrient supplementation on cancer incidence, recurrence, and mortality. Another priority is to translate the knowledge gained from these studies into useful information for the benefit of the American public. The NCI, working with other agencies and the private sector, designs and implements appropriate public information activities to help Americans adopt dietary patterns that may reduce cancer risk. This report describes the NCI's approach to quantifying the role of diet and specific dietary constituents in cancer prevention; presents current and new directions in diet and cancer prevention research initiatives; and concludes with a description of health promotion activities.

RESEARCH

Many of the NCI's diet and cancer research initiatives aim to strengthen and expand the scientific foundation upon which prevention strategies are designed. These efforts are

Nutrition and Biotechnology in Heart Disease and Cancer
Edited by J.B. Longenecker *et al.*, Plenum Press, New York, 1995

229

concentrated in two broad categories: diet and nutrition and chemopreventive approaches using dietary constituents. These areas of research are pursued through both extramural and intramural programs, covering a broad spectrum that includes cellular and molecular effects of nutrients and other dietary constituents; nutritional metabolism and physiology; epidemiologic studies; development of analytical methods for quantifying the nutrient and nonnutrient composition of foods; study of biomarkers of dietary intake and nutritional status; and the development of molecular and genetic biomarkers for cancer risk and intermediate endpoint markers for use in prevention clinical trials.

Complex questions in nutrition and cancer prevention often require investigative efforts that extend beyond the level practicable in a single research project. Accordingly, a new direction in the NCI research program is an increasing emphasis on interdisciplinary research and greater integration of the basic, clinical, and behavioral sciences. For example, a new initiative was recently implemented through investigator-initiated interactive research projects for the development of collaborative approaches to nutrition and cancer prevention. Another research project initiative focuses specifically on the development and conduct of clinical/metabolic studies for nutrition and breast cancer prevention. These interactive research projects aim to provide for building stronger research bridges between nutritional science and the other basic and clinical sciences. Special attention is encouraged for research on breast cancer, prostate cancer, and cancer in women and minorities.

The NCI-supported Clinical Nutrition Research Units (CNRUs) are yet another example of efforts to bring together investigators from relevant disciplines and to enhance and extend the effectiveness of research related to nutritional science and cancer prevention. The CNRUs provide a focus for increasing collaboration among groups of successful investigators at institutions with established comprehensive nutritional science and cancer research bases. Overall, the NCI research programs in cancer prevention and control are directed towards the reduction of cancer incidence, morbidity, and mortality through an orderly sequence from basic and clinical nutrition research to intervention clinical trials in defined populations and then to applications of the research results in the general population.

DIETARY MACROCONSTITUENTS AND CANCER

Dietary Fat

Associations of dietary fat intake with cancers of the breast, colon, and other sites have been demonstrated in both epidemiologic and laboratory research. International correlation studies have shown a strong association between the incidence of these cancers and total dietary fat intake, even after adjusting for total calorie intake (Armstrong and Doll, 1975; Hursting et al., 1990). However, the association of dietary macroconstituents with cancer risk from epidemiologic studies may be confounded by the inverse correlation of fat intake with consumption of fiber or other food constituents. There is substantial evidence that populations consuming a high-fat, low-fiber "Western" diet experience high risks of certain cancers, and that the high-fiber, low-fat "Asian" diet is associated with low risk of these cancers. The Finnish diet represents an exception to the usual dietary pattern in that it provides both high-fat and high-fiber intake (Rose, 1992). Finnish colon and breast cancer risks lie between those in the United States and Japan—perhaps suggesting that risk associated with dietary fat may be modulated by a protective effect of dietary fiber.

To test the hypothesis that a low-fat diet may be protective against several cancers, the NCI is sponsoring clinical intervention trials. One of these trials, the Women's Health Trial Vanguard Study, examined the feasibility of lowering fat intake from 37 to 20 percent of calories in postmenopausal women (Henderson et al., 1990; White et al., 1992). This study was among the first to demonstrate the feasibility of such a study design and success in lowering fat intake in free-living individuals. To ensure that these results would be applicable to populations other than the largely white, middle-class population first studied, the Women's Health Trial Feasibility Study in Minority Populations was initiated to test methods to enable black, Hispanic, and low-income women to lower their fat intake—from 37 to 20 percent of calories from fat—while increasing consumption of dietary fiber, fruits, vegetables, and grain products. This study is evaluating the effect of differences in culture and economic status on the ability to achieve and maintain the low-fat diet. In addition, the effect of the low-fat eating pattern on potential biomarkers, including blood lipids, lipoproteins, and hormones, is being investigated. The results and experiences from these two trials have been used in the design and implementation of the dietary intervention component of the NIH Women's Health Initiative. This national study examines the effects of dietary modification, calcium/vitamin D supplementation, and hormonal replacement therapy on the major causes of death and disability—breast and colon cancer, coronary heart disease, and osteoporosis—in postmenopausal women.

The Women's Intervention Nutrition Study (WINS) is a 5-year, randomized clinical trial to test whether dietary fat reduction will reduce breast cancer recurrence and increase survival in postmenopausal women with Stage II breast cancer (Chlebowski et al., 1987; Chlebowski et al., 1990). The women in the trial were randomly assigned to either an intensive intervention group (IIG) or a nonintensive intervention group (NIG). The IIG will reduce dietary fat intake from a baseline of over 25 percent to 15 percent of calories; the NIG will be instructed in the NCI's 1990 Dietary Guidelines to reduce fat intake to 30 percent of calories. Taken together, these different trials with similar study designs should contribute a great deal to our knowledge of the determinants of breast cancer.

Another dietary intervention study, the Polyp Prevention Trial, is investigating whether a low-fat, high-fiber, and high fruit and vegetable eating pattern will lower the occurrence of adenomatous polyps of the large bowel. The intervention diet is targeted toward achieving a dietary pattern consisting of 20 percent of calories from fat, 18 g of total dietary fiber/1000 kcal, and 5 to 8 daily servings of fruits and vegetables. This study also will determine how this dietary pattern affects biomarkers of large bowel epithelial cell proliferation, whether the proliferation markers predict neoplasia (polyp recurrence), and to what extent changes in proliferation indexes account for the observed intervention effect.

More recent studies of dietary fat and cancer have focused on defining the relationship between specific types of fat and cancer risk. These studies indicate that different fatty acids may have various site-specific tumor-promoting or tumor-inhibiting properties. In an international correlation study (Hursting et al., 1990), the age-standardized incidence rates for breast, colon, prostate, lung, and cervical cancer in 20 countries were correlated with each country's estimated intake of total fat, saturated fat, polyunsaturated fat, omega-3 polyunsaturated fat, omega-6 polyunsaturated fat, and monounsaturated fat. Total fat and saturated fat were both strongly and positively associated with the incidence of breast, colon, and prostate cancers. Polyunsaturated fat and omega-6 polyunsaturated fat were strongly and positively associated with breast and prostate cancers, but not with colon cancer, whereas omega-3 polyunsaturated fat intake showed weak, but insignificant, inverse associations with breast, colon, lung, and cervical cancers. Monounsaturated fat intake was not associated with any of the cancers studied.

Conversely, a large cohort study, the Nurses' Health Study, has shown no association between either total or saturated fat intake and breast cancer incidence or mortality (Willett et al., 1992). A possible association between dietary fat and breast cancer, however, may have been obscured by errors in the assessment of individuals' diets and the difficulties in separating the effect of highly correlated dietary factors inherent to such an observational cohort study (Byar and Freedman, 1989). Analysis of data from this cohort did indicate that animal, but not vegetable, fat consumption was strongly associated with risk of colon cancer

(Willett et al., 1990). A combined analysis of 12 diet and breast cancer case-control studies showed a consistent, statistically significant positive association between saturated fat intake and breast cancer risk in postmenopausal, but not premenopausal, women (Howe et al., 1990). These results strengthen the idea that positive associations between fat consumption and cancer incidence are site specific and that particular types of fats have different associations with site-specific cancer incidence.

Another direction in diet and cancer research is a focus on possible mechanisms of action of dietary fat in influencing cancer risk. A large body of evidence implicates sex hormones as having a role in carcinogenesis. Studies have been undertaken to evaluate the influence of a low-fat, high-fiber, vegetable, and fruit enriched diet on sex hormone levels in the women and men participating in the Polyp Prevention Trial, the Women's Health Trial Feasibility Study for Minority Populations, and the Women's Intervention Nutrition Study. In another ongoing dietary intervention trial, hormone levels among adolescent girls and boys eating a low-fat diet are being evaluated. The effect of diet on total concentrations and bioavailable fractions of hormones is to be determined in this NCI-sponsored ancillary study to the Diet Intervention Study in Children (DISC) of the National Heart, Lung, and Blood Institute.

Dietary Fiber

International, migrant, case-control, and time-trend epidemiologic studies conducted over the past two decades suggest that high consumption of fiber-rich foods is associated with lower colorectal and, perhaps, breast cancer risk. More recent studies of dietary fiber and cancer have focused on identifying the protective effects of specific types of fiber. Followup data from the Nurses' Health Study demonstrated a dose-related decrease in colon cancer risk as consumption of fruit (crude fiber) increased; however, total dietary fiber, total crude fiber, vegetable fiber, and cereal fiber were not as clearly associated with risk (Willett et al., 1990; Willett et al., 1992). A prospective study of men in the Health Professionals Followup Study found that total dietary and crude fiber from vegetables, fruits, and cereal grains were associated with decreased risk for colorectal adenomas, which frequently are precursors of colon cancer (Giovannucci et al., 1992). In a 4-year clinical trial of patients with familial adenomatous polyposis, the size and number of rectal polyps diminished in those patients who consumed more than 11 grams/day of wheat fiber supplements (DeCosse et al., 1989).

Better understanding of the effects of specific types of fiber has been impeded because the analytical methods available do not satisfactorily quantify all fiber components, and national food composition tables do not provide data for individual food fiber fractions. To fill this gap, the NCI has sponsored research to develop and improve analytical methods for detection and quantification of dietary fiber components. Foods that provide significant sources of fiber in the U.S. diet were selected and the total dietary fiber and major dietary fiber components quantified (Marlett, 1992). This food fiber composition database will be a very useful tool for estimating the dietary fiber intakes of individuals and population groups and provide a better understanding of the role of specific types of dietary fiber in cancer prevention and control.

Dietary fiber may play a role in preventing breast cancer by lowering circulating levels of estrogen. Several studies examining the connection between diet and systemic hormone patterns in women indicated that high total fiber intake and high intakes of fiber from vegetables, grains, fruits, and berries were associated with low levels of testosterone, estrone, and androstenedione; low levels of plasma free estradiol and free testosterone; and elevated levels of sex hormone-binding globulin—the protein to which estradiol and other hormones are bound (Adlercreutz, 1990). Overall reduction in the bioavailability of these hormones suggests that a fiber-enriched diet could theoretically reduce the risk of hormone-dependent cancers. In one study of women, normal diets were supplemented with wheat, oat, or corn bran. The wheat bran diet reduced serum estrone and estradiol levels, suggesting that a diet high in wheat bran may decrease breast cancer risk (Rose et al., 1991). Wheat bran consumption may modulate estrogen metabolism via enterohepatic circulation, in which estrogens are conjugated in the liver, excreted in the bile, then deconjugated by bacterial enzymes in the gut, and finally reabsorbed into the blood. Continued efforts to elucidate the possible mechanisms by which dietary fiber and fiber components may influence the prevention of cancer represent an important focus for future research.

Another research question receiving increased attention is whether dietary fiber, per se, or other components of fiber-rich foods are responsible for observed protective effects. A number of naturally occurring nonnutritive components present in fiber-rich foods—such as vegetables, fruits, and cereal grains—have been found to inhibit carcinogenesis in laboratory animals when fed in a purified form and at pharmacologic levels.

DIETARY MICROCONSTITUENTS AND CANCER

Epidemiologic studies generally have found inverse relationships between intake of a variety of fruits and vegetables and risk of cancer at multiple sites. In addition to fiber, numerous vitamins and other microconstituents found in fruits and vegetables may be responsible for the cancer-protective effects observed with high intake of these foods. Two recent comprehensive reviews found consistent, though not universal, inverse associations between consumption of vegetables, fruits, and/or their microconstituents and cancer risk at 15 or more sites (Steinmetz and Potter 1991a, 1991b; Block et al., 1992). Epidemiologic studies generally show an inverse relationship between cancer incidence and intake of foods high in several specific antioxidant nutrients such as beta-carotene, selenium, vitamin C (ascorbic acid), and vitamin E (alpha-tocopherol) (Hirayama, 1979; Steinmetz and Potter, 1991a; Block, 1991; Knekt et al., 1991; Kromhout, 1987). Low intakes of vegetables, fruits, and carotenoids have been associated with increased risk of lung cancer (Rogers and Longenecker, 1988; Steinmetz and Potter, 1991a; Ziegler, 1991). A few studies suggest that vegetable and fruit intake also may reduce the risk of cancer of the oral cavity, pharynx, larynx, esophagus, stomach, pancreas, colon, rectum, bladder, breast, ovary, and cervix (Block et al., 1992; Steinmetz and Potter, 1991a; Block, 1991; Ziegler, 1991; Weisburger, 1991). Selenium also may reduce cancer risk, but the evidence is not conclusive (Rogers and Longenecker, 1988). Several epidemiologic studies suggest that vitamin C-rich foods may reduce the risk of stomach and esophageal cancers (National Academy of Sciences, 1989; Block, 1991, 1992).

A number of microconstituents of fruits and vegetables have already been proposed as anticarcinogenic dietary factors and more are likely to be suggested. The selected examples for the following discussion are some of the antioxidants (vitamins E and C, carotenoids, and selenium), folate, and phytoestrogens.

Antioxidants

There is substantial epidemiologic evidence supporting the association of foods high in antioxidants—specifically, certain carotenoids, vitamin E, vitamin C, and the trace element selenium—with reduced cancer risk at several sites, and this is reinforced by animal and *in vitro* studies (Comstock et al., 1992; Henson et al., 1991). A recent NCI-collaborative study (Blot et al., 1993; Benner and Hong, 1993) demonstrated that a combination supplement of beta-carotene, vitamin E, and selenium reduced the stomach cancer mortality rate in Linxian, China, whose rural population has one of the highest rates of cancers of the esophagus and gastric cardia in the world.

Several large prospective trials sponsored by the NCI are assessing the potential chemopreventive effects of beta-carotene alone or with other microconstituents. Investigators conducting the Physicians' Health Trial are examining the impact of beta-carotene on overall cancer risk. The Women's Health Study, which began recruiting during 1992, will examine cancer and heart disease risk in women receiving combinations of beta-carotene, vitamin E, and aspirin. In the Alpha-Tocopherol, Beta-Carotene (ATBC) study, smokers have received either beta-carotene, vitamin E, or both in a randomized 2 x 2 factorial design in an effort to determine if these supplements can reduce the incidence of lung cancer. The Carotene and Retinoid Efficacy Trial (CARET), a double-blind dietary supplementation study, is investigating the effects of a daily combination supplement of beta-carotene and retinyl palmitate on the incidence of lung cancer in high risk populations. A randomized, double-blind, placebo-controlled trial to test the efficacy of beta-carotene and vitamin C and E supplementation in preventing neoplastic polyps of the large bowel in a population with a high risk of colon cancer is also underway.

These trials highlight the need for further research on the physiological and pharmacological effects of the micronutrient antioxidants, as well as their possible mechanism(s) of action in modifying the process of carcinogenesis. The NCI has sought to improve the ability to assess the chemopreventive effects of dietary microconstituents, especially the carotenoids. A database with specific carotenoid values of foods has recently been completed as a joint effort between the U.S. Department of Agriculture and the NCI. This database will be used to determine the carotenoid content of fruits and vegetables and to assess carotenoid status of individuals included in previous epidemiologic studies, thereby providing additional information on the relationship of diet and cancer. In collaboration with workers at the National Institute for Standards and Technology, a liquid chromatographic technique has been developed to optimize recovery and resolution of the major carotenoids in human serum and plasma. The method is reproducible and relatively rapid, enabling use in epidemiologic studies with large numbers of samples, as well as investigation of effects of individual carotenoids. It also will permit assessment of the use of circulating carotenoid levels as a biomarker of the consumption of fruits and vegetables in dietary intervention trials.

Folate

Folic acid, found in fresh vegetables and fruits, is another dietary microconstituent that may contribute to the protective role of fruit and vegetable consumption against cancers at multiple sites. Prospective epidemiologic analyses of data from two large ongoing cohorts—the Nurses' Health Study and the Health Professionals Followup Study—indicate that folate intake is inversely associated with risk of colorectal adenomas for both women and men (Giovannucci et al., 1993).

Folate deficiency recently was shown to be associated with an increased risk of cervical dysplasia in women who tested positive for human papillomavirus (Butterworth et al., 1992a). However, another study, in which patients with established lesions received oral supplements of folic acid, showed no observable beneficial effects over a period of 6 months (Butterworth et al., 1992b). These results raise the possibility that folate deficiency may be a cocarcinogenic factor during cervical cancer initiation and may act through facilitation of the papillomavirus genome's incorporation into the host DNA at specific folate-sensitive fragile sites (Butterworth, 1992).

Biochemical studies have shown that biologically active folates are essential to the normal control of gene expression. This control involves the methylation of specific DNA cytosine bases; folate is a cofactor in the production of S-adenosylmethionine, the major donor in the process that produces 5-methylcytosine at those specific DNA sites. Abnormalities in DNA methylation may contribute to loss of normal controls in proto-oncogene expression. A study comparing demethylation in normal and neoplastic tissue revealed 9.2 percent demethylation in a specific cytosine site in the c-*myc* gene in normal colonic mucosa, 24.8 percent in hyperplastic polyps, 50.5 percent in adenomatous polyps, 66.1 percent in the carcinoma proper, and 83.1 percent in metastatic deposits (Sharrard et al., 1992). These data suggest that demethylation may play a role in the progression of carcinogenesis and support further investigation of hypotheses about the involvement of folates in that process.

Phytoestrogens

Foods contain a diversity of naturally occurring, nonnutrient constituents—some of which may have important biological activities and functions. Phytoestrogens present in the fiber-containing foods are one class of nonnutrient dietary constituents which have recently received increasing attention as having chemopreventive potential for the hormone-dependent cancers (Adlercreutz, 1990). The phytoestrogens are nonsteroidal estrogens, chemically classified as isoflavones and structurally similar to the steroid estrogens (Price and Fenwick, 1985).

Legumes, especially soybeans, are the primary source of the isoflavone phytoestrogens of dietary origin. Soybeans and soy food products contain several major isoflavones, genistein and daidzein, and their corresponding glucosides, genistin and daidzin (Murphy, 1982). These isoflavones have been shown to have antioxidant, estrogenic, antiestrogenic, and anticarcinogenic activities in numerous *in vitro* studies, including human mammary and prostate cancer cell lines (Peterson and Barnes, 1991, 1993). However, the *in vivo* anticarcinogenic activity of these soy-derived isoflavones as purified compounds is unknown.

Very little is known about the impact of isoflavone phytoestrogens in humans, and their potential as anticarcinogens is only beginning to be characterized. The epidemiologic evidence that Asian countries consuming diets high in soy foods also have low breast cancer rates has led some investigators to postulate that isoflavone phytoestrogens may be protective. However, the availability of epidemiologic and animal evidence to support the hypothesis is limited. In a recent case-control study conducted in Singapore, high intake of soy protein was associated with a reduction in breast cancer in premenopausal, but not postmenopausal, women (Lee et al., 1991). Animal studies have shown that soy-containing diets reduced the number of mammary tumors in the NMU and DMBA rat mammary model systems (Barnes et al., 1990; Hawrylewicz et al., 1991). Further studies are required to clarify whether the isoflavone phytoestrogens or other constituents present in soybeans are responsible for the chemopreventive effect.

Current NCI-supported efforts include developing reliable analytical methodologies to identify and quantify specific phytoestrogens in foods and to determine their stability during food processing and storage; elucidating the mechanisms of action for specific phytoestrogens that may inhibit cancer initiation and promotion; investigating the chemopreventive potential of specific phytoestrogens in preclinical studies; and conducting clinical studies to elucidate the *in vivo* metabolism, absorption, and bioavailability of dietary phytoestrogens. The results of these studies will be used to develop food composition databases and to plan new directions for research on the role of dietary phytoestrogens in hormone metabolism and cancer prevention.

PUBLIC INFORMATION

A goal of diet and cancer prevention research is to provide quantitative data for developing dietary guidance that can reduce the risk of diet-related cancers. Based on the available scientific evidence, the NCI published Dietary Guidelines (Butrum et al., 1988), which consist of the following six recommendations: reduce fat intake to 30 percent or less of calories; increase fiber intake to 20 to 30 grams daily, with an upper limit of 35 grams; include a variety of vegetables and fruits in the daily diet; avoid obesity; consume alcoholic beverages in moderation, if at all; and minimize consumption of salt-cured, salt-pickled, and smoked foods. These guidelines provide a basis for two current NCI-sponsored public health information programs—the National 5 A Day for Better Health Program and the Ethnic Low Literacy Nutrition Education Project.

The National 5 A Day for Better Health Program

The 5 A Day for Better Health Program is a public/private partnership between the NCI and the Produce for Better Health Foundation, a nonprofit foundation formed by the fruit and vegetable industry. Initiated in October 1991, the program encourages Americans to eat five or more servings of fruits and vegetables every day and provides practical advice on how to improve dietary habits to meet this goal. Baseline survey data obtained during the summer of 1991 indicated that nearly 70 percent of respondents believed that one to two servings of fruits and vegetables daily were sufficient for good health. Only eight percent thought five or more servings were necessary for good health. A second survey a year later indicated that the percentage of respondents aware of the 5 A Day recommendation had risen from eight to 22 percent. There was also a decline from 34 to 15 percent in those who felt one or fewer servings daily to be adequate.

The 5 A Day Program is based on theories of behavior change, including social marketing, health belief, and social cognitive theories. Its four major components are interventions at the point of purchase, a national media campaign, community interventions, and research. The first component currently includes promotion by 35,000 retail grocers across the nation, and strategies for involving restaurants and cafeterias are under development. The media component, based on social marketing theory, uses messages derived from consumer research. In one type of message, for example, celebrities who consume large amounts of fruits and vegetables speak about their food habits. The third component involves 44 state health agencies to coordinate activities and develop coalitions of government, industry, and volunteer agencies to work synergistically in promoting the 5 A Day message. The applied research component of the 5 A Day for Better Health Program currently consists of nine projects funded in May 1993. These projects include school-based 5 A Day interventions, 5 A Day activities at the worksite, 5 A Day interventions in the African-American adult church population, and an intervention to target low-income women in the U.S. Department of Agriculture Special Supplemental Food Program for Women, Infants, and Children (WIC). Components of the Program are interwoven in an effort to create not only a change in awareness, but a change in behavior.

The Ethnic and Low Literacy Nutrition Education Project

The Ethnic and Low Literacy Nutrition Education Project is sponsored jointly by the NCI Division of Cancer Prevention and Control, the NCI Office of Cancer Communications, and the Office of Minority Health of the Department of Health and Human Services. The project aims to develop nutrition education materials that can be used by primary health care and other social service providers for counseling individuals in specific ethnic and low literate populations on how to achieve the NCI Dietary Guidelines.

This project will help meet one of the nutrition goals of the "Healthy People 2000: National Health Promotion and Disease Prevention Objectives"—to increase to at least 75 percent the proportion of primary care providers who offer nutritional assessment and counseling and/or referral to qualified nutritionists or dietitians.

Materials have been developed to reach underserved Americans in seven populations: American Indians, Alaska Natives, Native Hawaiians, Asians (Chinese, Filipinos, and Vietnamese), Hispanics, blacks, and whites. These culturally sensitive materials include booklets, posters, audio and video scripts, post cards, and a gameboard to be used interactively at health fairs and in clinics. Some of the materials will be used as part of a new initiative, "5 A Day in Primary Care." This initiative will evaluate patient outcomes as a function of the training primary care physicians have received for providing culturally appropriate nutrition counseling and is aimed at meeting the Healthy People 2000 objective.

SUMMARY

Through an orderly sequence incorporating epidemiologic and laboratory research, human clinical/metabolic studies, and clinical intervention trials, the NCI develops and maintains cancer prevention and control programs that are directed toward the overall goal to reduce cancer incidence, morbidity, and mortality. Epidemiologic studies have shown correlations between consumption of numerous dietary constituents and cancer risk. Results of *in vitro* and animal studies have reinforced many of these epidemiologic associations and data from clinical/metabolic studies are being used to evaluate the relevance of these associations in humans. Although much remains to be learned about the influence of specific dietary constituents and dietary patterns on cancer risk, it is clear that diet can have a significant impact in cancer prevention and control. Investigations on the cellular and molecular effects of dietary constituents, as well as their metabolic and physiologic effects should provide better insight on the mechanisms of action of these dietary constituents. There also is a need to develop biomarkers of dietary intake that could be used to monitor compliance in intervention studies, as well as biomarkers as clinical trial endpoints that can be used to predict the emergence and progression of cancer. Working with other agencies and the private sector, the NCI will continue to design and implement information programs that translate the knowledge gained from these diet and cancer prevention studies to help Americans adopt dietary patterns that may reduce cancer risk. Continued emphasis will be placed on intervention programs targeted toward high-risk and underserved segments of the U.S. population in an effort to reduce the high incidence of cancers in these groups.

REFERENCES

Adlercreutz, H., 1990, Western diet and Western diseases: Some hormonal and biochemical mechanisms and associations, Scand. J. Clin. Lab. Invest. 50:3.

Armstrong, B. and Doll, R., 1975, Environmental factors and cancer incidence and mortality in different countries, with special reference to dietary practices, Int. J. Cancer. 15:617.

Barnes, S., Grubbs, C., Setchell, K.D.R., and Carlson, J., 1990, Soybeans inhibit mammary tumors in models of breast cancer. In: Mutagens and Carcinogens in the Diet (ed., M. Pariza), Alan R. Liss, New York. pp. 239-253.

Benner, S.E. and Hong, W.K., 1993, Clinical chemoprevention: Developing a cancer prevention strategy [editorial], J. Natl. Cancer Inst. 85:1446.

Block, G., 1991, Vitamin C and cancer prevention: The epidemiologic evidence, Am. J. Clin. Nutr. 53:270s.

Block, G., Patterson, B., and Subar, A., 1992, Fruit, vegetables, and cancer prevention: A review of the epidemiological evidence, Nutr. Cancer. 18:1.

Blot, W.J., Li, J.-Y., Taylor, P.R., Guo, W., Dawsey, S., Wang, G.-Q., Yang, C. S., Zheng, S.-F., Gail, M., Li, G.-Y., Yu, Y., Liu, B.-Q., Tangrea, J., Sun, Y.-H., Liu, F., Fraumeni, J.F., Jr., Zhang, Y.-H., and Li, B., 1993, Nutrition intervention trials in Linxian, China: Supplementation with specific vitamin/mineral combinations, cancer incidence, and disease-specific mortality in the general population, J. Natl. Cancer Inst. 85:1483.

Butrum, R.R., Clifford, C.K., and Lanza, E., 1988, NCI dietary guidelines: Rationale, Am. J. Clin. Nutr. 48: 888.

Butterworth, C.E., Jr., 1992, Effect of folate on cervical cancer: Synergism among risk factors, Ann. N.Y. Acad. Sci. 669:293.

Butterworth, C.E., Jr., Hatch, K.D., Macaluso, M., Cole, P., Sauberlich, H.E., Soong, S.-J., Borst, M., and Baker, V.V., 1992a, Folate deficiency and cervical dysplasia, JAMA. 267:528.

Butterworth, C.E., Jr., Hatch, K.D., Soong, S.-J., Cole, P., Tamura, T., Sauberlich, H.E., Borst, M., Macaluso, M., and Baker, V., 1992b, Oral folic acid supplementation for cervical dysplasia: A clinical intervention trial, Am. J. Obstet. Gynecol. 166:803.

Byar, D.P. and Freedman, L.S., 1989, Clinical trials in diet and cancer, Prev. Med. 18:203.

Chlebowski, R.T., Nixon, D.W., Blackburn, G.L., Jochimsen, P., Scanlon, E.F., Insull, W., Jr., Buzzard, I.M., Elashoff, R., Butrum, R., Wynder, E.L., 1987, A breast cancer nutrition adjuvant study (NAS): Protocol design and initial patient adherence, Breast. Cancer. Res. Treat. 10:21.

Chlebowski, R.T., Blackburn, G.L., Buzzard, M., Grosvenor, M., Insull, W., Jr., Nixon, D., York, R.M., Khandekar, J., Elashoff, R., Wynder, E.L., 1990, Current status: Evaluation of dietary fat reduction as secondary breast cancer prevention, Prog. Clin. Biol. Res. 339:201.

Comstock, G.W., Bush, T.L., and Helzlsouer, K., 1992, Serum retinol, beta-carotene, vitamin E, and selenium as related to subsequent cancer of specific sites, Am. J. Epidemiol. 135:115.

DeCosse, J.J., Miller, H.H., Lesser, M.L., 1989, Effect of wheat fiber and vitamins C and E on rectal polyps in patients with familial adenomatous polyposis. J. Natl. Cancer Inst. 81:1290.

Doll, R. and Peto, R., 1981, The causes of cancer: Quantitative estimates of avoidable risks of cancer in the United States today, J. Natl. Cancer Inst. 66:1191.

Giovannucci, E., Stampfer, M.J., Colditz, G., Rimm, E.B., and Willett, W.C., 1992, Relationship of diet to risk of colorectal adenoma in men, J. Natl. Cancer Inst. 84:91.

Giovannucci, E., Stampfer, M.J., Colditz, G.A., Rimm, E.B., Trichopoulos, D., Rosner, B.A., Speizer, F.E., and Willett, W.C., 1993, Folate, methionine, and alcohol intake and risk of colorectal adenoma. J. Natl. Cancer Inst. 85:875.

Hawrylewicz, E.J., Huang, H.H., and Blair, W.H., 1991, Dietary soybean isolate and methionine supplementation affect mammary tumor progression in rats, J. Nutr. 121:1693.

Henderson, M.M., Kushi, L.H., Thompson, D.J., Gorbach, S.L., Clifford, C.K., Insull, W., Jr., Moskowitz, M., and Thompson, R.S., 1990, Feasibility of a randomized trial of a low-fat diet for the prevention of breast cancer: Dietary compliance in the women's health trial vanguard study, Prev. Med. 19:115

Henson, D.E., Block, G., and Levine, M., 1991, Ascorbic acid: Biologic functions and relation to cancer, J. Natl. Cancer Inst. 83:547.

Hirayama, T., 1979, Diet and cancer, Nutr. Cancer. 1:67.

Howe, G.R., Hirohata, T., Hislop, T.G., Iscovich, J.M., Yuan, J.-M., Katsouyanni, K., Lubin, F., Marubini, E., Modan, B., Rohan, T., Toniolo, P., and Shunzhang, Y., 1990, Dietary factors and risk of breast cancer: Combined analysis of 12 case-control studies, J. Natl. Cancer Inst. 82:561.

Hursting, S.D., Thornquist, M., and Henderson, M.M., 1990, Types of dietary fat and the incidence of cancer at five sites, Prev. Med. 19:242.

Knekt, P., Aromaa, A., Maatela, J. et al., 1991, Vitamin E and cancer prevention, Am. J. Clin. Nutr. 53:283s.

Kromhout, D., 1987, Essential micronutrients in relation to carcinogenesis, Am. J. Clin. Nutr. 45:1361.

Lee, H.P., Gourley, L., Duffy, S.W., Esteve, J., Lee, J., and Day, N.E., 1991, Dietary effects on breast cancer risk in Singapore. Lancet. 337:1197.

Lubin, F., Wax, Y., and Modan, B., 1986, Role of fat, animal protein, and dietary fiber in breast cancer etiology: A case-control study, J. Natl. Cancer Inst. 77:605.

Marlett, J.A., 1992, Content and composition of dietary fiber in 117 frequently consumed foods, J. Am. Diet. Assoc. 92:175.

Murphy, P., 1982, Phytoestrogen content of processed soybean products, Food Tech. 36:60. National Academy of Sciences. 1989. National Research Council, Commission on Life Sciences, Food and Nutrition Board. Diet and Health. Implications for Reducing Chronic Disease Risk. Washington, DC: National Academy Press.

Peterson, G. and Barnes, S., 1991, Genistein inhibition of the growth of human breast cancer cells: Independence from estrogen receptors and the multi-drug resistance gene, Biochem. Biophys. Res. Commun. 179:661.

238

Peterson, G. and Barnes, S., 1993, Genistein and biochanin-A inhibit the growth of human prostate cancer cell lines but not epidermal growth factor receptor tyrosine autophosphorylation, Prostate. 22:335.

Price, K.R. and Fenwick, G.R., 1985, Naturally occurring oestrogens in foods--A review, Food Additives Contam. 2(2):73.

Rogers, A.E., Longnecker, M.P., 1988, Biology of disease. Dietary and nutritional influences on cancer: A review of epidemiologic and experimental data. Lab. Invest. 59:729.

Rose, D.P., 1992, Dietary fiber, phytoestrogens, and breast cancer, Nutrition. 8:47.

Rose, D.P., Goldman, M., Connolly, J.M., and Strong, L.E., 1991, High-fiber diet reduces serum estrogen concentrations in premenopausal women, Am. J. Clin. Nutr. 54:520.

Shankar, S., and Lanza, E., 1991, Dietary fiber and cancer prevention, Nutr. Cancer. 5:25.

Sharrard, R.M., Royds, J.A., and Rogers, S., and Shorthouse, A.J., 1992, Patterns of methylation of the c-*myc* gene in human colorectal cancer progression, Br. J. Cancer. 65:667.

Steinmetz, K.A., and Potter, J.D., 1991a, Vegetables, fruit, and cancer. I. Epidemiology, Cancer Causes and Control. 2:325.

Steinmetz, K.A., and Potter, J.D., 1991b, Vegetables, fruit, and cancer. II. Mechanisms, Cancer Causes and Control. 2:427.

Trock, B., Lanza, E., and Greenwald, P., 1990, Dietary fiber, vegetables, and colon cancer: Critical review and meta-analyses of the epidemiologic evidence, J. Natl. Cancer Inst. 82:650.

Weisburger, J.H., 1991, Nutritional approach to cancer prevention with emphasis on vitamins, antioxidants, and carotenoids, Am. J. Clin. Nutr. 53:226s.

White, E., Shattuck, A.L., Kristal, A.R., Urban, N., Prentice, R.L., Henderson, M.M., Insull, W., Jr., Moskowitz, M., Goldman, S., and Woods, M.N., 1992, Maintenance of a low-fat diet: Follow-up of the women's health trial. Cancer Epidemiol. Biomark. Prev. 1:315.

Willett, W.C., Stampfer, M.J., Colditz, G.A., Rosner, B.A., and Speizer, F.E., 1990, Relation of meat, fat, and their fiber intake to the risk of colon cancer in a prospective study among women. The New England Journal of Medicine. 323:1664.

Willett, W.C., Hunter, D.J., Stampfer, M.J., Colditz, G., Manson, J.E., Spiegelman, D., Rosner, B.A., Hennekens, C.H., and Speizer, F.E., 1992, Dietary fat and fiber in relation to risk of breast cancer—an 8-year follow-up, JAMA. 268:2037.

Ziegler, R.G., 1991, Vegetables, fruits, and carotenoids, and the risk of cancer, Am. J. Clin. Nutr. 53:251s.

NEW DIRECTIONS IN DIETARY STUDIES AND HEART DISEASE: THE NATIONAL HEART, LUNG AND BLOOD INSTITUTE SPONSORED MULTICENTER STUDY OF DIET EFFECTS ON LIPOPROTEINS AND THROMBOGENIC ACTIVITY

Henry N. Ginsberg

Department of Medicine
Columbia University
College of Physicians and Surgeons
New York, N.Y. 10032

INTRODUCTION

Although many clinical studies of dietary effects on plasma lipids and lipoproteins have been conducted during the past 30 years, significant issues remain to be clarified. Previous studies have often lacked adequate sample size and or dietary control, and the diets used have either been extreme in composition or comprised of liquid formulas. Additionally, most of the subjects studied to date have been young white males; very little is known about the relative efficacies of prudent diets in African Americans or in women. The issue of differential responsiveness to diet in pre- and post-menopausal females must also be addressed. Finally, almost nothing is known about the effects of diet on thrombotic factors and activity in humans. To address these issues, a National Heart, Lung and Blood Institute-sponsored multi-center study of the effects of varying levels of dietary saturated fat on plasma lipids, lipoproteins and thrombogenic activities is underway. It is the first of several studies to be conducted under this initiative.

GENERAL STUDY DESIGN ISSUES

The overall objectives of DELTA (DIET EFFECTS ON LIPOPROTEINS AND THROMBOGENIC ACTIVITY) are to develop and carry out multicenter, collaborative investigations of the effects of diet on lipids, lipoproteins and thrombogenic activity. Participating centers include Columbia University, the Pennington Biomedical Research Center, Pennsylvania State University, and the University of Minnesota. The University of North Carolina is the site of the Coordinating Center *(Table 1)*.

Nutrition and Biotechnology in Heart Disease and Cancer
Edited by J.B. Longenecker *et al.*, Plenum Press, New York, 1995

Table 1. Delta investigations

FIELD CENTERS	
1.	Columbia University
2.	Penington Biomedical Research Center
3.	Pennsylvania State University
4.	University of Minnesota
COORDINATING CENTER	
1.	Univeristy of North Carolina
2.	Virginia Polytech
3.	Imogene Bassett Hospital
4.	University of Vermont

Studies will be carried out during the next three years and will focus on nutrition questions that will have a significant impact on public health recommendations. The ability to conduct centrally standardized studies in several centers should enable the investigators to study large numbers of subjects, allowing for separate analyses of results in males and females, in pre- and post-menopausal females, and in African Americans. In developing such protocols, to be carried out in several centers around the country, the DELTA investigators could draw on previous experience to design some aspects of the study, such as recruitment, laboratory standardization, and data transmission to the coordinating center. For other aspects of the study, such as standardization of food procurement and meal preparation, as well as compositional analysis of the experimental diets, there was no prior experience relevant to the proposed studies. In addition, development of standard approaches to maintaining subject compliance in differing populations and ethnic groups posed a significant challenge. The development and implementation of the first protocol will be used as a model for approaches to the issues listed above.

FIRST STUDY DESIGN

The objective of the first DELTA study was to determine the effect of reducing total dietary and saturated fat intake on plasma lipids and lipoproteins, and on hemostatic factors in healthy male and female, black and white adults. In particular, we were interested in determining if further reductions in total and saturated fat, beyond the levels recommended in the Step 1 diet, are efficacious in these gender and ethnic specific groups *(Table 2)*.

Table 2. Goals of Delta

1. Address Important Nutrition Questions
2. Study large Numbers of Subjects
3. Study Males, Females Over a Wide Range Age Range
4. Study Different Ethnic Groups
5. Determine Diet Effects on Thrombogenic Factors
6. Develop Methods to Conduct Multicenter Collaborative Studies of Diet Effects

Although several previous studies, beginning with the classic studies of Keys (1984); Keys et al. (1965a and 1965b) and Hegsted et al. (1965) have demonstrated that reductions in dietary saturated fatty acids are associated with reductions in plasma LDL cholesterol levels, most of these studies were carried out in white males. DELTA will study adequate numbers of women and African-Americans to determine if similar responses to saturated fat characterize these groups as well. In the diet: Present AHA and NIH guidelines focus on reducing total fat to 30% of calories and saturated fat to less than 10% of calories. Many public health officials, physicians, and scientists believe that further reductions are needed, while others are concerned that further lowering of dietary fat will not significantly reduce LDL cholesterol and may even lower HDL cholesterol. Finally, there are few well controlled studies that have focused on the effects of diet on thrombotic potential: DELTA will specifically look at fibrinogen, factor VII and plasminogen activator inhibitor levels during consumption of diets differing in total fat and saturated fatty acids.

The specific aims were: To compare the effects of three diets, differing in total and saturated fat content, on plasma lipids and lipoproteins in healthy adults; to compare the effects of these diets on plasma hemostatic factors in healthy adults; to determine if males and females, and individuals with different ethnic origins respond to these diets; and to evaluate the role of the menstrual cycle in modulating diet response in females.

The primary lipid-related outcomes included plasma concentrations of total cholesterol (TC) and triglycerides (TG), low density lipoprotein cholesterol (LDLP), high density lipoprotein cholesterol (HDLP), apoprotein B (APO B), apoprotein A-I (APO A-I), and lipoprotein (a) (LP (a)) . The primary hemostatic-related endpoints included plasma levels of fibrinogen, factor VII and plasminogen activator inhibitor I (PAI-I).

RECRUITMENT

Recruitment was carried out in three phases *(Table 3)*. Initial contact was via telephone in response to advertisement of the study. Personnel were trained in a standardized interview which was based on a computer-compatible form. Major exclusionary criteria were identified and a specific description of the study protocol,

Table 3. Recruitment

1.	Telephone Screen:	1,657
2.	Eligibility Visit 1:	546
3.	Eligibility Visit 2:	281
4.	Randomized:	114

Characteristics of Responders
Determinants of Ineligibility

including subject responsibilities, was provided. General subjects characteristics, including age and ethnicity, were obtained as well. Finally, a general diet and food history was obtained. If the subject was not disqualified, and showed further interest in the study, an appointment was made for a visit to the center. At that visit, a detailed medical, social and diet history was elicited, height and weight was obtained (and body mass index calculated), blood pressure was determined, and blood cholesterol was measured by fingerstick using a Cholestech device. Plasma cholesterol levels had to fall between the 25th and 85th percentile for age, gender and race. If subjects were not excluded at this visit, and were still interested, they were scheduled for a final eligibility visit. At this visit, fasting blood was obtained by venipuncture for complete blood count, chemistry screen and lipid profile. Subjects with triglyceride levels above the 90th% or HDL cholesterol levels below the 10th% were excluded, even if plasma cholesterol levels fell between the 25th and 85th percentile.

Recruitment was carried out during a three month period, between June and September, 1993. A major emphasis was placed on goals to recruit both males and females, African Americans, males aged below and above 40 years, and pre- and post-menopausal females. The total number of subjects interviewed by telephone was 1657; 546 subjects completed the first on-site screening, and 281 returned for the second on-site screening visit. From this group, 114 subjects were randomized into the study. Of this total group, 64 were female, 50 were males, African Americans made up 27% of the total group. Males over forty comprised 36% of all men, while postmenopausal females made up 26% of all the females.

PROTOCOL

The first protocol was designed as a randomized, double-blind, three-way crossover study *(Table 4)*. Each diet period was eight weeks in length, with breaks of 4-6 weeks

Table 4. Study Design

Three-way, Randomized Crossover Design
Eight Week Diet Records: Four-Six Week Breaks
Blood Samples During Weeks Five through Eight

Plasma Endpoints:

Triglycerides (TG), Total Cholesterol (TC),
Low Density Lipoprotein, Cholesterol (LDLC),
High Density Lipoprotein, Cholesterol (HDLC),
Apoprotein B (APO B), Apoprotein A-I (APO A-I),
Lipoprotein (a) (LP(a)), Fibrinogen, Factor VII,
Plasminogen Activator Inhibitor I (PAI-I)

between diet periods. Subjects ate two meals each day (either breakfast and dinner or lunch and dinner) in a cafeteria setting. All food provided had to be eaten on site at that meal. In addition, subjects were provided with a packaged third-meal as well as snacks. On weekends, all meals except Saturday dinner were packaged and provided at the Friday evening meal. Saturday evening dinner was self-chosen within detailed guidelines provided by the staff. Subjects filled out daily compliance forms as well as weekend food records. Weights were obtained twice weekly.

Blood samples were obtained once during weeks five, six, seven and eight. This design, with several weekly samples per subjects, was required because of a lack of any published data concerning diet effects on thrombogenic factors. Subjects fasted overnight prior to venipuncture. Strict, standardized blood drawing protocols were developed to meet the requirements for successfully determining thrombogenic factors.

DIET DEVELOPMENT AND IMPLEMENTATION

The goal of the first DELTA study was to determine the effects of reductions in total and saturated fat on plasma lipids, lipoproteins and thrombogenic factors. The diets chosen for study were: an average American diet that included 37% of calories from fat with 16% saturated, 14% monounsaturated and 7% polyunsaturated fats; an AHA Step 1 diet with 30% of calories from fat with 9% saturated, 14% monounsaturated and 7% polyunsaturated fats; and a low fat diet with 26% of calories from fat with 5% saturated, 14%

monounsaturated and 7% polyunsaturated fats. Dietary carbohydrate was 48%, 55%, and 59% of total calories on the average American, AHA and low fat diets, respectively. All of the diets provided 15% of calories as protein. Dietary cholesterol was 300 mg/day on all diets *(Table 5)*.

Table 5. Diet Composition

AVERAGE AMERICAN DIET:	
	37% Calories from Fat
	16% SF, 14% MF, 7% PF
AHA STEP 1 DIET:	
	30% Calories from Fat
	9% SF, 14% MF, 7% PF
LOW-FAT DIET:	
	26% Calories from Fat
	5% SF, 14% MF, 7% PF
TWO MEALS PER DAY EATEN ON-SITE	
THIRD MEAL AND SNACKS PROVIDED	
WEEKEND MEALS PROVIDED	

Numerous logistic problems faced the DELTA investigators. Diets had to be developed that would be palatable and acceptable to subjects living in very different areas of the nation and with widely varying ethnic and socio-economic backgrounds. Diets had to be identical in all sites, particularly with reference to saturated, monounsaturated and polyunsaturated fatty acid content. Each individual's meals had to be separately prepared, with weights accurate to the gram. Food safety had to be assured despite a need to package meals for as long as 2-3 days (in the case of the occasional holiday weekend). Indeed, development of the diets for the first DELTA study proved to be an enormously complex and time-consuming task. Many face-to-face meetings and teleconferences were required; numerous contacts had to be made with representatives of the food industry to obtain central supplies of most items; bulk delivery systems had to be developed; freezer and refrigerated storage space had to be identified at each site; staff training in the preparation, cooking and serving of meals had to be completed.

In parallel with the development of systems for the central procurement and local preparation of meals, DELTA had to develop systems for validating the composition of the experimental diets and to allow for ongoing monitoring of diets served in each of the centers. Protocols were developed and tested, and both validation of the diets for study 1 and the ongoing monitoring of those diets were achieved.

LABORATORY STANDARDIZATION

Central laboratories were chosen for determination of apoproteins and thrombogenic factors. This strategy was associated with the need for standardization of sample processing and storage at each site at the time of each blood drawing, and standardized protocols for later shipment to the central laboratories. Plasma lipids, including total cholesterol and triglycerides, LDL cholesterol and HDL cholesterol were to be determined at each site after completion of the entire study. The decision to use each center's laboratory for lipid measurements necessitated a standardization program that was conducted with the assistance of the CDC lipid laboratory. The focus of the standardization was precision, as the endpoints were differences in plasma lipids on the different diets. Limits for accuracy were also set.

PRELIMINARY REPORT OF STUDY 1

Study 1 began at the end of September, 1993 and the first eight week diet period was recently completed. One-hundred and eighteen subjects were initially randomized into the study after a one-week preliminary run-in period during which all of the diets were tested for palability. During the first eight weeks, 9 subjects have left the study; these drop-outs have usually resulted from illness or difficulty with scheduling. Of the 109 subjects who completed the first eight weeks, 48 were males and 61 were females. There were 28 African Americans. Thirty-one of the males were less than 40 years of age and 17 were over 40 years. There were 43 pre-menopausal and 18 post-menopausal females. Although recruitment goals of were met for total enrollment and number of African Americans, the goals for older men and post-menopausal females were not attained. In general, it soon became clear that recruiting older individuals was difficult, despite targeted recruitment. In particular, recruiting older black men was extremely difficult: only a single black male over 40 years of age was randomized.

Diet validation, which required training in standard preparation and cooking techniques, as well as development of a standardized protocol to homogenize diets and prepare uniform aliquots for analysis, was carried out during June, July and August. Goals for absolute levels of fat and fatty acids classes, and for differences between the diets were achieved for the majority of the menus. As additional menus had been developed to insure validation of an eight day menu cycle, the study was able to start on schedule. Preliminary analyses of the ongoing, weekly monitoring of the diets served, achieved by preparing an additional, blinded menu each day, indicates that the goals of 7% and 4% differences in total and saturated fats between the three diets are being achieved.

The initial blood sampling required protocol development and on-site training by representatives from the central laboratories. It should be noted that in typical epidemiologic studies or clinical trials, subjects are recruited on a "rolling-basis" and procedures such as venipuncture are conducted on one or two subjects each day. In our studies, all subjects are recruited and studied simultaneously, generating significant logistic problems for blood sampling and sample processing. This is particularly true when biologically fragile endpoints, such as thrombotic factors are being measured. DELTA developed protocols for handling 6-8 subjects each day, but this was very labor intensive. Sampling during the final four weeks of the first diet period was completed without significant problems.

FUTURE DIRECTIONS

DELTA is not only a series of specific protocols to test hypotheses related to dietary effects on lipids, lipoproteins and thrombogenesis, it is also a model for multicenter, controlled diet studies. The investigators and staff of DELTA have already learned a great deal about the unique problems inherent in such studies and will, it is certain, continue to learn during the next three years. There are numerous questions with major public health/nutrition implications that can only be answered by tightly controlled, standardized diet studies. However, the previous model of such "metabolic" studies, that is the small group of homogenous subjects studies on metabolic units, is no longer viable. Extrapolation of the results of many previous studies to the population-at-large has not produced the expected results and has prompted investigators such as those in DELTA to attempt larger studies with subjects more characteristic of the population-at-large. As described above, investigators attempting to conduct these studies will have to deal with many problems for which there is no prior experience. DELTA will be a model for some of those studies.

More specifically, DELTA will, after completion of the first study, embark on two studies dealing with major nutrition controversies. One study being planned will examine the question of whether diets that are very low in saturated fat, but high in total and monounsaturated fats are more efficacious than very low fat diets (low in both total and saturated fats) that are high in carbohydrates. Controversy about these diets focuses on individuals with insulin resistance and hypertriglyceridemia/low HDL cholesterol. A second study being planned will focus on the relative cholesterol-raising activity of the major saturated fatty acids: myristic, palmitic, and stearic. Results of this study should be of great interest to the food industry in their search for the optimal fat source.

SUMMARY

The National Heart, Blood and Lung Institute-sponsored DELTA study is the first collaborative, multicenter diet study to utilize standardized protocols to feed specific diets to study participants. In the first study, the investigators are focusing on the effects of reducing dietary saturated fat on plasma lipids, lipoproteins, and thrombogenic activity. Future studies will attempt to address important diet/public health questions dealing with high carbohydrate vs. high monounsaturated fat diets, and the cholesterol raising activity of specific saturated fatty acids. Moreover, DELTA will be an invaluable model for other multicenter diet studies.

REFERENCES:

Hegsted, D.M., McGandy, R.B., Myers, M.L. and Stare, F.J., 1965, Quantitative effects of dietary fat on serum cholesterol in Man, Am. J. Clin. Nutr. 17:281-295.

Keys, A., 1984, Serum cholesterol response to dietary cholesterol, Am. J. Clin. Nutr. 40:351-359.

Keys, A., Anderson, J.T. and Grande, 1965a, Serum cholesterol response to changes in the diet II. The effect of cholesterol in the diet, Metabolism 14:759-765.

Keys, A., Anderson, J.T. and Grande, F., 1965b, Serum cholesterol response to changes in the diet. IV. Particular saturated fatty acids in the diet, Metabolism 14:776-787.

APPENDIX

ENTEROVIRUS-INDUCED MURINE MYOCARDITIS: IMPACT OF A SELENIUM-DEFICIENCY. MA Beck, VC Morris and OA Levander, Department of Pediatrics, The University of North Carolina at Chapel Hill, 27599 and USDA, Human Nutrition Research Center, Beltsville, MD, 20705.

Keshan disease, a selenium (Se)-responsive cardiomyopathy, has been described in the People's Republic of China. Although adequate Se intake prevents Keshan disease, its seasonal and annual incidence suggest an infectious co-factor. Coxsackieviruses have been isolated from the blood and tissue of Keshan disease victims, and in the West, coxsackieviruses are implicated in some cases of myocarditis and dilated cardiomyopathy. Se-deficient mice inoculated with coxsackievirus B3 or B4 develop more severe inflammatory heart disease than Se-adequate mice. In order to understand how a Se-deficiency affects the pathogenesis of a coxsackievirus infection, we measured several immune functions of coxsackievirus B3 (CVB3)-infected Se-deficient and Se-adequate mice. We found that although neutralizing antibody titers were equivalent between Se-deficient and Se-adequate mice, T cell responses to both mitogen and specific CVB3 antigen were depressed in the Se-deficient mice. Production of interleukin 2 mRNA was decreased in lymphocytes infiltrating the heart taken from CBV3-infected Se-deficient mice when compared with Se-adequate mice. Natural killer cell activity was similar between both groups of mice. Viral titers were higher in heart, liver, spleen and serum from Se-deficient CVB3-infected mice at each time point measured. Thus, a Se-deficiency affects specific aspects of the immune system, which may in turn allow for increased viral replication and increased heart pathology. Thus, the interaction of host nutrition and a viral infection may play an important role in the outcome of certain types of cardiomyopathies, such as Keshan disease.

SERUM LIPID LEVELS OF CHILDREN AND ADOLESCENTS DURING WEIGHT REDUCTION. LK Carlisle, A Babst, E De Leon, MS Sothern, RP Farris, JN Udall and RM Suskind, Department of Pediatrics, LSU School of Medicine, New Orleans, LA 70112.

Obesity is recognized as a significant cardiovascular disease (CVD) risk factor. Prevention of obesity in childhood may decrease its incidence in adults and, therefore, decrease the risk of CVD. Subjects (ages 7-17) participated in an outpatient weight reduction program which incorporated a protein sparing modified fast, an exercise program, and behavior modification. Fasting total cholesterol (TC), lipoproteins (LDL-C, HDL-C) and apolipoproteins (Apo A-1, Apo B-100) were measured at baseline and ten weeks. Percent of ideal body weight (mean $\pm$ SD) was 190$\pm$45 initially and 171$\pm$37 at ten weeks. Matched pair analyses indicate significant lowering of TC, LDL-C and Apo B-100, while no significant change is noted in HDL or Apo A-1.

	N	Baseline	Ten Weeks	p value
TC (mg/dl)	17	189.6 $\pm$ 35.9	173.2 $\pm$ 24.6	<.04
LDL-C (mg/dl)	17	126.7 $\pm$ 26.0	115.3 $\pm$ 21.7	<.05
Apo-B100 (mg/dl)	17	120.2 $\pm$ 23.7	106.3 $\pm$ 16.3	.01

These results suggest that a multidisciplinary weight reduction program is successful in reducing CVD risk factors while maintaining protective levels of HDL-C and Apo A-1 in obese children and adolescents.

A STUDY OF COPPER (Cu) STATUS AND RELATED FACTORS OF ELDERLY MEN ENROLLED IN A CARDIOVASCULAR REHABILITATION PROGRAM.

M Kumar, AA Suleiman and WL Dodson, Department of Home Economics, Mississippi State University, Mississippi State, MS 39762.

Copper status has been shown to be related to zinc (Zn) status, Cu-Zn Superoxide Dismutase (Cu-Zn SOD) activity and serum cholesterol concentrations, either directly or otherwise. The various factors indicated above were determined for 25 patients and 15 controls. The serum Cu and Zn concentrations (determined by atomic absorption) and SOD activity (determined by method of Fridovich and McCord) were:

	Zinc µmol/L + SD	Copper µmol/L + SD	Cu-Zn SOD U/g Hb + SD
Patients	11.6 ± 1.6	23.7 + 1.9	191.4 + 26.9
Controls	11.9 ± 4.4	23.4 ± 1.1	209.5 ± 46.3

The similarity in copper status, was reflected in the SOD activity, but not in the serum cholesterol profile, which was significantly greater in the patients than the controls. Also, the Cu concentration was approximately twice the Zn concentration, for both the patients and the controls. Further investigation is planned to study the relative roles of the various factors involved in these findings.

VITAMIN E, DHEA & EXERCISE EFFECTS ON HEART OXIDATIVE STRESS,

AH GOLDFARB, MD McINTOSH and BT BOYER, Exercise & Sport Science, and Food & Nutrition Departments, University of North Carolina at Greensboro, Greensboro, NC 27412.

Exercise can increase lipid peroxidation and alter glutathione status which are indicators of oxidative stress. The purpose of this study was to determine the effects of vitamin E (E) and dehydroepiandrosterone (DHEA) on oxidative stress in heart. Sixty-four male rats received one of four treatments; half injected daily with either corn oil or DHEA (100 mg/kg), half received E (250 IU/day) for 5 weeks. Rats (N=32) were run for 1 hour at 21 m/min u p a 12% incline. An animal from each treatment group was killed/day by decapitation. Samples were frozen in liquid nitrogen & stored in a -70% freezer until analyzed. Run time was unaffected by treatment. Thiobarbituric acid reactive technique (TBARS) showed an increase with exercise ($1.94 \pm .13$ vs $2.40 \pm .16$ nmol/mg protein). DHEA increased TBARS and E tended to reduce the exercise-induced catalase activity response. DHEA increased catalase at rest but showed a decrease with exercise. Total glutathione peroxidase activity was unaffected. These results suggest that exercise & DHEA can increase oxidative stress in heart muscle whereas vitamin E can attenuate this response.

EPIDEMIOLOGIC EVIDENCE FOR PREVENTIVE ROLE OF DIETARY CALCIUM IN HYPERTENSION.

H Gruchow and K Hixson, Departments of Public Health Education and Exercise and Sport Science, University of North Carolina at Greensboro, Greensboro, NC 27412.

The interrelationships of dietary calcium (Ca), sodium (Na), potassium (K), and alcohol to blood pressure were studied using NHANES I baseline and follow-up mortality data. In baseline cross-sectional data, at low Ca intakes (<400 mg/d for men and <800 mg/d for women) the ratio of Na to K (Na:K) was significantly associated with blood pressures; at higher Ca intakes neither Na:K nor any other nutrient, except alcohol, was associated with blood pressures. Na:K was more strongly associated with blood pressures than either nutrient alone, and low Ca intakes were necessary for the Na:K-blood pressure association to be evident. Interaction of these three dietary factors in relation to blood pressures was evident in all race and gender groups, and was independent of age, body-mass-index, and alcohol consumption. Analyses of follow-up ten-year mortality indicated that low calcium intake was significantly associated with higher rates of mortality from all causes, with higher rates of initiation of hypertension medication, and with higher blood pressures among survivors not on hypertension medications. The analyses also revealed higher (but not statistically significant) rates of mortality from hypertensive disease associated with low calcium intakes.

INFLUENCE OF THE MENSTRUAL CYCLE AND PHYSICAL ACTIVITY LEVEL UPON DIETARY FAT INTAKE. SC Holliman and AC Hackney, Dept. of PE Exercise & Sport Sciences and Dept. of Nutrition, University of North Carolina, Chapel Hill, 27599.

Exercise programs (i.e., training) result in an increased utilization of lipid as an energy source; furthermore, the estrogen hormones have been shown to facilitate lipolysis. Little research, however, exits that has addressed whether the metabolic alterations/influences of physical activity and menstrual hormonal changes impact food selection. Therefore, the present study was undertaken to examine how menstrual cycle phase and physical activity level effected dietary fat intake. Our subjects were 23 healthy, eumenorrheic young adult women; 12 were classified physically active (PA) and 11 were classified physically inactive (PIA). Daily dietary records were kept by the subjects (via micro-tape recorders) for one complete menstrual cycle. Each subjects' menstrual cycle was divided into a follicular phase (low estrogen levels) and luteal phase (high estrogen levels) based upon daily basal body temperature records. A significant ($p<0.01$) main effect between groups was found for percentage fat intake as the PA group consumed less than the PIA group (means = 23.1% vs. 30.8%, respectively). Additionally, a main effect for menstrual cycle phase was also found. There was a greater fat intake in both groups during the luteal phase (28.2%) compared to the follicular phase (25.7%; p=0.02). These findings suggest that physical activity level and menstrual cycle phase do influence the level of dietary fat a eumenorrheic women self selects. These findings have direct implications for persons giving nutrition guidance and health-care to menstruating women.

OPIOID ANTAGONISM POTENTIATES THE COUNTERREGULATORY RESPONSE TO DYNAMIC EXERCISE IN MAN. MS Hickey*, SW Trappe, AC Blostein, MD Vukovich and BW Craig, *Human Performance Laboratory, East Carolina University, Greenville, NC 27834 and Ball State University, Munice, IN 47306.

Brief, high intensity exercise in both lean and obese individuals induces a post-exercise hyperglycemia and hyperinsulinemia accompanied by transient insulin resistance. In an attempt to investigate the role of the endogenous opioid peptides in this phenomenon, eight trained cyclists completed two exercise trials at 90% VO2max until exhaustion. Trials were conducted following the administration of the opiate antagonist naloxone (NAL) or volume matched saline (SAL). Serum glucose was significantly higher at all time points during exercise and at 30 and 60 minutes of recovery in the NAL trial. Serum insulin did not differ between trials. Serum C-peptide was significantly lower during exercise in the NAL trial. Plasma glucagon was significantly higher during exercise in the NAL trial. However, the glucagon:insulin molar ratio was not significantly altered by naloxone administration. Plasma norepinephrine was significantly higher at 10 minutes of exercise (19.1±2.8 vs. 11.6±1.7 nM) and at termination (21.0±1.7 vs. 12.2± 1.4 nM) in the NAL trial. Plasma epinephrine was higher at termination (4.9±2.1 vs. 2.7±1.7 nM) in the NAL trial, although this was not statistically significant. Thus, the normal role of the endogenous opiod peptides may be to limit the magnitude of the counterregulatory hormone response to exercise and prevent prolonged hyperglycemic excursions.

EFFECT OF EXERCISE AND LACTATION ON PLASMA LIPIDS. CA Lovelady, LA Rivers, MA McCrory and KG Dewey. Dept. of Nutrition, University North Carolina at Greensboro, Greensboro, NC 27412 and Dept. of Nutrition, UC, Davis, CA 95616.

To test the hypothesis that the secretion of lipid into breast milk enhances the normalization of plasma cholesterol (CHOL) levels after pregnancy and that exercise increases HDL-CHOL levels during lactation, exclusively breastfeeding women were randomly assigned to an exercise (E) or control group at 6-8 wk postpartum (pp). E subjects performed aerobic exercise 45 min/d, 5 d/wk, for 12 wk. Fasting blood samples were collected at 6-8 and 18-20 wk pp and analyzed for triglycerides (TG) and total-, HDL- and LDL-CHOL. Milk samples were analyzed for lipid and energy content. Maternal body composition and dietary intake were measured. Exercise significantly increased only HDL-CHOL levels. Total- and LDL-CHOL and TG significantly decreased in both groups. There were significant inverse correlations between TG, total-CHOL, LDL-CHOL and *total* lipid or energy output in milk at 18-20 wk pp. Women with a greater average energy output in milk had a significantly greater decline in total-, LDL- and HDL-CHOL between 6-8 and 18-20 wk, controlling for maternal age, body composition and dietary fat intake. These results suggest that a greater daily output of lipid or energy in milk may lower plasma CHOL levels with positive health consequences for lactating women.

DIETARY EFFECTS ON ATHEROSCLEROSIS IN APO E DEFICIENT MICE. SH Zhang, RL Reddick, LH Kester and N Maeda, Department of Pathology, University of North Carolina, Chapel Hill, NC 27599-7525.

The apolipoprotein E deficient mice created by gene targeting have spontaneous hyperlipidemia and develop atherosclerosis (Zhang et al., Science, 258, 468-471).

When mice were fed a high fat diet (15.8% fat, 1.25% cholesterol), we observed a 4 times increase in plasma cholesterol in homozygotes (3000mg/dl) and heterozygotes (330mg/dl), and 2-3 times increase in normal animals (240mg/dl). After 3 months of feeding with the high fat diet, homozygotes developed 13 times more atherosclerotic lesions than those on regular diet. More excitingly, the lesions developed in heterozygotes were 44 times larger than those of normal mice after high fat feeding, though neither developed lesions on regular mouse chow. Diet with saturated and polyunsaturated fatty acids increased cholesterol levels by 40% in homozygotes on both diets. Their triglyceride levels doubled on polyunsaturated but not on saturated fatty acid diet. The effects of these diets on lesions is currently being studied.

This set of data shows that apo E gene deficiency makes mice susceptible to atherosclerosis due to dietary fat and cholesterol, even when only one allele is affected. This is the direct evidence that dietary and genetic factors act together to induce atherosclerosis.

EFFECTS OF HIGH POLYUNSATURATED FAT DIETS ON LDL RECEPTOR ACTIVITY IN FRESHLY ISOLATED HUMAN MONONUCLEAR CELLS. Sunmin Park and JT Snook, The Ohio State University, Columbus, OH 43210.

The purpose of this research was to determine if a high intake of polyunsaturated fat increases LDL degradation through LDL receptors and binding to the receptors in freshly isolated mononuclear cells. Since apolipoprotein E phenotype (Apo E) may affect LDL receptor activity, the activity was determined according to Apo E. Healthy females consumed two different fat diets in two 28-day periods separated by a wash-out period of 8 weeks. Prior to each period subjects consumed a prescribed diet (a high saturated fat diet, P:S ratio=0.4) at home. Then, two groups were provided 40% of energy as fat. The main fats were either corn oil or butter (54% of total fats). LDL degradation through and binding to the LDL receptor were 83 and 173% higher on the corn oil than the butter diet (n=12, p=0.02, p=0.01, respectively). Specific degradation of LDL was 114 % higher on the corn oil diet than on a butter diet when only subjects with Apo E 3/3 were considered (n=9, p=0.004). Subjects with Apo E 3/2 and 2/2 had lower LDL degradation through LDL receptors than those with Apo E 3/3, and their LDL specific degradation was not altered by dietary modification. Specific binding was not significantly affected by Apo E. Unsaturated fatty acid contents in mononuclear cells were increased in the corn oil group, and saturated fatty acid contents were decreased, compared to the butter group. These results suggested that a high corn oil diet increases the LDL receptor activity, represented by LDL degradation and binding through the receptor-mediated pathway, which could be enhanced by increased incorporation of unsaturated fatty acids into mononuclear cells. Apo E 3/2 and 2/2 may be impaired in LDL receptor activity, and this defect may not be changed by dietary modification.

RELATIVE EFFECTS OF SATURATED FATTY ACIDS ON SERUM LIPOPROTEINS AND LDL RECEPTOR ACTIVITY. JT Snook, Sunmin Park, E Stasny, M Morocco, R VanVoorhis and Sonhee Park, The Ohio State University, Columbus, OH 43210.

To assess effects of different saturated fatty acids on cholesterol metabolism we conducted a study involving 18 men, 18 to 32 years of age, who ate 3 diets in three 28-day periods separated by washout periods of 4-6 weeks. The subjects consumed a baseline diet at home for one week. Total, saturated, monounsaturated and polyunsaturated fatty acid comprised 40, 15, 7, and 18 energy % of the diet, respectively. Test fatty acids, providing 2/3rd of the saturated fat, were 12:0/14:0 (tropical oil diet), 14:0/16:0 (butter diet), and 16:0/18:0 (meat diet). Despite some reports to the contrary, the 3 diets had similar effects on concentrations of serum total and LDL-cholesterol as well as apoproteins A-1 and B-100, LCAT, CETP, platelet aggregation, and cholesterol biosynthesis in vitro in freshly isolated mononuclear cells. Compared to baseline, diet 12:0/14:0 raised serum HDL-cholesterol concentrations 11% (p=0.009 for treatment effect) while diet 16:0/18:0 raised 125I-LDL degradation by mononuclear cells 30% (p=0.03 for treatment effect) when only subjects with apoprotein E phenotype 3/3 were considered. Mean degradation did not increase when subjects with the 3/2 phenotype ate 16:0/18:0 and did not differ among all subjects when only post-diet values were compared. The lack of effect of the treatments on many parameters may be a function of the dietary PUFA level or the form in which the saturated fatty acids were fed. For example, part of the 14:0 was fed as the free fatty acid. (Supported by a grant from the National Live Stock and Meat Board.)

IMPLEMENTING "REVERSAL" CARDIAC DIETS. KG Rosati and M Spencer, P.O. Box 61343, Durham, NC 27715.

Our goal was to establish and implement a more effective hospital cardiac diet, while maintaining good patient acceptance. When this effort began the cardiac diet contained: less than 30% fat, 300 mgs. of cholesterol (chol) and 4,000 mgs. of sodium (Na)/day. The new cardiac dietary guidelines approved were: less than 25% fat, 200 mgs. of chol and 3,300 mgs. of Na/day. Patient interviews and written meal service evaluations were used to assess patient acceptance, regarding taste, appearance, temperature and variety of food served.

Accuracy was assessed via test trays, which were evaluated for portion size accuracy and correct item on the tray. Manual calculations of the highest fat day revealed: 19% fat, 4.4% calories from saturated fat, 153 mgs. chol and 2009 mgs. of Na on the 1800 calorie house cardiac diet. The results showed that 94% of the patients found the new cardiac diet "great", "good" or "fair", which increased to 100% within five months of quality assurance intervention. This project demonstrated that cardiac diets with less than 20% fat can be successfully implemented, with good patient acceptance, in a population traditionally consuming very high fat intakes.

EVIDENCE THAT DIETARY ARACHIDONIC ACID ELEVATES PLASMA TRIGLYCERIDES. J Whelan and BY Li, Department of Nutrition, University of Tennessee, Knoxville, TN and ME Surette, Centre Hospitalier de i'Universite' Lavai, Ste. Foy, Quebec, Canada.

Dietary polyunsaturated fatty acids (PUFA) possess weak but inconsistent hypotriglyceridemic activity compared to saturated fatty acids. The notable exception is n-3 PUFA derived from marine oils, which are very effective in lowering plasma triglycerides. However, some evidence suggests that the effects of dietary arachidonic acid (AA) on plasma triglyceride concentrations may not be consistent with the effects observed with other PUFA. Therefore, two different studies were designed, in part, to evaluate the effect of dietary AA on plasma/serum triglyceride concentrations. The first study randomly divided 30 male Syrian hamsters into 5 dietary groups. The second study randomly divided 48 male CD-1 mice into 4 dietary groups. The animals, with respect to each study, were fed identical diets that were supplemented with various fatty acid ethyl esters (1.0 - 1.5% w/w), viz. 18:1 n-9 (OA), 18:2 n-6 (LA), 20:4 n-6 (AA), 20:5 n-3 (EPA) and AA+EPA. In both studies, increasing chain length and unsaturation of the supplemented fatty acids resulted in a progressive decline in circulating triglyceride levels with EPA exerting the greatest triglyceride-lowering effect. However, the circulating triglyceride concentrations were significantly elevated in the groups consuming AA compared to the OA-, LA- and EPA-supplemented groups. These data suggest that moderate intakes of dietary AA may have a triglyceride-elevating effect compared to monosaturated and other PUFA and may abrogate the effects of n-3 PUFA.

FAT AND FIBER INTAKE AND ADHERENCE TO DIETARY RECOMMENDATIONS AMONG COLLEGE STUDENTS. SM Wunderlich and KT Beutler, Department of Home Economics, Montclair State College, Upper Montclair, NJ 07043.

Dietary levels of fat and fiber are related to cardiovascular disease and cancer. This study investigates the relationship between dietary intake of fat and fiber (from fruits, vegetables, bread, and cereal sources), and the knowledge of diet-disease relationships, food sources and dietary recommendations among 151 college students. The knowledge of food sources of fat, cholesterol, and fiber are significantly related to adherence to dietary recommendations. Fat intake was highly related to the students knowledge about food sources of fat and cholesterol with a relationship coefficient of -0.8 based on the scale used in the study. Enhancing the nutrition knowledge among college students is therefore an effective way to encourage healthy eating habits and reducing the incidence of heart disease and cancer.

RELATIONSHIP OF FAT INTAKE AND SERUM TOTAL LIPID FATTY ACID COMPOSITION IN A NIGERIAN POPULATION. LLL Yeh, LH Kuller, FA Ukoli, CH Bunker, SL Huston, N Markovic and A Fabio, Department of Epidemiology, University of Pittsburgh, Pittsburgh, PA 15261

The prevalence of heart disease (CHD) appears to be much lower in Nigeria than in the U.S. Diet may make a major contribution to this difference. To assess Nigerian food intake and to determine whether serum total lipid fatty acids reflect dietary fat intake, we studied *serum total lipid fatty acid composition* of 184 Nigerians, along with two 24-hour food recalls. Daily meat intake was 47.5 grams and fish intake was 69.1 grams; whereas, daily meat intake in Americans is 208 grams and fish intake is 17 grams (based on U.S. survey data). Dairy food intake was too low to be quantified. We compared the quantified fatty acid composition (mg/dl) to that of a previously studied American population (N=104). The total fatty acids of Nigerians was 198.2 ± 57.7 mg/dl and 359.7 ± 106.9 mg/dl for Americans. Except for eicosapentaenoic acid (EPA) and docosahexaenoic acid (DHA), each fatty acid in the Nigerian population was lower than in the American population. The significant difference ($p<0.0001$) of the weight of major fatty acids were:

	Palmitic	Oleic	Linoleic	Arachidonic	EPA	DHA
Nigerian	53.4 ± 18.3	48.4 ± 15.6	50.2 ± 16.5	10.4 ± 4.4	3.8 ± 2.7	7.7 ± 3.2
American	85.6 ± 32.4	72.9 ± 27.5	122.7 ± 31.8	24.2 ± 8.3	1.5 ± 0.9	4.2 ± 2.2

We also found a positive correlation between meat intake and the ratio, arachidonic acid: sum of EPA and DHA ($r=0.27$, $p=0.0007$). The different pattern of fat intake and its apparent effect on the quantitative distribution of *serum total lipids* may partly explain the lower prevalence of CHD in the Nigerian population.

PROTECTIVE EFFECTS OF VITAMIN E ON DHEA- AND EXERCISE-INDUCED LIPID PEROXIDATION IN PLASMA AND SKELETAL MUSCLE OF RATS. M McIntosh, A Goldfarb, B Boyer and J Fatouros, Departments of Food and Nutrition, and Exercise and Sports Science, University of North Carolina at Greensboro, Greensboro, NC 27412.

The purpose of this study was to determine if dehydroepiandrosterone (DHEA) and/or acute exercise increased indices of lipid peroxidation in muscle and blood of rats and whether supplementation of vitamin E prevented the induction of lipid peroxidation by these treatments. Sixty-four male Sprague-Dawley rats were randomly assigned to one of eight treatment groups of six rats each in this 2 x 2 x 2 factorial design. The factors tested were DHEA treatment (+ or -), acute exercise (+ or -) and vitamin E supplementation (+ or -). Vitamin E supplementation and DHEA treatment lasted 5 weeks. Rats in the acute exercise group were run on a treadmill prior to being killed. Exercise increased both indices of lipid peroxidation (TBARS and lipid hydroperoxides) in plasma and TBARS in soleus and white fast-twitch muscle. DHEA treatment increased indices of lipid peroxidation in plasma and in red fast-twitch muscle. DHEA treatment reduced GPX activity in plasma but increased GPX activity in all three muscle fiber types. Vitamin E supplementation reduced the amount of lipid hydroperoxides in plasma and TBARS in plasma and in all three muscle fibers.

INFLUENCE OF DIET ON TUMORIGENESIS IN TRANSGENIC TG•AC MICE. J French*, R Tice+ and R Tennant*, *NIEHS and +ILS, Research Triangle Park, NC 27709.

The influence of diet on cancer is poorly understood. We have employed genetically initiated (ζ-globin promoted v-Ha-ras; TG•AC/FVB strain) mice to determine the influence of a choline deficient diet on induction of papillomas by TPA or depilation. TG•AC mice (Leder, et al., PNAS, 87:9178-82, 1990) exhibited a characteristic tumor phenotype when fed PicoChow ad libitum and treated with TPA (1.25 or 0.31 µg/mouse, 2X/wk for 20 wks; maximum average of 27 or 12 papillomas/mouse, respectively) or depilated (wax stripped 1X/wk for 6 weeks; 32 pap/mouse). Choline deficient diet (Shinozuka, et al., Cancer Res, 38:1093, 1978) was associated with a decreased response to 0.31 µg TPA/mouse (7 vs 20 pap/mouse) and depilation (8 vs 22 pap/mouse), but not to 1.25 µg TPA/mouse (29 vs 28 pap/mouse) when compared to the Lombardi diet supplemented with choline. Choline deficient vehicle (acetone) control treated mice exhibited an increase in papillomas (4 pap/mouse) compared to PicoChow or the supplemented Lombardi diet vehicle control mice (1.5 vs 0.5 pap/mouse, respectively). These results are not conclusive, but demonstrate the potential for a choline deficient diet to influence skin tumorigenesis in this mouse model. Transgenic mice with a defined genetic event critical to tumorigenesis may be used to investigate dietary and nutrient effects on tissue specific gene expression and tissue specific tumorigenesis.

EFFECT OF AMINO ACID MODIFICATION ON TUMOR AMINO ACID LEVELS AND TUMOR GROWTH. B. Grossie, PhD, Surgical Oncology, University of Texas MD Anderson Cancer Center, Houston, TX.

The effect of dietary arginine (Arg) on tumor growth is controversial. Our results (J. Surg. Oncol. 50:161, 1992) show that Arg, but not ornithine (Orn), in parenteral nutrition (PN) is associated with an enhanced growth of a Ward colon tumor. To evaluate the effect of substituting Orn for Arg in PN on the tumor and circulating levels of Arg and Orn, tumors which respond differently to an Arg-containing PN (JPEN 13:590, 1989) were compared. Rats with a transplantable colon tumor growth SC were given PN. PN contained essential and nonessential amino acids which had Arg (ENA) or with Orn substituted (ENO); glucose, electrolytes, and multivitamins were added. Controls were fed chow ad libitum. After 4 days, circulating Arg and Orn levels increased for ENA, tumor-bearing rats. For ENO rats, however, circulating Arg levels were decreased while Orn was increased. Arg levels of the colon tumor were increased for rats receiving ENA and decreased for those receiving ENO. For sarcoma rats, ENA did not affect Arg levels while ENO resulted in a slight decrease. The response of tumor Arg levels to PN may determine the growth response of the tumor to an amino acid-modified PN.

EFFECTS OF CALORIC RESTRICTION ON SPONTANEOUS TUMORIGENESIS AND CELL CYCLE KINETICS IN P53-KNOCKOUT MICE. SD Hursting, SN Perkins and JM Phang, Laboratory of Nutritional and Molecular Regulation, National Cancer Institute, Frederick, MD 21702.

Transgenic mice with both alleles of the p53 tumor suppressor gene (frequently mutated in human tumors) knocked out by homologous recombination are susceptible to early onset of spontaneous tumors. To determine if tumorigenesis in p53-knockout mice can be modulated by caloric restriction (CR), which inhibits a variety of rodent tumors, male p53-knockout mice end wild-type littermates were either ad libitum (AL)-fed or were restricted to 60% of AL animals' calorie intake (30/group). Tumor incidence after 28 weeks was 100% (median survival=16 weeks) for AL/p53 knockout mice and 57% (median survival=24 weeks) for CR/p53 knockout mice (p=0.0002). Tumor incidence in wild-type mice after 28 weeks was 6% with AL-feeding versus 3% with CR. Cell cycle analyses were performed on splenocytes from a separate group of p53-knockout mice and wild-type littermates following 6 and 12 weeks of AL-feeding or CR (5/group/time-point). The percentage of cells in S-phase of the cell cycle was 3-fold higher for p53-knockout mice than wild-type mice, and CR reduced the percentage of S-phase splenocytes in both p53-knockout mice and wild-type mice by 25%. These data demonstrate that CR inhibits tumor development in p53-deficient mice genetically predisposed to spontaneous neoplasia and suggest that a possible mechanism for this dietary effect involves cell cycle modulation.

RRR-α-TOCOPHERYL SUCCINATE INDUCTION OF BIOLOGICALLY ACTIVE TRANSFORMING GROWTH FACTOR-BETA (TGF-β). B Zhao, M Simmons-Menchaca, A Charpentier, K Israel, B Sanders, and K Kline. Division of Nutritional Sciences and Department of Zoology, University of Texas at Austin, Austin, TX 78712-1097.

RRR-α-tocopheryl succinate at 15, 10, 5 and 1 µg/ml inhibits the proliferation of estrogen receptor-positive and estrogen receptor-negative human breast cancer cell lines in a dose-dependent manner in vitro. Analyses of cell conditioned medium from RRR-α-tocopheryl succinate growth inhibited cells revealed the presence of a potent antiproliferative activity when tested in the Mv1Lu-CCL-64 mink lung bioassay cells. Characterization of the antiproliferative activity as transforming growth factor-β (TGF-β) was established by: (1) growth inhibition of the TGF-β-responsive Mv1Lu-CCL-64 mink lung, Mv3D9 mink lung and murine CTLL-2 cell lines; (2) combination of physical characteristics including heat stability, acid stability, and Bio-Gel P-60 column chromatography elution profile; and (3) neutralization of the antiproliferative activity in the conditioned media by antibodies specific for TGF-β.

MODEL SYSTEM FOR CHEMOPREVENTION OF ESTROGENIC CARCINOGENESIS. CA Rinehart and DG Kaufman, Department of Pathology, University of North Carolina, Chapel Hill, 27599-7295.

Estrogens are etiologic agents for cancers of the female reproductive tract. The purpose of this project was to establish an in vitro model system based on human cells that would be suitable for the study of prevention of estrogen-induced transformation by nutritional means. Estrogens induce anchorage independent proliferation (AIP) in SV40-immortalized human endometrial stromal cells in the rank order of DES> estradiol> beta-dienestrol. Progesterone did not induce AIP, and tamoxifen inhibited AIP. To study long term exposure to estrogens that may more accurately mimic in vivo conditions, immortal stromal cells were treated with DES for 11 months. During this time progressive increases in AIP occurred in DES treated cultures, but not in control cultures receiving vehicle (EtOH). Increases in AIP were accompanied by increased expression of the estrogen receptor gene, and the interleukin-1α and groα genes. The progressive nature of DES-induced transformation is ideally suited for studies of chemoprevention. Furthermore, expression of the estrogen receptor, IL-1α and groα genes may serve as biomarkers of prevention. Supported by NIH CA 31733 & The Institute of Nutrition.

VITAMIN B$_6$ MODULATION OF STEROID HORMONE REGULATED GENE EXPRESSION IN HUMAN BREAST CANCER CELL LINES. DB Tully, AB Scoltock, VE Allgood, and JA Cidlowski, Departments of Physiology and Nutrition, University of North Carolina, Chapel Hill, NC 27599.

In many cases, breast cancer growth is regulated by sex steroid hormones. The degree to which these cancers respond to steroid hormones correlates with their content of functional steroid receptors. We have now shown that changes in vitamin B$_6$ levels alter the extent of progesterone-induced gene expression in the T47D human breast cancer cell line. Cells cultured in the presence of elevated vitamin B$_6$ exhibit a decreased capacity to respond to progesterone stimulation. Conversely, cells maintained in a vitamin B$_6$-deficient state show an enhanced response to the steroid hormone. Work is currently in progress to evaluate the effects of changes in vitamin B$_6$ levels on estrogen-mediated gene expression in another human breast cancer cell line, MCF-7. These results may indicate clinical importance for monitoring the nutritional status of patients undergoing endocrine therapy for breast cancer.

BIOTECHNOLOGY IN HEART DISEASE AND CANCER

DESIGN AND RATIONALE FOR "BIOMARKERS OF BLACK-WHITE DIFFERENCES IN PROSTATE CANCER". F Ennever, R Fleming, S Swarts, E Paskett and D Zaccaro, Bowman Gray School of Medicine, Winston-Salem, NC.

Biotechnology has the potential to elucidate prostate cancer risk factors, including dietary fat intake, both for prostate cancer in general and for the 50% higher rate in Black men. This case-control study will use three biomarkers, of exposure, effect, and susceptibility. The biomarker of exposure is cadmium concentration in urine, a marker of lifetime intake of the suspect prostate carcinogen cadmium. The biomarker of effect is oxidative damage to DNA, analyzed in the context of overall oxidative stress (plasma levels of antioxidants and lipid oxidation products) to inferinherent DNA repair capacity as follows:

Oxidative stress	DNA repair	Oxidized DNA in cells	Oxidized DNA in urine
High	High	Intermediate	Highest
High	Low	Highest	Intermediate
Low	High	Lowest	Intermediate
Low	Low	Intermediate	Lowest

The biomarker of susceptibility is activity of two major xenobiotic metabolism pathways: oxidation and N-acetylation. All three biomarkers will be combined with dietary and lifestyle information to determine interactions among exposures and susceptibilities in prostate cancer risk.

THE ALPHA-AMINE CARBOXYBORANES, 2'-DEOXYRIBONUCLEOSIDE CYANOBORANES AND HETEROCYCLIC AMINE BORANES AS HYPOLIPIDEMIC AND ANTI-NEOPLASTIC AGENTS IN RODENTS. IH Hall, A Sood, BF Spielvogel, BS Burnham and M.C. Miller, III Medicinal Chemistry, School of Pharmacy, University of North Carolina, Chapel Hill, N.C. and Boron Biologicals, Inc. Raleigh, N.C.

Substituted alpha-amino acids, peptides, deoxy-nucleosides and deoxy-nucleotides with the boron atom have demonstrated potent pharmacological activities. All of these derivatives lowered serum total cholesterol and triglyceride, VLDL-cholesterol and LDL-cholesterol levels in normal and diet-induced-hyperlipidemic rats at 8-20 mg/kg/ day P.O. The HDL-cholesterol levels were significantly elevated in both types of rats. These compounds suppressed LDL-receptor binding and LDL-degradation in cultures cells of rat hepatocytes and aorta foam cells, mouse small intestinal mucosa cells and human hepatocytes and fibro-blasts. HDL receptor binding activity was increased in hepatocytes as was HDL uptake of fibroblast intracellular cholesterol. These agents were acyl CoA cholesterol acyl transferase inhibitors but stimulate neutral cholesterol ester hydrolase activity *in vivo* and *in vitro*.

These same agents proved to be potent anti-neoplastic agents reducing cell proliferation through a variety of mechanisms, i.e. inhibition of DNA polymerase alpha, purine *de novo* synthesis and Topoisomerase II activities. Selected agents were effective in blocking inflammation and osteo-porosis in rodents. Central mechanisms of the agents appears to be in the regulation of cytokine release and cytokine receptor binding on target cells.

THE EFFECT OF LIPOSOMAL AMIODARONE ON CARDIAC ARRHYTHMIAS. J Somberg, W Cao and V Ranade, Department of Medicine and Pharmacology, University of Health Sciences/The Chicago Medical School, North Chicago, IL 60064 and GL Jendrasiak and RL Smith, Department of Radiation Oncology, East Carolina University, Greenville, NC 27858.

Amiodarone was incorporated into small, phosphatidyl choline unilamellar vesicles. NMR and x-ray diffraction measurements suggest that the amiodarone is located near the water-polar surface of the vesicles. The effects of amiodarone in this preparation were evaluated in vivo in rats following coronary occlusion. The evaluation in the blinded fashion showed the placebo had no effect on arrhythmia frequency following coronary occlusion while amiodarone delivered in a liposome moiety showed a significant reduction in VPC frequency and complete abolition of VT and VF events. Amiodarone, thus delivered in this preparation is effective against acute cardiac arrhythmias in the rat model.

AN NMR STUDY OF LDL PARTICLES FROM NORMAL AND DIABETIC SUBJECTS. GL Jendrasiak, RL Smith and HA Barakat, Departments of Radiation Oncology and Biochemistry, ECU-School of Medicine, Greenville, NC 27858.

The purpose of this study was to determine if previously observed differences in chemical composition of LDL from normal and diabetic subjects would be manifested in changes in the NMR spectra. LDL were purified by ultracentrifugation and column chromatography and NMR spectra obtained for LDL from 6 normal and 6 diabetic subjects. The proton spectra exhibited no differences between the two classes. Chaotropic anions had no obvious effect on the resonance from the phospholipid (PL) head-group region. Mn(II) significantly affected the head group resonance from both classes as did Fe(III), however, no obvious differential effects were noted. Magnevist, an MRI contrast agent, did apparently show a differential effect on the NMR spectra. The use of ions to probe the surface of LDL may shed light on differences in binding characteristics of LDL with its receptor which may, in turn, be related to coronary disease.

DEVELOPMENT OF A STEROL CARRIER PROTEIN-2 GENE TARGETING VECTOR. M Raabe, U Seedorf, G Assmann and N Maeda, Institute for Arteriosclerosis Research, University of Muenster, Muenster 48149 (Germany) and Department of Pathology, University of North Carolina, Chapel Hill, NC 27599 (USA).

The sterol carrier protein-2 (SCP-2) is believed to play an important role in the intracellular cholesterol metabolism. This small protein shows strong stimulatory effects on cellular cholesterol movement and reveals a high degree of evolutionary conservation. Ancient fusion of the SCP-2 gene and an ancestral thiolase gene resulted in the formation of a gene encoding two related sterol carrier proteins (SCP-2 and SCP-x). To clarify the role of SCP-2 in cellular metabolism we have developed a gene replacement vector to generate mice lacking *Scp-2* gene function. The construct comprises 7 kb of the mouse *Scp-2* gene and both the neomycin resistance gene as well as the *herpes simplex* thymidine kinase gene for positive and negative selection. The vector was used for electroporation of E14TG2a embryo stem cells. Selection and analysis of targeted ES-cells is in progress. In addition we have isolated two *Scp-2* related genomic loci, one of which encodes a processed pseudogene (*Scp-2-ps1*) in a highly repetitive surrounding.

MURINE MODEL OF FAMILIAL DEFECTIVE APOLIPOPROTEIN B100 BY TARGETED MODIFICATION OF THE APOLIPOPROTEIN B GENE. L Toth and N Maeda, Department of Pathology and Curriculum in Genetics, School of Medicine, The University of North Carolina at Chapel Hill, Chapel Hill, NC 27599-7525.

A correlation between high levels of low density lipoprotein (LDL) in plasma and premature coronary heart disease is well established. LDL is cleared from plasma via receptor-mediated endocytosis by the LDL receptor. Apolipoprotein B100 (apoB100) is virtually the only protein component of LDL and functions as the ligand for the LDL receptor. Genetic defects in the apoB100 protein that interfere with or abolish receptor-ligand interaction cause the human disorder familial defective apolipoprotein B100 (FDB). FDB is characterized by increased plasma concentrations of apoB100, cholesterol, and LDL and high risk for developing atherosclerosis. We are using an "in-out" targeting method in murine embryonic stem cells to generate mice with subtle site-specific apoB100 mutations that result in protein unable to bind to the LDL receptor. The "in step", a homologous integration reaction, disrupts apoB coding sequence and generates a duplication of the target locus. The "in step" introduces a premature stop codon into exon sequence yielding a truncated apoB protein. We have chimeric mice harbouring this genetic change and are breeding for transmission of the modified gene. Mice with this mutation are expected to be models of familial hypobetalipoproteinemia. During the "out-step", recombination between the duplicated repeats generated in the integration reaction removes the duplication and all exogenous sequence except for modifications in the region of putative LDL receptor binding that abolish receptor-ligand interaction. We have obtained correctly modified embryonic stem cell lines and are in the process of generating mice with this altered apoB100 allele. These animals are expected to be deficient in receptor-mediated endocytosis of LDL particles from plasma resulting in accumulation of atherogenic LDL in the circulation. Such animals may be at high risk for developing atherosclerotic cardiovascular disease and model FDB in humans.

EFFECTS OF DEHYDROEPIANDROSTERONE (DHEA) ON GENE EXPRESSION OF MARKERS OF LIPID METABOLISM AND GLUCOSE TRANSPORTERS IN MUSCLE AND LIVER OF RATS. H Bao, M McIntosh, T Tagliaferro and SM Wu, Department of Food, Nutrition and Foodservice Management, University of North Carolina at Greensboro, Greensboro, NC and Department of Animal and Nutritional Sciences, University of New Hampshire, Durham, NH.

The purpose of this study was to identify a mechanism by which the adrenal steroid hormone DHEA reduces body fat and the levels of blood glucose and lipids in rats. Fourteen male Sprague-Dawley rats were fed a semi-purified, high fat diet (20% w/w) with or without DHEA (0.1%) for 3 weeks. At the termination of the study, total RNA was extracted from liver and muscle samples using the guanidinium thiocyanate method. Total RNA was checked for integrity (distinct bands at S28 & S18) using 5 ug/lane of RNA in an electrophoresis chamber. mRNA (10 ug RNA/lane) was separated from intact total RNA by Northern blotting electrophoresis techniques (denaturing with formaldehyde) and transferred to nylon membrane by standard procedures. mRNA detection for fatty acid synthase (FAS), peroxisomal fatty acyl oxidase (FAO), glucose transporters in muscle (GLUT4) and liver (GLUT2), and beta-acting were be measured via 32-P labeled cDNA probes. Levels of each transcript were quantified on an imaging densitometer and standardized relative to beta-acting message levels. The results of this study will be discussed in light of DHEA's antiobesity/antidiabetic properties.

DETAILED LIPOPROTEIN PROFILES PRODUCED AUTOMATICALLY AND RAPIDLY BY NMR SPECTROSCOPY. JD Otvos and EJ Jeyarajah, Department of Biochemistry, North Carolina StateUniversity, Raleigh, NC 27695 and RM Krauss and PJ Blanche, Lawrence Berkeley Laboratory, Berkley, CA 94720.

Knowledge of the particle size distributions of the subspecies within the VLDL, LDL, and HDL lipoprotein classes may significantly enhance heart disease risk assessment. However, this information is clinically inaccessible owing to the extreme difficulty of measuring subspecies levels. We have developed a powerful new method to overcome this problem that uses a single, easily-obtained NMR spectral fingerprint" of plasma as the source from which chylomicron VLDL, LDL, and HDL subspecies concentrations are derived simultaneously. Close correspondence is found between lipid levels and subspecies distributions determined by NMR and the "gold standard" methods (beta-quantification and gradient gel electrophoresis,' respectively). NMR also provides a particularly sensitive means of following postprandial chylomicron clearance rates and the accompanying changes in remnant particle sizes, which are thought to be related to cardiovascular disease.

CLONING AND SEQUENCING OF A POTENTIAL TRANSFORMING GENE FROM THE AORTA OF JAPANESE QUAIL SUSCEPTIBLE TO ATHEROSCLEROSIS. DW Kelemen, JCH Shih, Department of Poultry Science, North Carolina State University, Raleigh, NC 27695-7608 and C Walker, Chemical Industry Institute of Toxicology, Research Triangle Park, NC 27709.

Aortic smooth muscle cell (SMC) proliferation was found at the early stage of the development of atherosclerosis in the Japanese quail on a high cholesterol diet. Proliferative transformation of SMC may be involved in the process of atherogenesis. It is therefore of great interest to determine the transforming potential of the DNA isolated from the quail aorta and, if possible, to isolate and identify the gene with transforming activity. DNA was isolated from pooled aortas and tested by the standard NIH3T3 cells/nude mice assay. Tumors developed in 7-13 weeks. Secondary transfection using primary tumor DNA was again positive. In both primary and secondary tumors, quail DNA sequence was detected by Southern hybridization. The positive fragments were separated, cloned, partially amplified by PCR, and subsequently sequenced. This sequence, which has been carried over from quail by two cycles of transfection, is believed to be important in cell transformation. Its role in atherogenesis is yet to be determined. (Grant support from NC Biotechnology Center and UNC Institute of Nutrition.

IMPACT OF BIOTECHNOLOGY ON NEW DRUG DEVELOPMENT. JM Walenga and J Fareed, Loyola University Medical Center, Maywood, IL 60153.

Recombinant technology has added a new dimension to the development of antithrombotic and thrombolytic agents for cardiovascular disorders. Such drugs as plasminogen activators and their variants, protein C/C_a and derivatives, thrombomodulin and its variants, antibodies and peptide targeting platelet membrane glycoproteins, plasma inhibitors, e.g. antithrombin III and heparin cofactor II and anticoagulant proteins such as the tissue factor pathway inhibitor (TFPI) are currently under development. Similarly, targeting of platelets and subendothelial lesions has been accomplished using cell or cytoskeletal specific antibodies. Cloning of genes for protein C/C_a, protein S and thrombomodulin has resulted in the development of therapeutic grade products for the management of thrombotic disorders. Genes for antithrombin III, heparin cofactor II and TFPI have been cloned and the expressed proteins exhibited therapeutic properties. The introduction of recombinant hirudin and its variants has added a new dimension to the management of such indications as cardiovascular bypass surgery, unstable angina, prophylaxis of DVT and coating biomaterials. While the developments in this area are dramatic, only a few products have entered into clinical use at this time. Currently, protein Ca, hirudins and glycoprotein targeting antibodies and TFPI are undergoing clinical trials and will potentially have a major impact in the future management of thrombotic disorders. Thus, the recent developments in this area offer a challenge for both basic science as well as clinical investigators which must be met in an objective and stepwise fashion. These developments will have a major impact on the management of cardiovascular disorders in coming years.

INDEX